Atlantis Studies in Computational Finance and Financial Engineering

Volume 1

Series editor

Argimiro Arratia, Barcelona, Spain

For further volumes:
www.atlantis-press.com

Series Information

This book series aims at publishing textbooks aimed at the advanced undergraduate and graduate levels as well as monographs within the fields of Computational Finance and Financial Engineering.

For more information on this series and our other book series, please visit our website www.atlantis-press.com

Atlantis Press
29, avenue Laumière
75019 Paris, France

Argimiro Arratia

Computational Finance

An Introductory Course with R

ATLANTIS
PRESS

Argimiro Arratia
Department of Computer Science
Universitat Politécnica de Catalunya
Barcelona
Spain

ISSN 2352-3255   ISSN 2352-3115   (electronic)
ISBN 978-94-6239-069-0   ISBN 978-94-6239-070-6   (eBook)
DOI 10.2991/978-94-6239-070-6

Library of Congress Control Number: 2014936173

Printed on acid-free paper

Preface

Finance is about managing money. It is about making provisions and allocations of funds for a business project with an assessment of risks and benefits. There are several instruments for dealing with money. There are several places for trading these financial instruments. There are mathematical models for optimal selection of financial portfolios and for predicting their assets' future values. And there are computers to help us with all the necessary real-world data crunching to exploit the formal models to unprecedented possibilities.

Computational Finance includes all numerical methods, theories of algorithms and optimization heuristics geared to the solution of problems in economics and finance. The subject area is broad and requires knowledge in computational statistics, econometrics, mathematical finance and computer science, of course.

This book is the product of a seminar course taught every year since 2010 at the Computer Science Department of the Technical University of Catalonia (Universitat Politècnica de Catalunya), in which I have attempted to cover some of the material from every diverse area of knowledge required for the computational resolution of financial problems. To accomplish this task, I had to avoid lecturing in-depth on any one topic, and touch upon the surface of every topic I considered necessary for Computational Finance. This style of learning is reflected in the book, where the emphasis is on methods and not much on theorems and proofs, although I try to be as rigorous as possible and, in the absence of a proof, I give pointers to the literature from where to learn it. It is this aspect that makes the contents of this book an introductory course, with special attention to the software implementation of econometric models and computational exploration of financial instruments. All programming examples in this book are implemented in the free programming language R, with all the codes exhibited and commented in the text; there is also the possibility to download all programs from the book's website.[1] The Appendix contains a brief introduction to R, for those not acquainted with this software language.

For whom is this book intended? The truth is that one writes the books that one would like to read. I am a mathematician, working in a Computer Science Department and an amateur investor in stock markets. Therefore, this book was

---

[1] http://computationalfinance.lsi.upc.edu.

written for advanced undergraduate or graduate students in computer science, mathematics, business and non-academic financial investors, at least to some extent. For the business and non-academic class of readers, I suggest the following reading plan: Chapters 1, 6 and 8; at any point beyond Chap. 1 look back into Chap. 2 for the basic concepts in statistics, and afterwards jump to any chapter that attracts your attention. Computer science students should follow the plan of the book in order; after all the book was conceived from a seminar course in a Computer Science Department, intended mainly for computer science students. Mathematicians can follow either of the previously suggested reading paths.

I am indebted to several people who provided comments or corrections. I am grateful to the students in the different editions of the seminar in Computational Finance, first for their willingness to participate in this seminar, and for their inputs on the lessons that originated most of the chapters in this book. My greatest gratitude is for Alejandra Cabaña and Enrique Cabaña, who made profound corrections to a first draft of the chapters on statistics and time series analysis, obliging me to rewrite them in a decent and correct manner. Alejandra went further on reading succeeding versions and provided me with further corrections. Many thanks go also to my friend Jesus Carrero, who gave me helpful insights from the perspective of a mathematician in the finance industry; following his suggestions, I extended the presentation on Monte Carlo methods more than I had originally intended. Thanks to Joaquim Gabarró who kindly proofread final versions of some chapters, providing many comments that helped me improve their readability, and to my editor Keith Jones and the supporting team of Atlantis Press for their patience and cooperation throughout the stages of production of the book. Finally, I am grateful to all the members of my family and friends for their continuous support during the process of writing this book. To them I dedicate this work.

Barcelona   Argimiro Arratia

Contents

1 An Abridged Introduction to Finance 1
  1.1 Financial Securities 1
      1.1.1 Bonds and the Continuous Compounding
           of Interest Rates 2
      1.1.2 Stocks: Trade, Price and Indices 4
      1.1.3 Options and Other Derivatives 12
      1.1.4 Portfolios and Collective Investment 18
  1.2 Financial Engineering 19
      1.2.1 Trading Positions and Attitudes 19
      1.2.2 On Price and Value of Stocks. The Discounted Cash
           Flow model 22
      1.2.3 Arbitrage and Risk-Neutral Valuation Principle 25
      1.2.4 The Efficient Market Hypothesis and Computational
           Complexity 31
  1.3 Notes, Computer Lab and Problems. 33

2 Statistics of Financial Time Series 37
  2.1 Time Series of Returns. 37
  2.2 Distributions, Density Functions and Moments 43
      2.2.1 Distributions and Probability Density Functions 43
      2.2.2 Moments of a Random Variable 45
      2.2.3 The Normal Distribution 49
      2.2.4 Distributions of Financial Returns. 51
  2.3 Stationarity and Autocovariance 56
  2.4 Forecasting. 60
  2.5 Maximum Likelihood Methods 62
  2.6 Volatility 64
  2.7 Notes, Computer Lab and Problems. 67

3 Correlations, Causalities and Similarities 71
  3.1 Correlation as a Measure of Association. 72
      3.1.1 Linear Correlation. 72
      3.1.2 Properties of a Dependence Measure. 76
      3.1.3 Rank Correlation. 77

3.2   Causality   78
      3.2.1   Granger Causality   79
      3.2.2   Non Parametric Granger Causality   81
3.3   Grouping by Similarities   84
      3.3.1   Basics of Data Clustering.   85
      3.3.2   Clustering Methods   87
      3.3.3   Clustering Validation and a Summary of Clustering
           Analysis.   94
      3.3.4   Time Series Evolving Clusters Graph   95
3.4   Stylized Empirical Facts of Asset Returns   103
3.5   Notes, Computer Lab and Problems.   104

4   Time Series Models in Finance   109
4.1   On Trend and Seasonality.   110
4.2   Linear Processes and Autoregressive Moving
     Averages Models.   111
4.3   Nonlinear Models ARCH and GARCH   124
      4.3.1   The ARCH Model.   124
      4.3.2   The GARCH Model   127
4.4   Nonlinear Semiparametric Models.   130
      4.4.1   Neural Networks.   131
      4.4.2   Support Vector Machines.   134
4.5   Model Adequacy and Model Evaluation.   136
      4.5.1   Tests for Nonlinearity   137
      4.5.2   Tests of Model Performance.   138
4.6   Appendix: NNet and SVM Modeling in R   140
4.7   Notes, Computer Lab and Problems.   142

**5   Brownian Motion, Binomial Trees and Monte Carlo
Simulation**   145
5.1   Continuous Time Processes   145
      5.1.1   The Wiener Process   146
      5.1.2   Itô's Lemma and Geometric Brownian Motion   149
5.2   Option Pricing Models: Continuous and Discrete Time   153
      5.2.1   The Black-Scholes Formula for Valuing
           European Options   154
      5.2.2   The Binomial Tree Option Pricing Model   158
5.3   Monte Carlo Valuation of Derivatives   164
5.4   Notes, Computer Lab and Problems.   172

6   Trade on Pattern Mining or Value Estimation   177
  6.1   Technical Analysis   177
      6.1.1   Dow's Theory and Technical Analysis Basic
            Principles.   178
      6.1.2   Charts, Support and Resistance Levels, and Trends   180
      6.1.3   Technical Trading Rules   183
      6.1.4   A Mathematical Foundation for Technical Analysis   190
  6.2   Fundamental Analysis   196
      6.2.1   Fundamental Analysis Basic Principles   196
      6.2.2   Business Indicators   197
      6.2.3   Value Indicators   199
      6.2.4   Value Investing.   202
  6.3   Notes, Computer Lab and Problems.   204

7   Optimization Heuristics in Finance.   207
  7.1   Combinatorial Optimization Problems   207
  7.2   Simulated Annealing   209
      7.2.1   The Basics of Simulated Annealing.   210
      7.2.2   Estimating a $GARCH(1, 1)$ with Simulated Annealing. . .   211
  7.3   Genetic Programming   213
      7.3.1   The Basics of Genetic Programming   215
      7.3.2   Finding Profitable Trading Rules with Genetic
            Programming   218
  7.4   Ant Colony Optimization   226
      7.4.1   The Basics of Ant Colony Optimization   227
      7.4.2   Valuing Options with Ant Colony Optimization   229
  7.5   Hybrid Heuristics   233
  7.6   Practical Considerations on the Use of Optimization
      Heuristics.   234
  7.7   Notes, Computer Lab and Problems.   236

8   Portfolio Optimization   239
  8.1   The Mean-Variance Model   239
      8.1.1   The Mean-Variance Rule and Diversification   239
      8.1.2   Minimum Risk Mean-Variance Portfolio   241
      8.1.3   The Efficient Frontier and the Minimum
            Variance Portfolio.   243
      8.1.4   General Mean-Variance Model and the
            Maximum Return Portfolio   244
  8.2   Portfolios with a Risk-Free Asset   247
      8.2.1   The Capital Market Line and the Market Portfolio   249
      8.2.2   The Sharpe Ratio   250
      8.2.3   The Capital Asset Pricing Model and the Beta
            of a Security.   251

8.3 Optimization of Portfolios Under Different Constraint Sets. 256
  8.3.1 Portfolios with Upper and Lower Bounds in Holdings. . . 257
  8.3.2 Portfolios with Limited Number of Assets 258
  8.3.3 Simulated Annealing Optimization of Portfolios 259
8.4 Portfolio Selection. 260
8.5 Notes, Computer Lab and Problems. 263

9 **Online Finance** 267
9.1 Online Problems and Competitive Analysis 268
9.2 Online Price Search. 269
  9.2.1 Searching for the Best Price. 269
  9.2.2 Searching for a Price at Random 270
9.3 Online Trading 272
  9.3.1 One-Way Trading 272
9.4 Online Portfolio Selection. 273
  9.4.1 The Universal Online Portfolio. 274
  9.4.2 Efficient Universal Online Portfolio Strategies 279
9.5 Notes, Computer Lab and Problems. 281

Appendix A: The R Programming Environment. 283
A.1 R, What is it and How to Get it 283
A.2 Installing R Packages and Obtaining Financial Data 284
A.3 To Get You Started in R 285
A.4 References for R and Packages Used in This Book 286

References 289

Index 297

Chapter 1
An Abridged Introduction to Finance

This chapter is intended for giving the reader the minimum background on the fundamentals of finance. An outlook on the most common financial instruments and the places where these are traded. An introduction to investment strategies, portfolio management and basic asset pricing. In short, we give a succinct review of the what, how, and when of financial business: what can we buy or sell, how to buy it or sell it, when to buy it or sell it. These subjects constitute the sources of algorithmic research in Computational Finance. The review includes the notions of bonds, common stocks, options, forward and future contracts, and other derivatives. However, the main focus for the rest of the book will be on common stocks and their options. We shall describe the composition and regulations of some stock markets, part of the lingo used by investors, and summarize some important concepts of finance and paradigms for asset pricing, which will be dealt in more depth later in the book at a more mathematical and computational level.

1.1 Financial Securities

The financial instruments that are the objects of study of the methods presented in this book belong to the broad category known as *securities*. A *security* is a fungible, negotiable financial instrument representing financial value.[1] There are three main type of securities:

Debt: to this class belong, in general, those securities that are "secure", in the sense of been risk-free, such as bonds, commercial notes and bank deposits.
Equity: refers generally to a company's stocks or share value.
Derivatives: in this class are securities whose value depend on the values of other, more basic underlying variables. Examples of derivatives are: *futures* and *forward*

---

[1] For a more precise and lengthy description, albeit legalistic, of what constitutes a security see the USA Government *Securities Exchange Act* of 1934 in http://www.sec.gov/.

A. Arratia, *Computational Finance*, Atlantis Studies in Computational Finance
and Financial Engineering 1, DOI: 10.2991/978-94-6239-070-6_1,
© Atlantis Press and the authors 2014

contracts, *swaps*, and *options*. A stock option, for example, is a contract whose value depends on the price of a stock at a certain date.

Some debt contracts are established directly between two financial institutions, or a financial institution and an investor, but in general securities are traded at organized markets known as *exchange markets*. Examples of exchange markets of stocks and options in the United States are the New York Stock Exchange (NYSE), Chicago Board Options Exchange (CBOE) and Nasdaq; in Europe the London Stock Exchange (LSE), Deutsche Börse, and Spanish Exchange Markets (BME); in Asia the Tokyo Stock Exchange and Shanghai Stock Exchange, among others.[2]

Bonds are usually considered as the benchmark risk-free security in financial engineering. We adhere to this consideration and, consequently, we begin with a brief review on bonds and the computation of the profit obtained by these securities. Then we focus on stocks and options as our objects of study for financial and mathematical modeling.

1.1.1 Bonds and the Continuous Compounding of Interest Rates

A bond is a long-term loan contract between two parties, the *issuer* and the *holder*, in which the issuer receives a specified amount of money from the holder and is obliged to pay it back at a later date together with some interest. This interest could be paid at various fixed points during the life of the bond. The issuer of the bond is known as the *borrower* or *debtor*, the holder of the bond is the *lender*, the amount given by the lender is called the *principal*, the successive interest payments are called *coupons*,[3] and the date the bond expires is the *maturity* date. The value of a bond depends obviously on the time to maturity and the interest rate. Also it depends on the frequency of payment of the interests as a consequence of how interests are computed, as we shall explain in the next paragraphs.

Compounding the interest. We should all be aware that most of all risk–free securities that earn interest at a certain fixed rate have their value computed through time by *compounding* the interest. This means that the interest earned at a certain moment is added to the current value of the investment and the next interest is computed over the resulting sum. Therefore the value of the security with interest depends not only on the interest rate, but also on the frequency in which the interest is compounded. Consider for example that an investor acquires a bond with a principal of € 100, at an annual interest rate of 10 %, and for n years. If the interest is paid annually then by the end of the first year the holder of the bond gets €110 = 100(1 + 0.1). By the second year the interest is computed over the previous sum and added to it (here is where

---

[2] A full list of exchange markets, organized by countries, market capitalization or other economic parameters, can be found through the internet, e.g.,en.wikipedia.org/wiki/List_of_stock_exchanges.

[3] The term is a legacy from the past when bonds came with coupons attached, which had to be presented by the lender to the borrower in order to get the interest payment.

Table 1.1 The effects of compounding frequency on €100 over 1 year at the interest rate of 10% per annum

Frequency	Number of payments (m)	Interest rate per period (r/m)	Value at the end of year
Annual	1	0.1	€110.00
Semiannual	2	0.05	€110.25
Quarterly	4	0.025	€110.381
Monthly	12	0.0083	€110.471
Weekly	52	0.1/52	€110.506
Daily	365	0.1/365	€110.516

compounding takes place), and thus the investor gets $110(1+0.1) = 100(1+0.1)^2$. Continuing this way we see that by the end of n years the value of the bond is equal to $100(1+0.1)^n$. In general, if P_0 represents the amount of the principal at the initial time t_0, r the interest rate per annum and n the number of years, then the value of the bond by the nth year is given by

$$P_n = P_0(1 + r)^n \tag{1.1}$$

If the frequency of payment of the interest augments, hence increasing the frequency of compounding the interest, say by $m > 1$ times in a year, then the fraction of the annual interest rate that is compounded at each shorter period is r/m (the interest rate per period), and the value of P_n is

$$P_n = P_0(1 + r/m)^{nm} \tag{1.2}$$

Table 1.1 shows the effects of augmenting the frequency of compounding the interest for our debt security.

Now, for the purpose of modeling the price behavior of stocks or options, or any security whose value is frequently changing, it is convenient to consider that trading can be done continuously in time. To adapt the computing of value of risk-free securities to this frequent trading assume that the interests are computed infinitely often, and rated over a generic time period $\tau > 0$ which could be n years as well as an infinitesimal step. Mathematically this means to take the limit for $m \to \infty$ in Eq. (1.2) (with $\tau = n$), and by basic calculus we arrived at the formula for valuing a risk-free security with a constant interest rate r that is *continuously compounded* in a period $\tau > 0$

$$P_\tau = P_0 e^{r\tau} \tag{1.3}$$

where P_τ is the value of the investment that mature from time t_0 to time $t_0 + \tau$, and e^x is the exponential function. Thus, in our previous example, the value at the end of 1 year ($\tau = 1$) of the initial investment $P_0 = €100$ at the annual rate $r = 0.1$ with continuously compounding is $P_1 = 100e^{0.1} = €110.517$, a value that is very close to that obtained by daily compounding.

More generally, if P_t is the value of the bond at time t, to know the value at a later instant $t + \tau$ with continuous compounding we compound the period

$$P_{t+\tau} = P_t e^{r\tau} \tag{1.4}$$

and to recover the value at a previous instant $t - \tau$ we discount the period

$$P_{t-\tau} = P_t e^{-r\tau} \tag{1.5}$$

Although Eq. (1.3) is a theoretical valuation formula of risk-free securities, it can be used in practice to compute exactly the (discrete) m times per annum frequently compounded interest value Eq. (1.2), by making a correction on the deviation in the continuous compounded rate, as follows. If we have a debt security earning an interest rate of R per annum, to have the m times per annum compounding at this rate coincide with the continuous compounding, we just resolve for r in the equality

$$P_0 e^r = P_0 (1 + R/m)^m$$

to get $r = m \ln(1 + R/m)$. Thus, at the continuous compounding rate of $m \ln(1 + R/m)$ we obtain the same terminal value as compounding with a frequency of m times per annum at the annual rate of R.

Payoff and profit of bonds. The *payoff* of any security is its value at maturity. For a bond, its payoff is the principal plus all interests paid on the principal, and given by Eq. (1.2), or if continuously compounding of interest is assumed it is given by Eq. (1.3). The *profit* of a security is its risk-adjusted payoff discounting the initial investment, which includes contract fees or any other transaction costs. (The idea of risk-adjusting the payoff will be clear from the ways we'll settle each case of security to come.) For bonds there are usually no transactions costs or fees (or if there were, we can assume them to be included in the principal), and the risk is null; hence, the profit given by a bond is obtained by discounting the principal to the payoff; i.e., for a constant interest rate r through a period τ and m coupons on a principal of P_0, this is

$$P_\tau - P_0 = P_0((1 + r/m)^{m\tau} - 1)$$

For an example look back to Table 1.1: a €100 investment on a 1-year bond at an annual interest rate of 10% and quarterly payments (4 coupons) gives a payoff of €110.381 and a profit of €10.381.

1.1.2 Stocks: Trade, Price and Indices

A share of a company's stock is a claim on part of the company's assets and earnings. There are two main types of stock: *common* and *preferred*. Common stock usually entitles the owner to vote at shareholders' meetings and to receive dividends. Preferred stock generally does not give voting rights, but have priority over common stock regarding payment of dividends. As implied by the name, the prevalent of the two among investors is the common stock, and hence, wherever one hears or reads about stocks it is most likely referring to common stocks. Consequently, we will also use in this book the term stock as synonym of common stock, and will be most of the time focusing on this type of stocks.

A company sells shares or participations on its business in order to raise more capital. These shares are sold to the investors through stock exchange markets. The shares commercialized in a stock exchange market are the company's *shares outstanding*. The *market value* or *market capitalization* of a company is the number of shares outstanding times the price of each share. Formally

$$\text{market capitalization} = (\text{number of shares outstanding}) \times (\text{price of share}).$$

Of course this market value of a company varies through time as the price does. An owner of stock of certain corporation is known as a *shareholder*, and what this imply is that this person has partial ownership of the corporation. This ownership is determined by the number of shares a person owns relative to the number of shares outstanding. For example, if a company has 1,000 shares outstanding and one person owns 100 shares, that person would own and have claim to 10% of the company's assets. In the past a shareholder would get a title (or certificate) to testify for the amount of his participation. Figure 1.1 shows an 1880 certificate for 500 shares of Wells Fargo Mining Co. Presently, stocks are bought in person or electronically, through financial institutions with brokerage services. The financial institutions assume the custody, but not the obligations, of the shares. The brokerage firm or individual broker serves as a third-party to facilitate the transactions of stocks between buyers and sellers. They are the ones responsible for executing the buy or sell orders at the stock exchange, and as instructed by their clients. The buy or sell orders that investors can issue to their brokers are regulated by the market authority and have different forms to allow for different trading strategies. We list the three most common forms of buy or sell orders for stocks (which can also apply to any other security traded in exchange markets).[4]

Market order: An order that is send to the market demanding for the trade on the security to be executed immediately at the current market price. One would use this type of order when the execution of the trade is a priority over the price. The price obtained when the order is executed (or "filled") is the best possible at the time of execution, but could be different (for better or for worse) than the price quoted at the time the order was sent.

---

[4] For a full list of possible trading orders available at the NYSE see http://www.nyse.com/pdfs/fact_sheet_nyse_orders.pdf.

Fig. 1.1 Wells Fargo Mining stock certificate, 1880

Limit order: An order that is send to the market demanding for a given or better price at which to buy or sell the security. The trade is only executed when a matching offer for the security shows up. This order gives the trader control over the price but not over the time of execution.

Stop orders: An order that is held by the broker and only send to the market when the price of the security reaches a specified amount (the *stop price*). When that pre-set price barrier is reached the stop order can be turned into a market order or a limit order, as specified by the investor. This type of order is used to put a limit to possible losses due to market reversal, in which case it is called a *stop loss* order, or to take a pre-set percentage of profits, in which case it is a *stop gains* or *profit taking* order.

There is a subtle, yet very important difference between a stop order and a limit order. Although both demand a certain price, i.e. the stop price and the limit price, the former is not send to the market while the latter is send immediately, even though it might not get executed. But while the limit order is hanging in the bid/ask table the market might withhold from the investor the full euro amount (for buying) or freeze the volume of the asset written on the order (for selling) as a guarantee of fulfillment of the trade. This is not the case for stop orders because these are retained by the investor's broker.

Example 1.1 Here is an example of a stop loss order with a limit. An investor has 200 shares of a company bought at € 40 each. To protect his investment from possible market downfall in the coming month, he decides that he should sell if the price of the share goes below € 35, but restraining his losses to no more than € 7 per share. Hence he writes the following order:

> To sell the 200 shares at the limit price of €33, and activate this (limit) order when the price of the share is less than or equal to € 35. Keep this order active until the end of next month.

Thus, we have a stop order that is send to the market when the price touches € 35, and at that moment it turns into a limit order asking to sell the shares for € 33 or higher.   □

General Electric Company (GE) - NYSE

22.71 ↓0.02(0.09%) Sep 28, 4:00PM EDT | After Hours : **22.70** ↓0.01 (0.04%) Sep 28,

Prices

Date	Open	High	Low	Close	Volume	Adj Close*
Sep 28, 2012	22.77	22.96	22.62	22.71	70,926,100	22.71
Sep 27, 2012	22.24	22.86	22.13	22.73	67,300,400	22.73
Sep 26, 2012	22.16	22.25	22.07	22.10	41,123,300	22.10
Sep 25, 2012	22.39	22.67	22.30	22.31	47,039,200	22.31
Sep 24, 2012	22.40	22.45	22.30	22.36	36,799,800	22.36
Sep 21, 2012	22.55	22.69	22.46	22.53	66,551,700	22.53
Sep 20, 2012	22.19	22.47	22.12	22.43	43,677,000	22.43
Sep 20, 2012			0.17 Dividend			
Sep 19, 2012	22.30	22.49	22.25	22.43	40,462,600	22.26
Sep 18, 2012	21.99	22.24	21.96	22.24	38,831,200	22.07
Sep 17, 2012	21.93	22.05	21.90	22.05	79,283,200	21.88
Sep 14, 2012	22.20	22.37	21.98	22.11	100,148,400	21.94

Fig. 1.2 Historical price table of GE (NYSE)

An important topic of research is to establish the stop price in stop orders, in particular in stop loss orders, because we don't want to sell at a sudden drop of the price that immediately turns around and skyrockets leaving us out of the profits. We would like to be certain that when our stop order triggers we are really minimizing our losses, not missing on future gains.

Reading a stock quotes table. The information on the price of a stock is nowadays available as fast as clicking the mouse of a PC to connect to one of the many financial sites in the internet that freely provide stocks *quotes* or *historical price* tables. These stocks quotes tables present a date or time ordered list of various parameters that define the stock and the movement of its price and other related quantities, such as its traded volume. In Fig. 1.2 we have a partial view of the General Electric Co. (NYSE) stock's quotes table, as provided by Yahoo! (from *finance.yahoo.com*). The most common, and relevant, pieces of information found in a stock quotes table are the following:

Ticker: the company's id for the market (e.g. for General Electric Co. is GE).
Open: the price of the stock at the opening time of the market session
Close: the price of the stock at the current time, if the market is in session, or the last price traded when the market closes.
High: (or Max) is the maximum price reached by the stock between the Open and the Close price.
Low: (or Min) is the minimum price reached by the stock between the Open and the Close price.

Volume: is the number of shares traded between the times of Open and Close.

Adjusted Close: is the closing price adjusted to include any *dividend* payments plus any other corporate actions that affect the stock price (e.g. splits and rights offerings).

The dividend is a cash payment given periodically by some companies reflecting profits returned to shareholders. There is no obligation for a company to pay dividends, and the decision to do it or not is part of the company's business strategy. A company that gives no dividends usually reinvest all profits back into business.

As a warm-up for our computational explorations in finance, start your R programming environment and run the following example (see the Appendix for details on usage of R).

R Example 1.1 We shall use the `getSymbols` function in the `quantmod` package to retrieve financial data for General Electric (GE) from *Yahoo*. After loading the package into our workspace with `library("quantmod")` type:

```
> getSymbols('GE',src='yahoo', from="2000-01-01", to="2009-12-30")
```

This creates an `xts` object (extended time series) with the Open, High, Low, Close, Volume, and Adjusted Close prices of GE's stock in the range of dates indicated. Verify the data labels with: `names(GE)`. Next, to see the data from 1st of January 2000 to 20th of January 2000, type:

```
> GE["2000-01-01/2000-01-20"]
```

or to see only the Adjusted Close prices for that period:

```
> geAdj = GE$GE.Adjusted["2000-01-01/2000-01-20"] ; geAdj
```

Compute the maximum, minimum and mean value:

```
> max(geAdj); min(geAdj); mean(geAdj)
```

And finally, make a beautiful chart of the financial data with the instruction:

```
> chartSeries(GE)
```

or a chart of the data from January to February 2001, omitting the volume information (and by default as *candlesticks*; for other form specify it with the parameter `type`[5]):

```
> chartSeries(GE,TA=NULL,subset='2001-01::2001-02')
```

Finally, save the data in your local disk with `saveRDS(GE,file="GE.rds")`. For future use, you can upload it with `GE=readRDS("GE.rds")`   □

There are some important observations on the price and volume quoted on a historical price table that are worth an explanation and our attention. First, the Open price is usually not equal to the previous session Close price (compare the Close with the Open of consecutive sessions in Fig. 1.2). The reason for this mismatch is that

---

[5] The different type of charts will be studied in Chap. 6.

some particular trades may significantly alter the flow of supply or demand and are thus put on hold by the market authority and executed after the market closes to the public. This is the case, for example, of orders whose quantity and price sellers and buyers had previously agreed upon, or orders of very large volume. Second, observe that volume is defined as the total number of shares that are flowing in the market, regardless of the direction of the flow as to whether the shares are been bought or sold. This detailed information is usually unknown through the cost-free channels of stock quotes, but at some exchange floors and through private brokerage services, investors can have access to information on the direction of flow of volume, and furthermore to the id of the broker making the trade. Yet another fact of interest about volume is the general observed relationship of volume to price, in that the former confirms the tendency of the latter. In Chap. 6 we make precise this connection of volume to price in the context of Technical Analysis.

Payoff and profit of stocks. Let us first begin with a simple situation where we trade with stocks that pay no dividends, and that there are no transaction costs. In this scenario, if we buy one share of a stock at a time $t = t_0$ for a price of S_0, and sell the share at a later time $t = T$ for a price of S_T, the payoff is S_T. (For m shares the payoff is obviously $m S_T$.) However, the profit obtained can not be just the result of discounting the initial investment to the payoff, because the equation $S_T - S_0$ involves cash flows at different points in time, and there is some risk involved in the investment which has to be taken into consideration. Any real investor could have instead put his money in a risk-free security (e.g. a bond) earning a risk-free interest rate r for the period $\tau = T - t_0$. Thus, assuming continuous compounding at a constant interest rate r, the profit for one share of a stock bought at time t_0 for S_0 a share, and sold at time T for S_T is

$$S_T - S_0 e^{r\tau} \qquad (1.6)$$

To include dividends and transaction costs on the profit equation, denote by D_τ the total amount of dividends per share received through the period we held to the stock, and $C(S_0)$ the sum of S_0 plus all transaction costs (i.e. the total cost of the investment). Then the profit, including dividends and transaction costs is

$$S_T + D_\tau - C(S_0) e^{r\tau} \qquad (1.7)$$

Stock indices. A measure of the value of the market, or of a sector, is given by the stock indices. Formally, an index is a mathematical function that measures the changes in a representative group of data points. In the stock market the data points are the prices of stocks, and a stock index tracks the changes in the value of a selected group of stocks that supposedly represent the market or an industrial sector. The purpose of a stock index is to serve as general reference of the market's value trend, and the general state of health of the economy. For the majority of investors it serves as a benchmark for the performance of their investments, as it also quotes in the market and investors can thus follow its performance through time. However, since it does not represents one company, but a hypothetical basket of companies without any economic endorsement, it cannot be traded by the investors; it is only a reference. Stock indices in used worldwide today are composed by either one of the following two methods:

Price weighted: only the price of each component stock is consider. For example, the Dow Jones Industrial Average (DJIA), which (as of today) takes the average of the prices of 30 of the largest publicly owned companies based in the U.S. Other price-weighted indices are: the NYSE ARCA Tech 100 and the Amex Major Market.

Capitalization-weighted: considers the market capitalization of each stock composing the index; that is, the stock price times the number of shares outstanding. Examples are: the NASDAQ Composite, FTSE 100, Russell 2000, CAC 40 and IBEX 35.

In a price-weighted index a significant change in the price of a single component may heavily influence the value of the index, regardless of the size of the company (i.e., number of shares outstanding). On the other hand, in a capitalization-weighted index a small shift in price of a big company will heavily influence the value of the index. These facts, together with other subjective assumptions underlying the composition of an index (e.g. the rules for admitting a company's stock to be part of the index) motivates loads of research on innovation of stock indices (e.g. see Note 1.3.5).

Example 1.2 (**A price weighted index**) The Dow Jones Industrial Average, conceived by Charles Dow and Edward Jones (1896), is computed by the following formula

$$\text{DJIA}_t = \frac{\sum_{i=1}^{30} S_{i,t}}{D} \tag{1.8}$$

where $S_{i,t}$ is the price of stock i at time t and D is the Dow Divisor, a constant included in the equation since 1928 to adjust, or rather stabilized, the average in case of stock splits, spinoffs or similar structural changes; as of July 2 of 2010 the divisor is 0.132129493.[6]   □

Example 1.3 (**A capitalization weighted index**) The Madrid Stock Exchange principal index, IBEX35, is computed by the following formula[7]

$$\text{IBEX}_t = \text{IBEX}_{t-1} \times \frac{\sum_{i=1}^{35} Cap_{i,t}}{\sum_{i=1}^{35} Cap_{i,t-1} \pm J} \tag{1.9}$$

where IBEX_t is the value of the index at time t, $Cap_{i,t}$ is the free float market capitalization of company i at time t, and J a coefficient used to adjust the index, similar in nature as the D in the DJIA equation.   □

---

[6] Updates on this number are periodically posted at http://www.cmegroup.com/trading/equity-index/files/djia-history-divisor.pdf.

[7] *source* Sociedad de Bolsas 1992.

Example 1.4 (**Value Line Stock Index**) One of the most reputed market index, used as benchmark for American and Canadian stock markets, is the Value Line Index. This is a composite index of about 1,700 companies from the New York Stock Exchange, NASDAQ, AMEX, and the Toronto Stock Exchange. Originally launched in 1961, it is calculated as an equally weighted geometric average, the index VLIC. Then in 1988, an arithmetic average version for the same composition of stocks, the VLIA, was released. The formula for calculating the Value Line Geometric Index (VLIC) is[8]

$$\text{VLIC}_t = \text{VLIC}_{t-1} \times \left(\prod_{i=1}^{n} \frac{C_{i,t}}{C_{i,t-1}} \right)^{1/n} \tag{1.10}$$

where VLIC_t stands for the value of the index at time t, $C_{i,t}$ is the current (or Close) price of stock i at time t, and n is the number of stocks. For the Value Line Arithmetic Index (VLIA), the formula with similar nomenclature is given by

$$\text{VLIA}_t = \text{VLIA}_{t-1} \times \frac{1}{n} \sum_{i=1}^{n} \frac{C_{i,t}}{C_{i,t-1}} \tag{1.11}$$

Although capitalization is not considered in the formula, changes in the geometric version of the Value Line index (VLIC) are more a reflection of changes in the price of medium to big capitalization companies, due to the composition of the index. On the other hand, just as with the DJIA, changes in the arithmetic Value Line index (VLIA) can be consequence of considerable changes in price of any company, regardless of capitalization. □

Comparing the performance of two or more stocks. As mentioned before, a way that investors have to gauge the performance of their stock investments is to compare their price behavior with another reference stock's price history, or usually with the history of the market index. More precisely, the comparison is made between the cumulative *rate of benefits* that one obtains throughout the investment period with the cumulative rate of benefits that one would have obtained in the same time period if the investment would have been made in the reference stock, or (hypothetically) in the market index. Note that it is important to consider the successive sums of rates of benefits because under this factor of price variation all stocks are measured at a same scale, so comparisons make sense. For any given price series $\{P_t : t \geq 0\}$ the rate of benefit (also known as *return*), from one time instant to the next, is given by $R_t = P_t/P_{t-1} - 1$. In Chap. 2 we analyze in depth the statistical properties of an asset's returns time series.

Example 1.5 As an example of using an index as benchmark for our investments, Fig. 1.3 shows the performance in percentage terms through 1 year of four U.S. companies, Coca-Cola (KO), Apple (AAPL), Mc Donalds (MCD), General Electric

---

[8] *source* Kansas City Board of Trade's webpage.

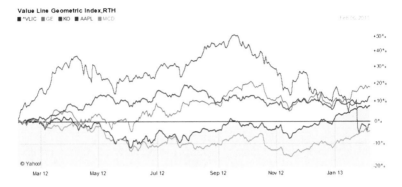

Fig. 1.3 VLIC vs KO, AAPL, MCD, GE (*source* Yahoo!)

(GE), and the Value Line Geometric Index. From the graph it is clear that GE and KO have outperformed the U.S. market by a median of approximately 10%; AAPL was over by 20% most of the year 2012, to drop below the market line in the first quarter of 2013; MCD has underperformed the whole year considered (from February 2012 to February 2013). The picture was produced with Yahoo! financial chart services, but one can build similar comparison charts in R. We will show how to do this in Note 1.3.4.   □

1.1.3 Options and Other Derivatives

Options. An option is a contract that investors can buy for a fee in order to have the opportunity, but not the obligation, to trade an asset in an exchange market at a future date, and at a given price. The date for making the trade, as established in the option, is known as the *exercise date*, *expiration date* or *maturity*. The price of the asset written in the option is known as the *exercise price* or *strike price*. If the trade in the option is to *buy* the asset, the option is known as a *call* (a *call option*); whereas if the option is giving the right to *sell* the asset, then it is known as a *put* (a *put option*). By *exercising the option* it is meant that the trade specified in the option (i.e. a call or a put) is to be carried out by the exercise date and at the exercise price. Depending on the way of exercising the option we have different *styles of options*. We list only those that will be treated later in this book.

European: these options may only be exercised on expiration date.

American: may be exercised on any trading day on or before the expiration date.

Bermudan: may be exercised only on predetermined dates before expiration; so this option is somewhat between European and American.

Asian: the payoff of these type of options is determined by the average of the price of the underlying asset during some period of time within the life of the option.

Barrier: these options may only be exercised if the price of the underlying asset passes a certain level (i.e. a barrier) during a certain period of time.

European and American options are classified as *vanilla*, a term that applies to options with direct and simple payoff conditions; other options with more involved conditions for exercising are termed *exotic*. Furthermore, the Asian and Barrier are *path dependent* options because for calculating their payoffs it is necessary to know their past price history. This characteristic adds a further difficulty for solving the problem of option pricing (a problem that will be defined precisely later in this section), and we shall see how to approach this problem with Monte Carlo methods in Chap. 5.

Reading an option quotes. The price of an option, also known as the *premium*, is given on a per share basis. An option on a stock corresponds to 100 shares. Thus, for example, if an option on stocks for GE (NYSE) has a quoted premium of 2.64, the total amount to pay for that option would be $264 (=2.64 × 100), and this will entitle the holder to buy or sell 100 shares of GE at maturity. The writer of an option on a stock presents for negotiation different possibilities of premiums which depends on the stock's current price, the time to maturity and the type. The options are then organized in series of different premiums, for a fixed maturity date, and quoted at the exchange market using the following standard nomenclature:

$$\begin{matrix} \text{Security} & & \text{Expiration} & & \text{Type} & & \text{Strike} \\ \text{Symbol} & + & \text{Date} & + & \text{Call/Put} & + & \text{Price} \end{matrix}$$

Figure 1.4 shows a series of options for GE, offered on the 2nd of November of 2012 and expiring on the 18th of January of 2014. For example, the symbol GE140118C00013000 stands for an option on GE, expiring on 2014-01-18, of type C (call), with strike price of 13. The last quoted premium for this option is $8.30. Observe the premium changes the further away the strike is from the current price. Calls are generally more expensive than puts. Options with maturity over a year, as presented in the table, are called LEAPS. For details on the factors affecting the price of stock's options read Hull (2009, Ch. 9).

Payoff and profit of options. The payoff of an option depends on the possibility of exercising it or not. If the option is exercised its payoff is the difference between the exercise price and the price of the asset at maturity, but if it is not exercised the payoff is 0. We can define precisely the payoff functions for the call and put contracts of vanilla options. If K is the strike price and P_T is the price of the underlying asset at the exercise date T, then the payoff for a call option is

$$\max(P_T - K, 0) \tag{1.12}$$

since we would only exercise the option if $P_T > K$. By analogous reasoning the payoff for a put option is

$$\max(K - P_T, 0) \tag{1.13}$$

The payoff functions for path dependent options depend on the conditions tying the strike price to past prices of the asset. For example, for an Asian call option, the payoff can be

General Electric Company (GE) NYSE   ⊕ Add to Portfolio   ▣ Like 812

21.31 ↓0.03(0.14%) Nov 2, 4:01PM EDT | After Hours : 21.33 ↑0.02 (0.09%) Nov 2, 7:34PM EDT

Options

View By Expiration: Nov 12 | Dec 12 | Jan 13 | Feb 13 | Mar 13 | Jun 13 | **Jan 14**
Options Expiring Friday, January 17, 2014

Calls							**Strike Price**	**Puts**						
Symbol	Last	Change	Bid	Ask	Volume	Open Int		Symbol	Last	Change	Bid	Ask	Volume	Open Int
GE140118C00003000	19.40	0.00	16.05	20.60	0	0	3.00	GE140118P00003000	0.02	0.00	N/A	0.03	0	30
GE140118C00005000	12.80	0.00	14.10	18.70	0	0	5.00	GE140118P00005000	0.08	0.00	0.01	0.05	100	1,471
GE140118C00008000	14.84	0.00	11.05	15.60	1	93	8.00	GE140118P00008000	0.15	0.00	0.07	0.11	0	909
GE140118C00010000	11.20	0.00	10.75	12.40	7	1,135	10.00	GE140118P00010000	0.18	0.00	0.13	0.16	1	13,136
GE140118C00013000	8.30	0.00	8.25	8.80	1	910	13.00	GE140118P00013000	0.36	0.00	0.33	0.36	18	22,115
GE140118C00015000	6.45	↑0.33	6.40	6.50	25	9,850	15.00	GE140118P00015000	0.57	↓0.01	0.58	0.60	145	28,667
GE140118C00017000	4.75	↑0.15	4.65	4.75	15	11,409	17.00	GE140118P00017000	0.96	↓0.07	0.98	1.00	22	56,357
GE140118C00020000	2.64	↑0.12	2.60	2.65	230	34,665	20.00	GE140118P00020000	1.98	0.00	2.00	2.04	47	26,293
GE140118C00022000	1.63	↑0.12	1.60	1.62	349	23,000	22.00	GE140118P00022000	3.05	0.00	3.00	3.10	945	5,108
GE140118C00025000	0.71	↑0.10	0.68	0.71	4,092	28,256	25.00	GE140118P00025000	5.15	0.00	5.10	5.20	3	4,778
GE140118C00027000	0.36	↑0.06	0.35	0.38	300	8,126	27.00	GE140118P00027000	5.80	0.00	6.50	6.90	1	781
GE140118C00030000	0.11	0.00	0.12	0.15	392	9,339	30.00	GE140118P00030000	9.45	↓0.25	9.55	9.70	10	1,880
GE140118C00035000	0.04	0.00	0.02	0.04	60	3,410	35.00	GE140118P00035000	12.65	↓1.55	14.30	14.65	1	310

Fig. 1.4 Options on GE expiring January 2014 (*source* Yahoo!)

$$\max(A(T_o, T) - K, 0) \tag{1.14}$$

where $A(T_o, T)$ is the average price of the asset from date T_o to date T. Other variants can be obtained by putting weights to this average $A(T_o, T)$ or different ways of computing it (e.g. as a geometric average instead of arithmetic.)

To compute the profit on options we must take into consideration the fee or commission given at the time of settling the agreement (the *price* of the option), which depends on the price P_0 of the underlying asset at that initial time t_0 and the time T set for maturity. Hence, we denote by $C(P_0, T)$ the price of the option, which is always a positive amount. Then again, as with stocks, we should deduct the price paid for the option from the payoff, but considering that these amounts are given and received at different points in time, and that we could have done a safe investment instead, we must include the possible gains given by a risk-free asset earning a constant interest rate r for the period $\tau = T - t_0$ and continuously compounded. Therefore, under all these considerations, the profit of a call option is

$$\max(P_T - K, 0) - C(P_0, T)e^{r\tau} \tag{1.15}$$

The equations for computing the profits of a put and other options are obtained with similar reasoning, and are left as exercises.

There are at least two advantages from using options on stocks as opposed to buying common stocks directly. One is that we can buy (or reserve to buy at later date) more shares with less money, since we only pay a premium which is always less than the price of the underlying stock. This fact should be clear from common sense (why would one want to buy an option on a stock for a price greater or equal to the price of the stock?); but this can be proven formally with arguments based on

Table 1.2 Profits for the call option and for holding 100 shares of the stock

Stock price	Profit for 100 shares	Call	
		Price	Profit
5	−1500	0	−200
10	−1000	0	−200
15	−500	0	−200
20	0	0	−200
22	200	2	0
25	500	5	300
30	1000	10	800
40	2000	20	1800

no arbitrage (Sect. 1.2.3 below and Problem 1.3.6). The second advantage of using options on stocks is that we have in advanced a limit on losses, while we can still make theoretically unlimited profits. To best appreciate the benefits of options it is good to build different investment scenarios and look at the resulting *profit graphs*. (To simplify calculations it is common practice to deduct only the initial investment from the payoff to obtain the profit; or equivalently, to assume that the risk-free interest rate $r = 0$.)

Example 1.6 Suppose that common stock of company XYZ is trading today at €20, and a call option on XYZ with strike $K = 20$ and a month of maturity is selling for $C_0 = 2$, or €200 (= 2 × 100). Buying one call option would have at expiration a loss on the downside of the price of the stock limited to −200, since we would not exercise the option. On the contrary, buying the stock could have a future loss of −2000, in the worst scenario where the stock losses 100 % of its value. Table 1.2 shows a comparison of profits for the call versus holding directly 100 shares of the stock, for different prices of the stock. Figure 1.5 shows the corresponding profit graphs for the call (solid line) and the stock (dashed line) with their evolution of value as described in the table. □

Options are also useful for creating different trading strategies with a desired profit diagram, in combination with other securities or alone.

Example 1.7 The *straddle* strategy consists on buying a call and a put options on a stock with same strike price and expiration date. Figure 1.6 presents the profit graph (solid line) of this strategy resulting from the sum of the two profit graphs (dashed lines), one for the call (in blue) with premium C_0 and the other for the put (in red) with premium P_0, and common strike price K. If the price of the stock at maturity, S_T, is close to the strike price K (i.e $|S_T − K| < C_0 + P_0$) then the straddle strategy produces a loss, whereas if S_T moves far away from K, in either positive or negative direction, then there could be significant profits. Thus, a straddle strategy should be used if one expects in the future a significant variation of the price of the stock but is uncertain about the direction it will move. □

Other popular options' strategies are reviewed in the Notes.

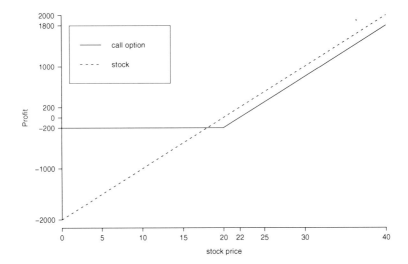

Fig. 1.5 Profit graphs for the call and the stock

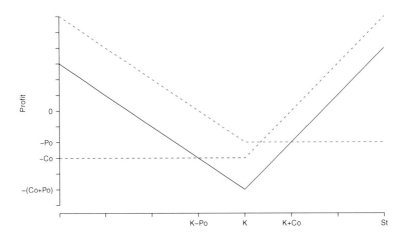

Fig. 1.6 The straddle strategy

The option pricing problem. In entering an option contract both parties involved assume a financial risk. The holder of the option assumes the payment of the premium which could be a financial lost if the option is not exercised. On the other hand, if the option is exercised, the writer incurs in a lost for financing the underlying asset at a price worse than the market price at the time of delivery. Consider, for example, an European call option on a stock. If at exercise time T, the price of the stock S_T is less than or equal to the strike price K, the holder of the option would not exercised it and loses the premium. If $S_T > K$ the option would be exercised and the writer must deliver a share of the stock for K units of cash, loosing the difference with the real price S_T. Thus, the problem of pricing an option is to determine a price for the option contract that is deemed fair for the writer and the holder; that is, an amount of cash that the buyer should be willing to pay and the seller should charge and not systematically lose.

The solution to this problem seems at first to depend on the possibly different attitudes to risk held by each party of the option contract, and therefore, we can not pretend to have a "universal" solution: one formula good for all buyers and sellers of options. However, we would see in Chap. 4 that such universal formulas are possible, under certain assumptions pertaining to market equilibrium that will be reviewed in Sect. 1.2.3.

Forward and futures contracts. Forward and futures contracts are derivatives that differ from options essentially in their terms of liability. In these contracts two parties subscribe an obligation to trade a specific asset (one party to buy it and the other to sell it), for a specified price (the *delivery* price) and at a specified future date (for the forward contract) or time period (e.g. a month, for the futures contract). The asset specified in the contract is usually any commodity (i. e. agricultural products or raw materials), currencies, or bonds, but could also be any stock.

Forward contracts are not traded on exchange markets while futures contracts are traded on exchanges. This imply that the delivery price in forward contracts is settled by the two parties involved, while in futures contracts is marked to the market on a regular basis. A forward contract can be used by an investor (the buyer) to lock in a future purchase of some asset from a supplier (the seller) at a price settled today, without actually paying the full amount up until the date of expiration or maturity on the contract. Futures contracts are used as investment instruments and traded at the Chicago Mercantile Exchange, the London International Financial Futures Exchange, the Tokyo Commodity Exchange, and about one hundred other futures and futures options exchanges worldwide.

Payoff and profit. If K is the delivery price of the asset written in the contract and P_T is the price at exercise date T, the payoff for the buyer in the contract on one unit of the asset is $P_T - K$, while the payoff for the seller on one unit of the asset is $K - P_T$. Since there are no fees for writing a forward or futures contract, nor initial investment, the profit function for either party is the same as the payoff, and this can be positive or negative at maturity.

Now, as with options, one important problem is to assign a value to these contracts that should be considered fair for both parties involved. At the time of entering any of these contracts the obvious fair value for both parties is zero, which means that the delivery price is equal to the price of the underlying asset. This is because neither party, with the current information at hand, is assuming any risk. But as time progresses, while the delivery price remains the same, the price of the asset may change producing an imbalance in the odds of profits for both parties. The fair situation for both parties is that at maturity the price of the asset is in line with the delivery price. Therefore, for a forward (resp. future) contract one defines its *forward price* (resp. *future price*), for any time t, as *the delivery price that gives the contract a value of zero*. We later show how to compute the forward price of a forward contract, and leave to the reader the task of doing the bibliographic research on the problem of settling future prices.

1.1.4 Portfolios and Collective Investment

A *portfolio* is a collection of one or more securities owned by an investor or an investment company. The elements of a portfolio are also referred to as the *positions*. The value of a portfolio at any given time is the sum of the values of its positions at that point in time, and the profit at a certain time is the sum of the profits given by its positions at that time.

Portfolios are designed to fit certain investment objectives and aiming at accomplishing the investor's expected rewards which are in continuous confrontation with his tolerance to risk. The management of portfolio risk involves *opening* new positions (i.e. adding positions) with better expected reward, and closing positions in the portfolio. To *close out* a position can mean to sell it in the market, or to add to the portfolio the inverse of the position so that its future value performance does not affect the future value of the portfolio. For example, a forward contract to buy an asset for a certain delivery price and given maturity, can be close out by another forward contract to sell the same asset (and same amount of it) for the same delivery price and maturity.

This dynamic of opening and closing out positions in a portfolio is known as *asset allocation*, and in determining a proper asset allocation one aims to minimize risk and maximize profits. This optimization problem will be treated in detail in Chap. 8.

Mutual funds and ETF. An alternative to personally manage your own portfolio is to put your money, alongside other investors money, in a collective investment vehicle managed by a professional company that does the selection of securities, monitors their performance, and the subsequently asset allocation of the portfolio. The most common type of these collective investment instruments are the *mutual funds*, characterize by being regulated and registered in the exchange market, publicly available and either closed-end (meaning that the number of shares in the portfolio is fixed and hence, once all shares are sold, no new investor can enter the fund) or open-end (meaning that there are no restrictions on the amount of shares that can be issued, and hence investors can enter the fund anytime, or exit by selling their shares back to the fund).

A further variety is the Exchange-Traded Fund (ETF) which is like a closed-end fund with the added feature that it is traded in the stock exchange. Thus investors can buy or sell shares of an ETF, that is representing a large collection of stocks, just as buying or selling shares of a single stock directly in the market. There are many ETFs build with the purpose of tracking an specific index (i.e. a portfolio of stocks that replicates the index's price behavior), so this is a way for an investor to trade on the index.

1.2 Financial Engineering

We have so far acquired knowledge on various securities and how to trade them at exchange markets, so our next immediate concern should be to form some investment criteria. What attitudes should we have towards investment? What is a fair price to pay for a security? How to control risk and secure some benefits? We introduce these matters in this section and expand them further in later chapters.

1.2.1 Trading Positions and Attitudes

Long and short positions. In the financial parlance, when an investor owns a security it is said that she is *long* in that security, and when she owes the value of a security she is *short* in that security. Moreover, an investor who buys a security is assuming a *long position* on the security, and that who sells a borrowed security is assuming a *short position*. For example, the buyer of a call option is long on the option, while the writer of that option is short. In a forward contract, one of the parties assumes a long position by agreeing to buy the underlying asset while the other party assumes a short position by agreeing to sell it. The buyer of a stock opens a long position on the stock. But selling a stock one owns does not means to become short on that stock, it just means to close the long position. To have a short position on a stock one has to borrow shares of the stock from some broker and sells them in the exchange market. Then at a later time one buys the shares back to return them to the lender. There will be profit if by the time one has to return the shares the price is less than it was at the time the shares were borrowed and sold in the market. This operation is known as *short selling*, and is highly regulated and sometimes banned by governments because it is believe to contribute to the downfall of market prices.[9]

The general rule for profiting from a long or short position is the following:

> Being long in any security implies to have a profit if the price increases, and a lost if the price decreases; whereas being short carries a profit if the price of the security decreases, and a lost if it increases.

Combining long and short positions is a way of reducing losses and possibly increasing benefits.

Example 1.8 (**A portfolio insurance strategy**) Suppose an investor is long on stocks from XYZ. Then to cover himself from possible losses, in the event that the price of the stock goes down, he can simultaneously enter into a long position on a put option (i.e. a *long put*) for similar amount of XYZ shares. In this way the investor is sure to

---

[9] cf. Note 1.3.3.

sell part of the stock at the strike price of the put. For this strategy it is preferable to have the flexibility of an American put option.   □

Hedgers, speculators and arbitrageurs. Hull (2009, §1.5) identifies three types of traders: hedgers, speculators and arbitrageurs. Hedgers trade so as to reduce or eliminate the risk in taking a position on a security; their main goal is to protect the portfolio from loosing value at the expense of lowering the possible benefits. This attitude (or trading strategy) is called *hedging*. An hedging strategy usually involves taking contrarian positions in two or more securities. For example, taking a long position on a stock and a short position on another stock with inverse price behavior; or the portfolio insurance strategy described in Example 1.8. The easiest, and for most investors realistically possible, hedging strategy is to buy an *Inverse ETF*[10] on the market index or industry where their portfolios have taken most of their positions. Thus, a fall (respectively, rise) on the value of the portfolio is compensated by a proportional rise (fall) of the Inverse ETF.

Speculators take risk on purpose by betting on future movements of the security's price. For example, an speculator buys a stock under the conviction that its price will rise.

Arbitrageurs are traders that take advantage of a price difference between two or more markets. They look for an imbalance in the pricing of a security in two different markets, and buy it at the cheap price in one market to immediately sell it at the higher price in the other market, making a profit on the difference. This procedure is called *arbitrage*. For example, suppose that the exchange rates of EUR/USD in Frankfurt quotes at $1.25, while in New York it quotes at $1.3 at some instant of time[11]; hence, the euro is undervalued in Frankfurt and overvalued in New York. An arbitrageur would exploit this imbalance by exchanging dollars for euros in Frankfurt and immediately exchange the euros for dollars in New York. The profit for each dollar in this double exchange is 0.8–0.769 = 0.031 or a 3 % on the amount invested. In real life, there are transaction costs that would probably reduce considerably the profit, but moreover, it is natural to believe that this opportunity to profit from the different valuations on the exchange rates EUR/USD would come to the attention of more than one investor, and once a few of these alerted investors act upon this chance of easy profiting, the exchange rates EUR/USD will equal at both sides of the Atlantic, and hence terminating the arbitrage opportunity. In fact, by the same argument that there will always be investors ready to act upon this or any other arbitrage opportunities, and that by doing so the chance of profiting from it will be immediately diluted, we must admit that the possibility of arbitrage is in practice non existent, as common belief holds. This assumption of the impossibility of arbitrage opportunities (or no arbitrage) is the basis for many arguments for valuing derivatives.

---

[10] This is a ETF that simulates short positions on various stocks or indices.

[11] Exchange rates of (Currency A)/(Currency B) gives the amount of Currency B that exchanges for one unit of Currency A. Thus, we are saying that €1 = $1.25 in Frankfurt while €1 = $1.3 in New York.

Bulls and bears. When the market trend is upward, as reflected by the main index or the majority of its composite stocks, it is said to be *bullish*, or that the bulls are in charge. On the contrary if the market trend is downward then it is *bearish* or the bears have taken over. Thus, *bulls* are those investors that take long positions in stocks whose prices are in uptrend, while *bears* are investors that go short in stocks whose prices are in downtrend. So here is your first trading principle:

> When the market is bullish you should assume long positions; and when bearish you should assume short positions.

...or in doubt, go both long and short, which is a form of hedging. Nonetheless, determining if the market is bullish or bearish, which boils down to determining the expected slope of the trend and its sustainability in time, involves loads of subjective judgement. Here is what the Vanguard group thinks a signal of bearish market should be:

> While there's no agreed-upon definition of a bear market, one generally accepted measure is a price decline of 20% or more over at least a two-month period.[12]

Market timing or buy-and-hold. As for possible trading strategies, at one end of the wide range of possibilities is the passive attitude, followed by the majority of investors, consisting on buying various securities and maintaining them in the portfolio for a long time. What sustains this static strategy is the observed tendency of almost all securities to increase their value if sufficient time is given to them to mature. This type of strategy is known as *buy-and-hold*. On the other hand, while waiting patiently for his investments to reach the desired rate of benefits, an investor will surely observe how the price roller-coasters through time, and may wonder if returns would be greater (and obtained quicker) by selling at the peaks and re-buying at the troughs. There are many empirical reports sustaining the increase in benefits of applying this timely sequence of entries and exits in the shareholding of stocks; provided, of course, that the times of trading are rightly chosen.[13] Various methods exist for devising these *market timing strategies*, which include using models for forecasting returns, analysis of structural properties of prices and of the business economic fundamentals, and others. These subjects are treated in the following chapters. In any case, whatever the methodology chosen for constructing a market timing trading strategy, one should expect that at least it should beat the results obtained by doing no intermediate trade between the first buy and the final sell. In other words, the general accepted test of performance of a market timing strategy is to compare its results with the buy-and-hold strategy.

---

[12] Vanguard Group,"Staying calm during a bear market". https://retirementplans.vanguard.com/VGApp/pe/PubVgiNews?ArticleName=Stayingcalmbearmkt.

[13] Shilling (1992) describes a hypothetical situation where a smart investor applying an optimal market-timing strategy could have earned a 26.9% annual gain from 1946 to 1991 in the Dow Jones, as opposed to the 11.2% yearly gains given by buy-and-hold.

1.2.2 On Price and Value of Stocks. The Discounted Cash Flow model

Now that we have learned the basic tricks of the trade at exchange markets, it is appropriate to stop for a moment and reflect upon what is it that we are paying for when we buy shares of a stock at the market. Today it has become so easy and fast to buy stocks that the sense of partnership in a business, intrinsic in being a shareholder, is almost lost among investors and replaced by the thrill of participating of the sudden gains or losses registered in the quotes table, and nothing more. The company behind the stock is not as much considered as the figures quoted at the market, when it comes to buying, or selling, by many investors. Consequently, the price of a stock at the exchange market, more often than not, does not reflects the worth of its business and is completely determined by what people are willing to pay for it. Thus, keep well framed in your mind this basic fact about the price of a stock quoted at a exchange market:

> The price of a stock is determined by the forces of supply and demand from investors.

But, is the quoted price of a stock what it really worths? For an investor the value of a stock depends on the possible future capital gains that can be made from investing in it. In principle, the future gains come from the expected dividend per share and the expected price appreciation per share. Then to have an estimate of the *present value* S_0 of a share of a stock, say 1 year ahead, taking into account these sources of capital gain, the most common and simple model is to estimate the expected future payoff (future price plus dividend), and disregarding transaction costs and considering a constant discount rate r per year, discount this cash flow back to present to obtain

$$S_0 = \frac{D_1 + S_1}{1 + r} \tag{1.16}$$

where D_1 and S_1 are the expected dividend and expected price per share at the end of a year. The next year price per share can be forecasted also by applying the same discount formula, so that $S_1 = (D_2 + S_2)/(1 + r)$, where D_2 and S_2 are the expected dividend and price per share at the end of 2 years. Putting this forecast for S_1 in Eq. (1.16), one gets

$$S_0 = \frac{D_1}{1 + r} + \frac{D_2}{(1 + r)^2} + \frac{S_2}{(1 + r)^2}$$

which determines the present value of the stock in terms of expected yearly dividends, for the next 2 years, and expected price at the end of second year, everything discounted by a constant rate. The argument can be repeated for an arbitrary number T of years, and doing so one gets

$$S_0 = \sum_{t=1}^{T} \frac{D_t}{(1+r)^t} + \frac{S_T}{(1+r)^T} \tag{1.17}$$

where D_t is the expected dividend for year $t = 1, \ldots, T$, and S_T is the expected price at the end of the investment horizon. At this point, one should think that prices could not grow for ever while the discount rate keeps accumulating; hence, it is reasonable to assume that the expected discounted future price approaches zero as the investment horizon T goes to infinity: $\lim_{T \to \infty} S_T/(1+r)^T = 0$. Then, taking limits to infinity in Eq. (1.17) one obtains

$$S_0 = \sum_{t=1}^{\infty} \frac{D_t}{(1+r)^t} \tag{1.18}$$

This is known as the *discounted cash flow* (DCF) formula for the present value of a stock, which represents the actual price as a perpetual stream of expected cash dividends. In order to make estimations with it, or with the partial sum given by Eq. (1.17), we should first make precise the nature of the discount rate r. In principle, it could be taken as the risk-free interest rate of a high grade bond, but that would be an underestimation and does not reflects the investors acceptance of risk. Consider instead resolving r from Eq. (1.16):

$$r = \frac{D_1 + S_1}{S_0} - 1$$

This says that r is the expected return for the stock, or it could also be the expected return of any other asset with similar risk as the one we are analyzing. The latter is known as the *market capitalization rate* or *cost of equity capital*, and it is a more accurate interpretation for the discount rate r. Thus, with an estimation of future gains at hand, knowing the cost of equity capital we can calculate the present value of the stock; or knowing the present (intrinsic) value of the stock we can calculate its expected return. The modeling of expected returns and forecasting of price are not simple matters, and are subjects of a large corpora of research papers; both are main themes of study in this book.

Example 1.9 Suppose Exxon Mobile (XOM:NYSE) is selling for $90.72 today. Let's take that as its present value. The expected dividend over the next year is $2.52 and the expected 1 year future price is estimated to be $95.17. Then the expected return for Exxon is $r = (2.52 + 95.17)/90.72 - 1 = 0.076$, or 7.6%.

On the other hand, if the market capitalization have been estimated to be of 11% (e.g. as the average expected return of stocks from companies in the major integrated oil and gas industry), take $r = 0.11$ and as before $D_1 = 2.52$ and $S_1 = 95.17$. Then today's price of Exxon should be $S_0 = (2.52 + 95.17)/(0.11 + 1) = \$88$. This is a fair price to pay with respect to the market capitalization rate of the oil and gas industry. $\square$

A simpler closed formula for estimating present value can be obtained from Eq. (1.18) with the following argument: one just does an estimation of next year dividend D_1, and for the following years assumes that this dividend grows at a constant rate g. Thus, $D_2 = D_1(1+g)$, $D_3 = D_1(1+g)^2$, and in general, $D_t = D_1(1+g)^{t-1}$. Then, Eq. (1.18) becomes

$$S_0 = \sum_{t=1}^{\infty} \frac{D_1(1+g)^{t-1}}{(1+r)^t} \tag{1.19}$$

Now, this growth rate of dividends can not exceed the market capitalization rate for too long, otherwise the company's stock becomes a riskless investment ...perpetually!.[14] Therefore, it is fair to assume $g < r$, and the right hand side of Eq. (1.19) is a geometric series with finite sum[15]:

$$S_0 = \frac{D_1}{1+g} \sum_{t=1}^{\infty} \left(\frac{1+g}{1+r} \right)^t = \frac{D_1}{1+g} \cdot \frac{1+g}{r-g} = \frac{D_1}{r-g} \tag{1.20}$$

This formula, based on growing perpetually, expresses the present value as the first year expected dividend divided by the difference between the perpetual market capitalization rate and the perpetual dividend growth rate. In practice, financial analysts compute some fixed long term return and dividend growth rate, usually on a 5 years period, and make a leap of faith into perpetuity.

Example 1.10   Let us apply the perpetual growth formula (1.20) to compute present value of Exxon. We have, from Example 1.9, that $D_1 = 2.52$ and the market capitalization rate $r = 0.11$. Using data from Yahoo finance services we find a 5 years dividend growth rate[16] of 8.33 % for Exxon, and use this as the company's perpetual dividend growth rate, so that $g = 0.083$. Then $S_0 = 2.52/(0.11-0.083) = \$93.3$.

Indeed, the reader should be amused with the discrepancies between this and previous calculations. But let's not forget that Eqs. (1.16) and (1.20) are based on different time horizons: while the former assumes a 1 year discount rates, the latter is based on an endless discount rate accumulation. Nevertheless, all these present value estimations should be taken as rough approximations, as they are highly dependent on the long term forecasts of growth and market capitalization on which they are based. We emphasize this fact in the remark that follows.   □

Remark 1.1   Be aware of the sensitivity of the perpetual growth formula (1.20) to the parameters g and r, and its possible prone to error. If the estimations of g or r in Example 1.10 differ from the ones considered by just 1 %, then the present value

---

[14] It becomes an endless *arbitrage opportunity* against the cost of capital determined by the market. This is too good to be true, and as we shall also argue against such possibility in the next section.

[15] Recall that $\sum_{t=1}^{\infty} a^t = \frac{a}{1-a}$, for $0 < a < 1$.

[16] A way that some analysts have to compute this long term dividend growth rate is to multiply the long term return on equity (ROE) by the ratio of reinvesment after payment of dividends per share's earnings. We shall give details on this formula in Chap. 6.

can change for as much as 35%. Take, for example, $r = 0.12$ and g as before, then $S_0 = 2.52/(0.12-0.083) = 68.1$. Now take $g = 0.09$ and $r = 0.11$, then $S_0 = 2.52/(0.11-0.09) = 126$. The figures said all: results of constant discount (in perpetuity or long term) cash flow models should be treated as approximations, often overshot, to the true value.   □

For a probably more exact estimate of present value of a stock one must introduce into the equation considerations pertaining the economy in general, and the profitability of the company's business in particular. The latter suggest to equate the value of the stock with the value of the company behind the shares, measured in terms of its earnings, debt and other financial figures. This is a valuation based on a company's financial fundamentals, a topic that we will treat in Chap. 6. For a survey of more refine forms of *discounted factor models* (for which the DCF is a particular case) see Campbell (2000).

1.2.3 Arbitrage and Risk-Neutral Valuation Principle

To obtain closed solutions to the option pricing problem, we need to make yet some other economic assumptions and upon these economic scenarios develop some mathematical models for the price behavior of the underlying stock. A discounted cash flow model as Eq. (1.18) for options, where a rate of possible returns is discounted, is not possible because it is very hard to estimate expected returns on the option. For options the expected benefits depends on the underlying stock price relative to the exercise price. When the stock price falls below the option's exercise price, investors will deem the option riskier (and less profitable), and if the price of the stock were over the exercise price then it will be regarded as safer (and more profitable); hence, investors expected returns on options can be hardly considered as evolving at constant rate through any period of time. We must elaborate our models upon other economic hypotheses.

A fundamental economic assumption underlying many mathematical models of price is that it is not possible to make a profit without a risk of loosing money; that is not possible to get something from nothing; or as Milton Freedman stated bluntly: *"There is no such a thing as a free lunch"*.[17] This assumption is formally summarized as the *principle of no arbitrage*.

Principle of No Arbitrage: there are no arbitrage opportunities.

The no arbitrage principle is usually accompanied with other mild assumptions to create a much controlled economic situation, where the wheat can be isolated from

---

[17] This is the title used by Milton Freedman for one of his economic books, published in 1975, and which helped popularized the "no free lunch" adage.

the chaff, making possible some very clean pricing models. We simply named this set of assumptions expanding no arbitrage the *extended no arbitrage*.

Definition 1.1 *(Extended no arbitrage)* The *extended no arbitrage* is the following list of assumptions about the market of securities:

- Arbitrage is not possible.
- There are no transaction costs, no taxes, and no restrictions on short selling.
- It is possible to borrow and lend at equal risk-free interest rates.
- All securities are perfectly divisible.

A consequence of working under the assumption that extended no arbitrage holds is that we can determine precisely when a portfolio is replicating the value through time (or the time value) of another portfolio. Given a portfolio $\mathscr{A}$ and a time instant t, let $v(\mathscr{A}, t)$ be the value of $\mathscr{A}$ at time t. We have the following result.

Proposition 1.1 *Assuming extended no arbitrage, two portfolios with equal value at time T must have equal value at all time $t \leq T$.*

Proof Let $\mathscr{A}$ and $\mathscr{B}$ be two portfolios such that $v(\mathscr{A}, T) = v(\mathscr{B}, T)$, and assume that for some $t' < T$, $v(\mathscr{A}, t') < v(\mathscr{B}, t')$. Then at that time t' an investor buys all positions of $\mathscr{A}$, short sells every position of $\mathscr{B}$ and invest the difference $D(t') = v(\mathscr{B}, t') - v(\mathscr{A}, t')$ in some bond at the fixed rate r. This combined portfolio is summarized in the following table:

Portfolio
Long on A
Short on B
Bond with principal $D(t')$ at interest r

The new portfolio has at time t' a value of

$$v(\mathscr{A}, t') - v(\mathscr{B}, t') + D(t') = 0$$

which means that this investment has no initial cost. At time T, the bond in the portfolio has matured its value to $D(T) = D(t')e^{r(T-t')}$. Then the value of the portfolio at time T is

$$v(\mathscr{A}, T) - v(\mathscr{B}, T) + D(T) = D(t')e^{r(T-t')} > 0$$

Thus, the investor got something from nothing! This contradicts the no arbitrage hypothesis. Therefore it must be that $D(t') = 0$ and that $v(\mathscr{A}, t) = v(\mathscr{B}, t)$ for all $t \leq T$ □

By a similar proof one can show the following generalization.

Proposition 1.2 *Assuming extended no arbitrage, if $\mathscr{A}$ and $\mathscr{B}$ are two portfolios with $v(\mathscr{A}, T) \geq v(\mathscr{B}, T)$ then, for all time $t \leq T$, $v(\mathscr{A}, t) \geq v(\mathscr{B}, T)$.* □

Proposition 1.1 provides us with the following *replicating portfolio* technique to price a derivative. Given a derivative whose value at a given time we want to know, build a portfolio $\mathscr{A}$ containing the derivative and possibly other securities. Build another portfolio $\mathscr{B}$ containing securities with known values and such that the total value of $\mathscr{B}$ is the same as the value of $\mathscr{A}$ at a future time T. Then, assuming (extended) no arbitrage, the values of $\mathscr{A}$ and $\mathscr{B}$ must be equal at all time $t \leq T$. We can then deduce the unknown value of the security from the equation of values of $\mathscr{A}$ and $\mathscr{B}$. We say that portfolio $\mathscr{B}$ *replicates* portfolio $\mathscr{A}$. The next example shows how to use the replicating portfolio method to price a call option on a stock.

Example 1.11 We want to price a call option on stock XYZ, with 6 months to maturity and exercise price $K = €115$. Assume the current price S_0 of a share of XYZ is equal to the option's exercise price, that is, $S_0 = €115$ today.[18] Assume that at the time of delivery, the price of XYZ must be either $S_d = €92$ (i.e. 20% lower) or $S_u = €149.5$ (30% higher). (We can think of S_d and S_u as the lower and upper bounds in a range of expected values that we have estimated using some mathematical model of price, as the ones we will study in Chap. 4.) Assume the risk-free interest rate for lending and borrowing is 3% per year, which amounts to 1.5% for a 6 months period. So, if we let $r' = 0.03$ be the yearly interest rate, we are considering the interest rate for only half of the period, that is $r = r'/2 = 0.015$.

We then have a portfolio $\mathscr{A}$ consisting of one call option on the stock XYZ, whose current price C we want to compute. The payoff of the option depends on the two possible prices S_d and S_u of the stock at maturity, and given by $\max(S - K, 0)$. The payoff can be either,

$$C_u = S_u - K = 149.5 - 115 = 34.5 \quad \text{or} \tag{1.21}$$
$$C_d = 0 \quad (\text{since } S_d < K).$$

Build another portfolio $\mathscr{B}$ with

- Δ shares of the stock, and
- a loan from the bank for the amount of $\widehat{B}$ euros,

where

$$\Delta = \frac{C_u - C_d}{S_u - S_d} \quad \text{and} \quad \widehat{B} = \frac{\Delta S_d - C_d}{1 + r} = \frac{\Delta S_u - C_u}{1 + r} \tag{1.22}$$

(The formula for Δ will be explained in Chap. 4, Sect. 5.2.2. At the moment take it as a recipe for building the appropriate replicating portfolio $\mathscr{B}$. The formula for $\widehat{B}$ is

---

[18] In financial parlance when such equality occurs one would say that the option is *at the money*. If it were the case that $S_0 > K$ then one says that the call option is *in the money*, and when $S_0 < K$ it is *out of the money*. What these expressions indicate is the possibility of making money or not if one were to exercise the option.

because we want to leveraged the position in Δ shares, so we borrow the difference between the payoffs from Δ shares and the payoffs from the option, discounting the interests.[19])

The payoff of $\mathscr{B}$ at the time of expiration of the call has two possible values, depending on the final price of the stock, which is either $S = S_d$ or $S = S_u$, and is given by subtracting the loan plus interests from the payoff of Δ shares of the stock. Formally, the payoff function for $\mathscr{B}$ is

$$\Delta S - (1+r)^t \widehat{B}$$

where $t = 1$ at the end of the period, and $t = 0$ at the beginning. In our scenario $\Delta = 34.5/57.5 = 0.6$ and $\widehat{B} = (0.6)92/1.015 = ((0.6)149.5 - 34.5)/1.015 = 55.2/1.015 = €54.38$. Hence, at the expiration time of the call, the payoff of $\mathscr{B}$ is either

$$\Delta S_u - (1+r)\widehat{B} = (0.6)149.5 - 55.2 = 34.5 \quad \text{or}$$
$$\Delta S_d - (1+r)\widehat{B} = (0.6)92 - 55.2 = 0$$

which are the same payoffs as obtained in Eq. (1.21) for portfolio $\mathscr{A}$. By Proposition 1.1 both portfolios must have the same value at all times. This implies

$$C = \Delta S - (1+r)^t \widehat{B}$$

In particular, at the time of writing the call ($t = 0$), when the stock's price is $S_0 = K = €115$, we have $C = (0.6)115 - 54.38 = €14.62$. This is the price of the call option on XYZ that we wanted to know, obtained under the extended no arbitrage assumption. □

Using the replicating portfolio method we can compute the price for a forward contract.

Theorem 1.1 (Pricing a forward contract) *At any time t the forward price F_t of a forward contract to buy an asset which has a price S_t at time t, with delivery price K, maturity date T, and considering a constant interest rate r through the life of the contract is either:*

(1) $F_t = S_t e^{r(T-t)}$, if the asset pays no dividends and has no costs.
(2) $F_t = (S_t - D_t)e^{r(T-t)}$, if the asset pays dividends or has some costs whose discounted total value is D_t.
(3) $F_t = S_t e^{(r-q)(T-t)}$, if the asset pays interests at a continuous rate of q.

Proof We prove case (1) and leave the other cases as exercises. We consider the asset to be a stock. We enter into a forward contract to buy a stock whose price is S_t,

---

[19] We use $\widehat{B}$ to distinguish it from the B that the reader will encounter in Chap. 4, in our discussion of the Binomial model. The relation is $B = -\widehat{B}$, because $\widehat{B}$ is for borrowing while B is for lending.

at some time t; the delivery price in the contract is K, and maturity date T. At this delivery time T, our portfolio $\mathscr{A}$ consisting of this forward contract has as value, $v(\mathscr{A}, T)$, the payoff

$$v(\mathscr{A}, T) = S_T - K$$

This is the same value of a portfolio $\mathscr{B}$ with a long position on one share of the stock, and a short position on some bond with principal value K (which have to be paid back at maturity). By Prop. 1.1, for all $t \leq T$, $v(\mathscr{A}, t) = v(\mathscr{B}, t)$. At any time $t \leq T$, we have $v(\mathscr{B}, t) = S_t - Ke^{-r(T-t)}$, since we have to apply the discounting factor for the value of the bond at time t. On the other hand, by definition, the forward price F_t is at any time t the delivery price such that $v(\mathscr{A}, t) = 0$. Therefore,

$$0 = v(\mathscr{A}, t) = v(\mathscr{B}, t) = S_t - F_t e^{-r(T-t)},$$

and the result follows. $\square$

Another important consequence of no arbitrage, where again the method of replicating portfolios is employed for its proof, is the put-call parity for European options.

Theorem 1.2 (Put-Call parity for European options) *Let $C(S_t, \tau)$ and $P(S_t, \tau)$ denote respectively the value of a call and a put options at time instant t, with same strike price K, and maturity date T; $\tau = T - t$ is the time to maturity and S_t the price of the underlying asset at the instant t. Let r be the fixed continuous interest rate.*

(1) If the asset pays no dividends and has no costs, then at all times $t \leq T$,

$$P(S_t, \tau) + S_t = C(S_t, \tau) + Ke^{-r\tau} \tag{1.23}$$

(2) If the asset pays dividends or has some costs with discounted total value D_t, then at all times $t \leq T$,

$$P(S_t, \tau) + S_t = C(S_t, \tau) + D_t + Ke^{-r\tau} \tag{1.24}$$

(3) If the asset pays interests at a continuous rate of q, then for all $t \leq T$,

$$P(S_t, \tau) + S_t e^{(q-r)\tau} = C(S_t, \tau) + Ke^{-r\tau} \tag{1.25}$$

Proof Once again we assume for simplicity that the underlying asset is a stock. And once again, we prove (1) and leave (2) and (3) as exercises.

On the left hand side of Eq. (1.23) we have the value at time $t \leq T$ of a a portfolio $\mathscr{A}$ containing a put and one share on the same stock. Let $\gamma_t = C(S_t, \tau)$ and $\pi_t = P(S_t, \tau)$. At expiration date, portfolio $\mathscr{A}$ has a payoff

$$\pi_T + S_T = \max(K - S_T, 0) + S_T$$
$$= \max(S_T - K, 0) + K = \gamma_T + K$$

We have that at time T, portfolio $\mathscr{A}$ has same value as a portfolio $\mathscr{B}$ containing a call on the same stock and some cash amount K (or a bond with principal of K). By no arbitrage and Proposition 1.1, both portfolios must have same value at any prior time $t \le T$. The value $v(\mathscr{B}, t)$ of portfolio $\mathscr{B}$ at $t \le T$ must take into account the continuous discount rate to time t for the cash K. This is e^{-rt}. So we have that for $t \le T$, $\pi_t + S_t = \gamma_t + Ke^{-rt}$.

Example 1.12 We use the pull-call parity relation to compute the price of a put option on the stock XYZ with the conditions presented in Example 1.11. Note that implicitly we have assumed the option as European since we want to know the price when we exercise the option at the maturity date. We had that $S_0 = K = €115$, and the interest rate for the 6 months is $r = 0.015$. The value of the call computed was $C = €14.62$. Then, by Eq. (1.23), the value P of the corresponding put is

$$P = C + Ke^{-r} - S_0 = 14.62 + 115(e^{-0.015} - 1) = €12.91 \qquad \square$$

Risk-neutral valuation. Another way to value an option can be derived from the assumption that investors are indifferent about risk. This is a strong hypothesis but make for doing pretty valuation. Observe that in a risk-neutral world all securities should behave much like a bond. Therefore, on the one hand, the rate of benefit that investors can expect from a stock should be equal to the risk-free rate, since they don't care about risk. And on the other hand, the present value of a derivative can be obtained by calculating its expected future value, and discounting it back to present at the risk-free rate (cf. Eq. (1.7)). Let us see how these facts allow us to compute the price of the call option in the situation considered in Example 1.11.

Example 1.13 Recall from Example 1.11 that we wanted to write a call option on stock XYZ, with six months to maturity and exercise price $K = €115$. We assume that the current price of the stock is equal to K. The risk-free interest rate for the life of the call is $r = 0.015$, and it is assume that by the end of the period the price of the stock may go up 30% or down 20%. With this information, and assuming investors are neutral to risk, we can compute the probability p of the stock going up, since the expected rate of benefit can be estimated as the sum of all the outcomes' probabilities weighted by their expectations, and all this is assumed equal to r. We have,

Expected rate of benefit $= (p \times 0.3) + (1 - p) \times (-0.2) = 0.015$.

This gives $p = 0.43$. On the other hand, the payoff of the option is $C_u = 34.5$ (if stock goes up), or $C_d = 0$ (if stock goes down). Then the expected future value of the option (EC) is the sum of these possible values weighed by their probabilities of occurring,

$$EC = (p \times C_u) + (1 - p) \times C_d$$
$$= 0.43 \times 34.5 + 0.57 \times 0 = 14.84$$

The present value C of the call is $C = \frac{EC}{1+r} = \frac{14.84}{1.015} = €14.62$, which coincide with the value computed by no arbitrage in Example 1.11 (as should be expected, of course!). □

Thus, both arbitrage and risk-neutral attitude of investors are two important and most used economic hypothesis for pricing derivatives. We will see their relevance in deriving the option pricing models to be discussed in Chap. 4.

1.2.4 The Efficient Market Hypothesis and Computational Complexity

A more general paradigm for market equilibrium arises from the assumption that markets are "informationally efficient". This means:

> The information available at the time of making an investment is already reflected in the prices of the securities, and in consequence market participants can not take advantage of this information to make a profit over the average market returns.

Such a belief of price behavior has been termed the *Efficient Market Hypothesis* (EMH), and an earlier version can be traced back to a paper by Working,[20] but a formal formulation, that became the standard definition, was given later by Fama (1970).

There is a large number of research papers devoted to testing the EMH (see the extensive surveys by Fama (1991) and Park and Irwin (2007)). The general methodology underlying much of these empirical tests of the EMH is summarized in Campbell et al. (1997) as a two part process: first, the design of a trading strategy based on an specific set of information and second, to measure the excess return over average returns obtained by the trading strategy. For the first part one must specify the information set, and in this regard the general accepted forms of efficiency sum up to the following three[21]:

Weak: only the price history of the securities constitutes the available information.
Semi–strong: all *public* information known up to the present time is available.
Strong: all *public and private* information (i.e., all possible information) known up to the present time is available.

Thus, for example, the Weak Efficient Market Hypothesis (Weak EMH) refers to the impossibility of making a profit above average when the only market information

---

[20] Working, H., "The investigation of economic expectations", *American Economic Review*, 39, 150–166, 1949.

[21] Campbell et al. (1997) attributes this classification to Roberts, H. (1967), *Statistical versus Clinical Prediction of the Stock Market*, unpublished manuscript, University of Chicago. A publicly available presentation and discussion of this taxonomy is given by Jensen (1978).

available is the history of prices; and similarly are defined the other two forms of EMH with respect to the available information, Semi-Strong EMH and Strong EMH.

For the second part of the market efficiency test one must specify a model of normal security returns, in order to have a precise measure of the surplus of benefits that can be attributed to the trading strategy (i.e., that can be forecasted using the given information).

In this context, trading strategies based on Technical Analysis are tests for the weak EMH, since their basic paradigm is to rely solely on the past history of the price, while those strategies based on Fundamental Analysis are tests for the semi-strong EMH. We shall study Technical and Fundamental analysis in Chap. 6. On the other hand, the autoregressive linear models, the random walks, and other models of the behavior of returns, all to be studied in Chap. 4, can be used for measuring the possible excess return given by trading strategies using the history of prices, and to some extend the public information represented as some random noise. As for the strong form of the EMH, there are several arguments for its impossibility in practice. For example, Grossman and Stiglitz (1980) developed a *noisy rational expectations* model that includes a variable that reflects the cost for obtaining information, besides the observable price. Then they showed that, in a competitive market, informed traders (those who pay the cost for the information) do have an advantage over uninformed traders (those who only observe prices), since prices cannot reflect all possible information, for otherwise no one would have an incentive to obtain costly information and the competitive market breaks down. Thus, while their model supports the weak EMH, it does not supports the strong form of the EMH.

Notwithstanding the soundness of the empirical methodology and variety of tools for testing the efficient market hypothesis, at least in its weak or semi-strong forms, the results of such tests are always difficult to interpreted. As Campbell et al. (1997) observe: "If efficiency is rejected, this could be because the market is truly inefficient or because an incorrect equilibrium model has been assumed. This joint hypothesis problem means that market efficiency as such can never be rejected". Therefore these authors advised to not preoccupy ourselves with testing efficiency, but focus on developing statistical tools for measuring it. This is a sensible advise that we subscribe to completely, although we believe that a recently developed conception of market efficiency in terms of computational complexity might better explain the experienced inefficiency of markets by some financial traders, who have made profits with their own strategies and models.

From a computational complexity point of view, a market is defined to be efficient with respect to computational resources R (e.g. time or memory), if no strategy using resources R can generate a substancial profit. This definition allows for markets to be efficient for some investors but not for others. Those who have at their disposal powerful computational facilities (let these be machines or human expert programmers) should have an advantage over those who haven't. Note that this definition is in the same spirit of noise rational expectation models mentioned above, but the relativization of efficiency is not to a monetary cost but rather to a computational cost.

Within this computational complexity framework for market efficiency, Hasan-hodzic et al. (2011) studied strategies that depend only on a given number of the most recent observations (so called memory bounded models), and that have an impact in the market by their decision to buy or sell (e.g., imagine the effects of a large fund that trades big amounts of shares of a stock at one time). Their conclusion was that such optimal strategies using memory m can lead to market bubbles, and that a strategy capable of using longer memory $m' > m$ can possible make larger profits.

In this same computational complexity context, Philip Maymin (2011) goes a step further and explores the computational difficulty of finding a profitable strategy among all possible strategies with long memory, and shows that under the assumption that the weak EMH holds, all possible trading strategies can be efficiently evaluated (i.e., using computing time bounded by some polynomial in the size of the informa-tion set, or in other words, using small computational resources), and that this fact implies that $\mathbf{P} = \mathbf{NP}$. Here $\mathbf{P}$ is the class of problems having deterministic algorithmic solutions with running time bounded by some polynomial in the size of the inputs, while in $\mathbf{NP}$ are those problems that can be *verified* by deterministic procedures polynomially bounded in time, but *solved* nondeterministically in polynomial time. Obviously $\mathbf{P} \subseteq \mathbf{NP}$, but there are many problems in $\mathbf{NP}$ with only known exponential deterministic solutions. The question whether $\mathbf{P} = \mathbf{NP}$ or not, is *the* most important problem in the theory of Computational Complexity, declared as one of the New Millennium Problems with a US\$ 1 million prize for its resolution.[22]

1.3 Notes, Computer Lab and Problems

1.3.1 Bibliographic remarks: For a global view of modern corporate finance, theory and practice, read Brealey et al. (2011), a renown textbook in business schools. For a deeper mathematical analysis and detailed account of derivatives see Hull (2009), Cox and Rubinstein (1985) and Neftci (2000). A more practical view of options, their history and a long list of trading strategies is in The Options Institute (1995). A review of the Efficient Market Hypothesis is given by Fama (1991) and Park and Irwin (2007), and can be complemented with the discussion on the subject found in Campbell et al. (1997, §1.5). For an in-depth treatment of the theory of Computa-tional Complexity see Papadimitriou (1994). To complement the topics treated here see Rubinstein (2006) chronological review of the most influential research papers in the financial theory of investments. From this source one learns, for example, the historical development of the DCF models (Eqs. (1.17) and (1.20)) and gets the authorship record straight: the formulas are often attributed to Gordon and Shapiro,[23] but they were first developed in 1938 by Williams in his PhD thesis (Williams 1938). For more on the history of derivatives and asset pricing models read Bernstein (1992,

---

[22] see http://www.claymath.org/millennium-problems.

[23] M. Gordon and E. Shapiro, "Capital equipment analysis: the require rate of profit", *Management Science* 3, 102–110, 1956.

Ch. 11). It is from this source where we learned about the possibly first option contract in history (see following note). Finally, for your bedtime reading we recommend the financial chronicle of Paulos (2007).

1.3.2 Possibly the First Option Contract in History: Aristotle reports in his book of *Politics* that the Greek mathematician and philosopher Thales the Milesian, incited by their common fellow's belief of the inutility of his maths, set out to prove them all wrong and worked on a financial scheme based on his knowledge of astronomy. Reading the stars one winter, Thales was able to foresee that by the coming year the olive harvest would be larger than usual; so he took his savings and paid all the owners of olive presses in Chios and Miletus for the right to use their presses in the upcoming harvest-time. He was able to negotiate this reservation of presses at a low price since no one could know what the harvest would be that much time ahead, and hence no one bid against his offer. Aristotle ends this anecdote as follows: "When the harvest-time came, and many [presses] were wanted all at once and of a sudden, he let them out at any rate which he pleased, and made a quantity of money. Thus he showed the world that philosophers can easily be rich if they like, but that their ambition is of another sort".[24]

1.3.3 On Short Selling: Short selling a stock is a bet for the future decrease of the price. If the short seller is a well-known investment corporation, and since these short positions are communicated to the public by the market authorities, the knowledge of this news can induce the stampede from the stock by many small investors, which would indeed knock down the price. The European Commission has adopted a series of regulatory technical standards on short selling that can be read at http://ec.europa.eu/internal_market/securities/short_selling_en.htm. See also what the US Securities and Exchange Commission writes about short sales at http://www.sec.gov/answers/shortsale.htm.

1.3.4 R Lab: We can reproduce in R Fig. 1.3 showing the different performances for four US stocks and the Value Line Index. To compute rate of benefits or returns from a given time series x use quantmod's function Delt(x). To express the resulting return series in percentage terms just multiply the series by 100. The function cumsum(x) does cumulative sum (for each time t, sum of terms up to time t). After loading quantmod, write the following commands in your R console:

```
> symbols=c('^VLIC','GE','KO','AAPL','MCD')}
> getSymbols(symbols,src='yahoo',from="2012-02-01",to="2013-02-01")
> #obtain adjusted closed
> VLICad = VLIC$VLIC.Adjusted; GEad= GE$GE.Adjusted;
+ KOad=KO$KO.Adjusted; AAPLad=AAPL$AAPL.Adjusted;
+ MCDad = MCD$MCD.Adjusted
> #compute cumulative sum (cumsum) of daily returns (Delt)
+ #Remove first term of the series, with [-1,],
+ #since cumsum is not defined for it.
```

---

[24] *source* Aristotle, *Politics*, Book I §11. In collection: *Encyclopedia Britannica, Great Books of the Western World*, 9, p. 453 (1982). See also Bernstein (1992, Ch. 11).

```
> vl = cumsum((Delt(VLICad)*100)[-1,])
> ge = cumsum((Delt(GEad)*100)[-1,])
> ko = cumsum((Delt(KOad)*100)[-1,])
> ap = cumsum((Delt(AAPLad)*100)[-1,])
> md = cumsum((Delt(MCDad)*100)[-1,])
> ###range of values for the plot
> lim = c(min(vl,ge,ko,ap,md),max(vl,ge,ko,ap,md))
> ###the plot
> plot(vl,main="",ylim=lim,xlab="dates",ylab="% benefits")
> lines(ge,col="green"); lines(ko,col="red")
> lines(ap,col="violet"); lines(md,col="yellow")
> legend(x="topleft",cex=0.4,c("VLIC","GE","KO","AAPL","MCD"),
+    lty=1, col=c("black","green","red","violet","yellow"))
```

1.3.5 Research Problem: Which market index is best? No market index has shown to be optimal for the purposes of (long term) asset allocation with highest return and lowest risk. Thus it is important to compare and statistically test which index (or measure of indexing) has historically come closer to this optimal goal. Furthermore, to research new ways of computing an index that may realize these optimality requirements. This is important because investors typically perceive an index to be a neutral choice of long term risk factor exposure. Some papers on the subject: Amenc, N., F. Goltz, and V. Le Sourd, 2006, Assessing the Quality of Stock Market Indices, *EDHEC Pub.*; and Arnott, R.D., J. Hsu, and P. Moore, 2005, Fundamental Indexation, *Financial Analysts Journal* 60 (2), 83–99.

1.3.6 Problem (Simple Bounds on Calls and Puts): Using no arbitrage arguments show that for options on stocks: (i) the stock's price is an upper bound for the price of a call; and (ii) the strike price is an upper bound for the price of a put.

1.3.7 Problem (More Strategies with Options): The following is a list of well-known investment strategies obtained by different combinations of put and call options on the same underlying asset. For each one of these strategies compute the payoff function and draw the profit graph. Additionally argue about the situations where the strategy is profitable.

The strangle: Similar to the straddle (see Example 1.7). It consists on buying a put and a call with the same expiration date but different strike prices. If K_c is the strike price for the call, and K_p is the strike price for the put, then the strategy requires $K_c > K_p$.

The strip: This strategy consists of long positions in one call and two puts, all with the same strike price and expiration date.

The strap: This one consists of long positions in two calls and one put, all with the same strike price and expiration date.

The butterfly spread: This is made with options of the same type. Suppose we use calls and that the underlying asset is a stock. Then a butterfly spread of calls consist on short selling two calls with strike price K_0 close to the current stock price, and buying two calls, one with strike price K_0-c and the other with strike price $K_0 + c$, where $2c > 0$ is the length of the spread chosen by the investor.

1.3.8 Problem: Use the replicating portfolio method (Example 1.11) to estimate the current value of a six-month call option on a stock with the following characteristics. The current price is €25, the exercise price is €20, at exercise time it is estimated that the stock will be in the range of €15 to €40, and the six-month risk-free interest rate is 5 %. Also, compute the value of the corresponding put option using the put-call parity relation.

1.3.9 Problem: In Example 1.11 we encounter the *option delta* (for a call), which is computed as the quotient of the spread of possible option prices over the spread of possible share prices Eq. (1.22). Denote this quantity Δ_C, and we have seen that it gives the number of shares needed to replicate a call option. Define analogously Δ_P to be the option delta for a put, and show that for a call and a put European options on the same stock and with the same strike price, $\Delta_P = \Delta_C - 1$. (Hint: use the put-call parity for European options.)

Chapter 2
Statistics of Financial Time Series

The price of a stock as a function of time constitutes a financial time series, and as such it contains an element of uncertainty which demands the use of statistical methods for its analysis. However, the plot of the price history of a stock in general resembles an exponential curve, and a time series of exponential terms is mathematically hard to manipulate and has little information to give even from classical functional transformations (e.g., the derivative of an exponential is again exponential). Campbell et al. (1997 Sect. 1.4) give two solid reasons for preferring to focus the analysis on returns rather than directly on prices. One reason is that financial markets are almost perfectly competitive, which implies that price is not affected by the size of the investment; hence, what is left for the investor to gauge is the rate of benefit that could be derived from his investment, and for that matter it is best to work with a size–independent summary of the investment opportunity, which is what the return represents. The second reason is that returns have more manageable statistical properties than prices, such as stationarity and ergodicity, being the case more often than not of dynamic general equilibrium models that give non stationary prices but stationary returns. Therefore, this chapter's main concern is the study of returns and their fundamental statistical properties. We briefly review some of the fundamental concepts of statistics, such as: moments of a distribution, distribution and density functions, likelihood methods, and other tools that are necessary for the analysis of returns and, in general, financial time series.

2.1 Time Series of Returns

Begin by considering the curve drawn by the price history of the Dow Jones Industrial Average (DJIA) from 1960 to 2010 . This can be obtained with the following R commands.

```
> require(quantmod)
> getSymbols("DJIA",src="FRED") ##DJIA from 1896-May
> plot(DJIA['1960/2010'],main="DJIA")
```

A. Arratia, *Computational Finance,* Atlantis Studies in Computational Finance and Financial Engineering 1, DOI: 10.2991/978-94-6239-070-6_2, © Atlantis Press and the authors 2014

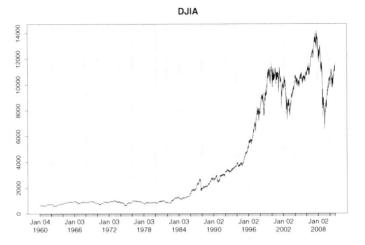

Fig. 2.1   Dow Jones Industrial Average from 1960 to 2010

The resulting picture can be seen in Fig. 2.1. One can observe that from 1960 to about 1999 the DJIA has an exponential shape; from 2000 to 2002 it has an exponential decay, to rise up exponentially again until 2008, and so on. This might suggest to fit an exponential function to this curve to forecast its future values. We show you how to do this in the R Lab 2.7.2. However, as argued in the introduction to this chapter it is best to turn to returns to analyze a stock's behavior through time. There are many variants of the definition of returns, according to whether we allow some extra parameters to be considered in their calculation, like dividends or costs of transactions. Thus, we first look at the most simple definition of returns where only the price is considered.

Definition 2.1   *(Simple return)* Let P_t be the price of an asset at time t. Given a time scale τ, the *τ-period simple return at time t*, $R_t(\tau)$, is the rate of change in the price obtained from holding the asset from time $t - \tau$ to time t:

$$R_t(\tau) = \frac{P_t - P_{t-\tau}}{P_{t-\tau}} = \frac{P_t}{P_{t-\tau}} - 1 \tag{2.1}$$

The *τ-period simple gross return at time t* is $R_t(\tau) + 1$. If $\tau = 1$ we have a one-period simple return (respectively, a simple gross return), and denote it R_t (resp., $R_t + 1$).
□

There is a practical reason for defining returns backwards (i.e. from time $t - \tau$ to t, as opposed to from t to $t + \tau$), and it is that more often than not we want to know the return obtained today for an asset bought some time in the past. Note that return values range from -1 to ∞; so, in principle, you can not loose more than what you've invested, but you can have unlimited profits.

Example 2.1 Consider the daily closing prices of Apple Inc. (AAPL:Nasdaq) throughout the week of July 16 to July 20, 2012[1]:

	1	2	3	4	5
Date	2012/07/16	2012/07/17	2012/07/18	2012/07/19	2012/07/20
Price	606.91	606.94	606.26	614.32	604.30

Let us refer to the dates in the table by their positions from left to right, i.e., as 1, 2, 3, 4 and 5. Then, the simple return from date 3 to date 4 is

$$R_4 = (614.32 - 606.26)/606.26 = 0.0133.$$

The return from day 1 to day 5 (a 4-period return) is $R_5(4) = (604.30 - 606.91)/606.91 = -0.0043$. The reader should verify that

$$1 + R_5(4) = (1 + R_2)(1 + R_3)(1 + R_4)(1 + R_5) \qquad (2.2)$$

Equation (2.2) is true in general:

Proposition 2.1 *The τ-period simple gross return at time t equals the product of τ one-period simple gross returns at times $t - \tau + 1$ to t.*

Proof

$$R_t(\tau) + 1 = \frac{P_t}{P_{t-\tau}} = \frac{P_t}{P_{t-1}} \cdot \frac{P_{t-1}}{P_{t-2}} \cdots \frac{P_{t-\tau+1}}{P_{t-\tau}}$$
$$= (1 + R_t) \cdot (1 + R_{t-1}) \cdots (1 + R_{t-\tau+1}) \qquad (2.3)$$

For this reason these multiperiod returns are known also as *compounded* returns. □

Returns are independent from the magnitude of the price, but they depend on the time period τ, which can be minutes, days, weeks or any time scale, and always expressed in units. Thus, a return of 0.013, or in percentage terms of 1.3%, is an incomplete description of the investment opportunity if the return period is not specified. One must add to the numerical information the time span considered, if daily, weekly, monthly and so on. If the time scale is not given explicitly then it is customary to assumed to be of one year, and that we are talking about an *annual rate of return*. For τ years returns these are re-scale to give a comparable one-year return. This is the *annualized* (or average) return, defined as

$$\text{Annualized } (R_t(\tau)) = \left(\prod_{j=0}^{\tau-1} (1 + R_{t-j}) \right)^{1/\tau} - 1 \qquad (2.4)$$

---

[1] *source* http://finance.yahoo.com/q/hp?s=AAPL+Historical+Prices

It is the geometric mean of the τ one-period simple gross returns. This can also be computed as an arithmetic average by applying the exponential function together with its inverse, the natural logarithm, to get:

$$Annualized\ (R_t(\tau)) = \exp\left(\frac{1}{\tau}\sum_{j=0}^{\tau-1}\ln(1 + R_{t-j})\right) - 1 \qquad (2.5)$$

This equation expresses the annualized return as the exponential of a sum of logarithms of gross returns. The logarithm of a gross return is equivalent, in financial terms, to the continuous compounding of interest rates. To see the link, recall the formula for pricing a risk–free asset with an annual interest rate r which is continuously compounded (Chap. 1, Eq. (1.3)): $P_n = P_0 e^{rn}$, where P_0 is the initial amount of the investment, P_n the final net asset value, and n the number of years. Setting $n = 1$, then $r = \ln(P_1/P_0) = \ln(R_1 + 1)$. For this reason, the logarithm of (gross) returns is also known as *continuously compounded returns*.

Definition 2.2 *(log returns)* The *continuously compounded return* or *log return* r_t of an asset is defined as the natural logarithm of its simple gross return:

$$r_t = \ln(1 + R_t) = \ln\left(\frac{P_t}{P_{t-1}}\right) = \ln P_t - \ln P_{t-1} \qquad (2.6)$$

Then, for a τ-period log return, we have

$$\begin{aligned} r_t(\tau) &= \ln(1 + R_t(\tau)) = \ln((1 + R_t)(1 + R_{t-1})\cdots(1 + R_{t-\tau+1})) \\ &= \ln(1 + R_t) + \ln(1 + R_{t-1}) + \cdots + \ln(1 + R_{t-\tau+1}) \\ &= r_t + r_{t-1} + \cdots + r_{t-\tau+1} \end{aligned} \qquad (2.7)$$

which says that the continuously compounded τ-period return is simply the sum of τ many continuously compounded one-period returns.

Besides turning products into sums, and thus making arithmetic easier, the log returns are more amenable to statistical analysis than the simple gross return because it is easier to derive the time series properties of additive processes than those of multiplicative processes. This will become more clear in Sect. 2.2.

Example 2.2 Consider the same data from Example 2.1 and now, for your amazement, compute the continuously compounded, or log return, from date 3 to date 4: $r_4 = \ln(614.32) - \ln(606.26) = 0.0132$. And the 4-period log return

$$r_5(4) = \ln(604.3) - \ln(606.91) = -0.0043.$$

Observe how similar these values are to the simple returns computed on same periods. Can you explain why? (Hint: recall from Calculus that for small x, i.e., $|x| < 1$, $\ln(1 + x)$ behaves as x.) □

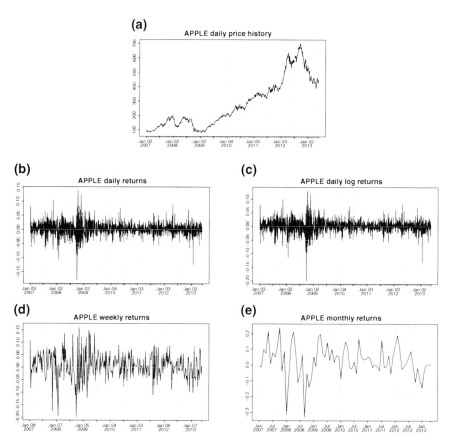

Fig. 2.2 Apple (Nasdaq) 2007–2013. **a** daily price; **b** daily return; **c** daily log return; **d** weekly return; **e** monthly return

Figure 2.2 shows different plots of Apple shares historical data from year 2007 to 2013. Again we can see an exponential behavior in the plot of price (a). Note the similarities between plots (b) and (c) corresponding to daily simple returns and log returns. Also observe that at any time scale the return series shows a high degree of variability, although this variability seems similar over several time periods.

The different period returns are easily computed with R's quantmod package. For example Fig. 2.2b can be obtained with

```
> getSymbols("AAPL",src="yahoo")  ##data starts from 2007
> aplRd = periodReturn(AAPL,period="daily")
> plot(aplRd, main="APPLE daily returns")
```

Returns with dividends. If the asset pays dividends periodically then the simple return must be redefined as follows

$$R_t = \frac{P_t + D_t}{P_{t-1}} - 1 \tag{2.8}$$

where D_t is the dividend payment of the asset between dates $t - 1$ and t. For multi-period and continuously compounded returns the modifications of the corresponding formulas to include dividends are similar and are left as exercises.

Remark 2.1 (**On adjusted close prices and dividends**) Some providers of financial data include in their stock quotes the adjusted close price (cf. Chap. 1, Sect. 1.1.2). Since this price is adjusted to dividend payments and other corporate actions on the stock, it is then more accurate to work with the adjusted close price when analyzing historical returns.   □

Excess return. It is the difference between the return of an asset A and the return of a reference asset O, usually at a risk-free rate. The simple excess return on asset A would be then $Z_t^A = R_t^A - R_t^O$; and the logarithmic excess return is $z_t^A = r_t^A - r_t^O$. The excess return can be thought of as the payoff of a portfolio going long in the asset and short on the reference.

Portfolio return. Let $\mathscr{P}$ be a portfolio of N assets, and let n_i be the number of shares of asset i in $\mathscr{P}$, for $i = 1, \ldots, N$. Then, at a certain time t, the net value of $\mathscr{P}$ is $P_t^{\mathscr{P}} = \sum_{i=1}^{N} n_i P_t^i$, where P_t^i is the price of asset i at time t. Applying Eq. (2.3), we get that the τ-period simple return of $\mathscr{P}$ at time t, denoted $R_t^{\mathscr{P}}(\tau)$, is given by

$$R_t^{\mathscr{P}}(\tau) = \sum_{i=1}^{N} w_i R_t^i(\tau) \tag{2.9}$$

where $R_t^i(\tau)$ is the τ-period return of asset i at time t, and $w_i = (n_i P_{t-\tau}^i)/(\sum_{j=1}^{N} n_j P_{t-\tau}^j)$ is the weight, or proportion, of asset i in $\mathscr{P}$. Therefore, the simple return of a portfolio is a weighted sum of the simple returns of its constituent assets. This nice property does not holds for the continuously compounded return of $\mathscr{P}$, since the logarithm of a sum is not the sum of logarithms. Hence, when dealing with portfolios we should prefer returns to log returns, even though empirically when returns are measured over short intervals of time, the continuously compounded return on a portfolio is close to the weighted sum of the continuously compounded returns on the individual assets (e.g., look back to Example 2.2 where the log returns almost coincide with the simple returns). We shall be dealing more in depth with the financial analysis of portfolios in Chap. 8. In the remaining of this book we shall refer to simple returns just as returns (and use R_t to denote this variable); and refer to continuously compounded return just as log returns (denoted r_t).

2.2 Distributions, Density Functions and Moments

We depart from the fact that a security's time series of returns are random variables evolving over time, i.e., a *random process*, and that usually the only information we have about them is some sample observations. Therefore, in order to build some statistical model of returns, we must begin by specifying some probability distribution or, stepping further, by treating returns as continuous random variables, specify a probability density function, and from this function obtain a quantitative description of the shape of the distribution of the random values by means of its different moments. This will give us a greater amount of information about the behavior of returns from which to develop a model.

2.2.1 Distributions and Probability Density Functions

The *cumulative distribution function* (CDF) of a random variable X is defined, for all $x \in \mathbb{R}$, as $F_X(x) = \mathbb{P}(X \leq x)$. F_X is said to be continuous if its derivative $f_X(x) = F'_X(x)$ exists; in which case f_X is the *probability density function* of X, and F_X is the integral of its derivative by the fundamental theorem of calculus, that is

$$F_X(x) = \mathbb{P}(X \leq x) = \int_{-\infty}^{x} f_X(t)dt \qquad (2.10)$$

When F_X is continuous then one says that the random variable X is continuous. On the other hand, X is discrete if F_X is a step function, which then can be specified, using the *probability mass function* $p_X(x) = \mathbb{P}(X = x)$, as

$$F_X(x) = \mathbb{P}(X \leq x) = \sum_{\{k \leq x : p_X(k) > 0\}} p_X(k).$$

The CDF F_X has the following properties: is non-decreasing; $\lim_{x \to -\infty} F_X(x) = 0$ and $\lim_{x \to \infty} F_X(x) = 1$. We can then estimate $F_X(x)$ from an observed sample $x_1, \ldots, x_n$ by first ordering the sample as $x_{(1)} \leq x_{(2)} \leq \cdots \leq x_{(n)}$ and then defining the *empirical cumulative distribution function* ECDF:

$$F_n(x) = (1/n) \sum_{i=1}^{n} \mathbf{1}\{x_{(i)} \leq x\}$$

where $\mathbf{1}\{A\}$ is the indicator function, whose value is 1 if event A holds or 0 otherwise. Observe that the ECDF give the proportion of sample points in the interval $(-\infty, x]$. If it were the case that the CDF F_X is strictly increasing (and continuous) then it is invertible, and we can turn around the problem of estimating $F_X(x)$ and find, for a given $q \in [0, 1]$, the real number x_q such that $q = F_X(x_q) = \mathbb{P}(X \leq x_q)$; that is, for a given proportion q find the value x_q of X such that the proportion of observations below x_q is exactly q. This x_q is called the *q-quantile* of the random variable X with distribution F_X. However, in general, F_X is not invertible so it is best to define the q quantile of the random variable X as the smallest real number x_q such that $q \leq F_X(x_q)$, that is, $x_q = \inf\{x: q \leq F_X(x)\}$.

When it comes to estimating quantiles one works with the inverse ECDF of a sample of X or other cumulative distribution of the order statistics. It is not a trivial task, apart from the 0.5 quantile which is estimated as the *median* of the distribution. But in R we have a good sample estimation of quantiles implemented by the function `quantile`.

R Example 2.1 Consider the series of returns for the period 06/01/2009–30/12/2009 of Allianz (ALV), a German company listed in Frankfurt's stock market main index DAX. Assume the data is already in a table named `daxR` under a column labelled `alvR`. The instructions to build this table are given in the R Lab 2.7.6. After loading the table in your work space, run the commands:

```
> alv=na.omit(daxR$alvR)
> quantile(alv,probs=c(0,1,0.25,0.5,0.75))
```

This outputs the extreme values of the series (the 0 and 100% quantiles) and the *quartiles*, which divide the data into four equal parts. The output is:

```
        0%          100%          25%           50%           75%
-0.098800569  0.124268348 -0.013133724  0.001246883  0.017879936
```

The first quartile (at 25%) indicates the highest value of the first quarter of the observations in a non-decreasing order. The second quartile (at 50%) is the median, or central value of the distribution, and so on.   □

Now, if we want to deal with two continuous random variables jointly, we look at their joint distribution function $F_{X,Y}(x, y) = \mathbb{P}(X \leq x; Y \leq y)$, which can be computed using the joint density function $f_{X,Y}(x, y)$ if this exists (i.e. the derivative of $F_{X,Y}(x, y)$ exists) by

$$F_{X,Y}(x, y) = \mathbb{P}(X \leq x; Y \leq y) = \int_{-\infty}^{y} \int_{-\infty}^{x} f_{X,Y}(s, t)dsdt \qquad (2.11)$$

Also with the joint density function $f_{X,Y}$ we can single handle either X or Y with their *marginal probability densities*, which are obtained for each variable by integrating out the other. Thus, the marginal probability density of X is

$$f_X(x) = \int_{-\infty}^{\infty} f_{X,Y}(x, y)dy \qquad (2.12)$$

and similarly, the marginal probability density of Y is $f_Y(y) = \int_{-\infty}^{\infty} f_{X,Y}(x, y)dx$.

The conditional distribution of X given $Y \leq y$ is given by

$$F_{X|Y \leq y}(x) = \frac{\mathbb{P}(X \leq x; Y \leq y)}{\mathbb{P}(Y \leq y)}$$

Again, if the corresponding probability density functions exist, we have the *conditional density of X given* $Y = y$, $f_{X|Y=y}(x)$, obtained by

$$f_{X|Y=y}(x) = \frac{f_{X,Y}(x, y)}{f_Y(y)} \tag{2.13}$$

where the marginal density $f_Y(y)$ is as in Eq. (2.12). From (2.13), we can express the joint density of X and Y in terms of their conditional densities and marginal densities as follows:

$$f_{X,Y}(x, y) = f_{X|Y=y}(x)f_Y(y) = f_{Y|X=x}(y)f_X(x) \tag{2.14}$$

Equation (2.14) is a very important tool in the analysis of random variables and we shall soon see some of its applications. Right now, observe that X and Y are independent if and only if $f_{X|Y=y}(x) = f_X(x)$ and $f_{Y|X=x}(y) = f_Y(y)$. Therefore,

Two random variables X and Y are *independent* if and only if their joint density is the product of their marginal densities; i.e., for all x and y,

$$f_{X,Y}(x, y) = f_X(x)f_Y(y) \tag{2.15}$$

A more general notion of independence is the following:

X and Y are independent if and only if their joint distribution is the product of the CDF's of each variable; i.e., for all x and y,

$$F_{X,Y}(x, y) = F_X(x)F_Y(y) \tag{2.16}$$

Eq. (2.16) is stronger than Eq. (2.15) since it holds even if densities are not defined.

2.2.2 Moments of a Random Variable

Given a continuous random variable X with density function f, the *expected value* of X, denoted $E(X)$, is

$$E(X) = \int_{-\infty}^{\infty} xf(x)dx. \tag{2.17}$$

If X is discrete then (2.17) reduces to $E(X) = \sum_{\{x:f(x)>0\}} xf(x)$.

Let $\mu_X = E(X)$. Then μ_X is also called the mean or first moment of X. The geometric intuition is that the first moment measures the central location of the distribution.

More general, the *n-th moment* of a continuous random variable X is

$$E(X^n) = \int_{-\infty}^{\infty} x^n f(x)dx, \tag{2.18}$$

and the *n-th central moment* of X is defined as

$$E((X - \mu_X)^n) = \int_{-\infty}^{\infty} (x - \mu_X)^n f(x)dx \tag{2.19}$$

The second central moment of X is called the *variance* of X,

$$Var(X) = E((X - \mu_X)^2) \tag{2.20}$$

and most frequently denoted by σ_X^2. The variance measures the variability of X (quantified as the distance from the mean). Observe that $\sigma_X^2 = Var(X) = E(X^2) - \mu_X^2$. The term $\sigma_X = \sqrt{E((X - \mu_X)^2)}$ is the *standard deviation* of X.

The third and fourth central moments are known respectively as the *skewness* (denoted $S(X)$) and the *kurtosis* ($K(X)$), which measure respectively the extent of asymmetry and tail thickness (amount of mass in the tails) of the distribution. These are defined as

$$S(X) = E\left(\frac{(X - \mu_X)^3}{\sigma_X^3}\right) \quad \text{and} \quad K(X) = E\left(\frac{(X - \mu_X)^4}{\sigma_X^4}\right) \tag{2.21}$$

Estimation of moments. In practice one has observations of a random variable X and from these one does an estimation of X's moments of distribution. Given sample data $x = \{x_1, \ldots, x_m\}$ of random variable X, the *sample mean* of X is

$$\widehat{\mu}_x = \frac{1}{m} \sum_{t=1}^{m} x_t, \tag{2.22}$$

and the *sample variance* is

$$\widehat{\sigma}_x^2 = \frac{1}{m-1} \sum_{t=1}^{m} (x_t - \widehat{\mu}_x)^2. \tag{2.23}$$

The *sample standard deviation* is $\widehat{\sigma}_x = \sqrt{\widehat{\sigma}_x^2}$. The *sample skewness* is

$$\widehat{S}_x = \frac{1}{(m-1)\widehat{\sigma}_x^3} \sum_{t=1}^{m} (x_t - \widehat{\mu}_x)^3, \tag{2.24}$$

and the *sample kurtosis*

$$\widehat{K}_x = \frac{1}{(m-1)\widehat{\sigma}_x^4} \sum_{t=1}^{m} (x_t - \widehat{\mu}_x)^4 \tag{2.25}$$

Remark 2.2 The above estimators constitute *the basic statistics* for X. The sample mean and variance are *unbiased* estimators of their corresponding moments, in the sense that for each estimator $\widehat{\theta}_x$ and corresponding moment θ_X, we have $E(\widehat{\theta}_x) = \theta_X$. This does not hold for sample skewness and kurtosis, and so these are said to be biased estimators. From now on, whenever X or its sample x are clear from context we omit them as subscripts.   □

R Example 2.2 In R Lab 2.7.6 we guide the reader through the R instructions to compute some basic statistics for a group of German stocks. We chose to analyze the four companies Allianz (ALV), Bayerische Motoren Werke (BMW), Commerzbank (CBK) and Thyssenkrupp (TKA). The returns from dates 06/01/2009 to 30/12/2009 for these stocks are placed in a table labelled daxR. Using the basicStats() command, from the package fBasics we get the following results. (A note of caution: the kurtosis computed by basicStats() is the *excess kurtosis*, that is, $K(X) - 3$. We will explain later what this means. Hence, to get the real kurtosis add 3 to the values given in the table.)

```
> basicStats(na.omit(daxR[,2:5]))
                   alvR        bmwR        cbkR        tkaR
nobs         251.000000 251.000000 251.000000 251.000000
NAs            0.000000   0.000000   0.000000   0.000000
Minimum       -0.098801  -0.077849  -0.169265  -0.081459
Maximum        0.124268   0.148430   0.187359   0.162539
1. Quartile   -0.013134  -0.015040  -0.025762  -0.015535
3. Quartile    0.017880   0.019219   0.020863   0.020753
Mean           0.001339   0.001988   0.000844   0.002272
Median         0.001247  -0.000389  -0.001709   0.000437
Sum            0.336189   0.498987   0.211965   0.570381
SE Mean        0.001880   0.001866   0.003071   0.002162
LCL Mean      -0.002363  -0.001687  -0.005204  -0.001986
UCL Mean       0.005041   0.005663   0.006893   0.006531
Variance       0.000887   0.000874   0.002367   0.001173
Stdev          0.029780   0.029563   0.048652   0.034253
Skewness       0.236224   0.663122   0.392493   0.618779
Kurtosis       2.060527   2.911924   2.126179   2.038302
```

You can also get the particular statistics for each individual return series (and here the function kurtosis() gives the real kurtosis, not the excess).

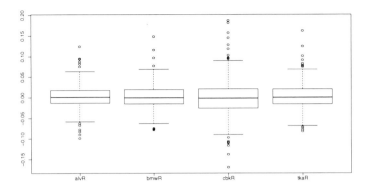

Fig. 2.3 Box plots for ALV, BMW, CBK and TKA (stocks listed in DAX)

```
> alvR <- na.omit(daxR$alvR)
> mean(alvR) ## the mean
> var(alvR) ## variance
> sd(alvR) ## standard deviation
> kurtosis(alvR, method="moment") #gives real kurtosis
> skewness(alvR)
```

We explain the terms in the table of basicStats that are not immediately clear: nobs is the number of observations; Sum is the sum of all nobs observations (hence, for example, note that Mean = Sum/nobs); SE Mean is the standard error for the mean, which is computed as the standard deviation (Stdev) divided by the square root of nobs; LCL Mean and UCL Mean are the Lower and Upper Control Limits for sample means, computed by the formulas:

$$LCL = Mean - 1.96 \cdot SE \quad and \quad UCL = Mean + 1.96 \cdot SE \qquad (2.26)$$

and represent the lower and upper limit of a *confidence band*.[2]

A better way to "see" the results is through a *box plot*:

```
> boxplot(daxR[,2:5])
```

The resulting picture is in Fig. 2.3.

A box plot (proposed by John Tukey in 1977) gives a description of numerical data through their quartiles. Specifically, for the R function boxplot( ), the bottom and top of the box are the first quartile (Q_1) and the third quartile (Q_3), and the line across the box represents the second quartile (Q_2), i.e., the median. The top end of the whisker (represented by the dotted line) is the largest value $M < Q_3 + 1.5(Q_3 - Q_1)$; the low end of the whisker is the smallest value $m > Q_1 - 1.5(Q_3 - Q_1)$. The dots above and below the whisker (i.e. outside $[m, M]$) represent outliers. Inside the box

---

[2] Confidence refers to the following: if the probability law of the sample mean is approximately *normal*, then within these LCL and UCL bounds lie approximately 95 % of sample means taken over nobs observations. We shall discuss normality in the next pages.

there is 50 % of the data split evenly across the median; hence if, for example, the median is slightly up this is indication of major concentration of data in that region.

Viewing the box plots and the numeric values we can infer some characteristics of the behavior of stock's daily returns. We can see that the daily return of a stock presents:

- small (sample) median;
- more smaller (in value) returns than larger returns;
- some non null skewness, hence some gross symmetry;
- a considerable number of outliers (approx. 5 % shown in the box plots), and a high kurtosis. In fact, a positive excess kurtosis, for recall that basicStats gives $\widehat{K}_x - 3$ for the Kurtosis entry.

These observations suggest that stocks' returns are distributed more or less evenly around the median, with decaying concentration of values as we move away from the median, and almost symmetrical. We can corroborate this suggestion by plotting a histogram of one of these return series and observing that the rectangles approximate a bell-shaped area (using the data from R Example 2.2):

```
> alvR <- na.omit(daxR$alvR)
> hist(alvR,probability=T,xlab="ALV.DE returns",main=NULL)
```

Consequently, as a first mathematical approximation to model the distribution of stock returns we shall study the benchmark of all bell-shaped distributions.

2.2.3 The Normal Distribution

The most important and recurred distribution is the *normal* or *Gaussian* distribution. The normal distribution has probability density function

$$f(x) = \frac{1}{\sigma\sqrt{2\pi}} \exp(-(x - \mu)^2/2\sigma^2), \tag{2.27}$$

for $-\infty < x < \infty$, and with $\sigma > 0$ and $-\infty < \mu < \infty$. It is completely determined by the parameters μ and σ^2, that is by the mean and the variance, and so, we could write $f(x)$ as $f(x; \mu, \sigma^2)$ to make explicit this parameter dependance. Here the terminology acquires a clear geometrical meaning. If we plot $f(x) = f(x; \mu, \sigma^2)$ we obtain a bell shaped curve, centered around the mean μ, symmetrical and with tails going to infinity in both directions. Thus, the mean is the most likely or expected value of a random variable under this distribution. For the rest of the values we find that approximately 2/3 of the probability mass lies within one standard deviation σ from μ; that approximately 95 % of the probability mass lies within 1.96σ from μ; and that the probability of being far from the mean μ decreases rapidly. Check this by yourself in R, with the commands:

```
> x = seq(-4,4,0.01)
> plot(x,dnorm(x,mean=0.5,sd=1.3),type='l')
```

which plots the normal distribution of a sequence x of real values ranging from -4 to 4, and equally spaced by a 0.01 increment, with $\mu = 0.5$ and $\sigma = 1.3$.

The normal distribution with mean μ and variance σ^2 is denoted by $N(\mu, \sigma^2)$; and to indicate that a random variable X has normal distribution with such a mean and variance we write $X \sim N(\mu, \sigma^2)$.

Remark 2.3 In general, $X \sim Y$ denotes that the two random variables X and Y have the same distribution. $X|A \sim Y|B$ means X conditioned to information set A has same distribution of Y conditioned to information set B.   □

The *standard normal distribution* is the normal distribution with zero mean and unit variance, $N(0, 1)$, and the standard normal CDF is

$$\Phi(x) = \int_{-\infty}^{x} \frac{1}{\sqrt{2\pi}} \exp(-t^2/2) dt, \quad -\infty < x < \infty. \tag{2.28}$$

The normal distribution has skewness equal to 0 and kurtosis equal to 3. In view of this fact, for any other distribution the difference of its kurtosis with 3, namely $K(X) - 3$, is a measure of *excess of kurtosis* (i.e., by how much the kurtosis of the distribution deviates from the normal kurtosis), and the sign gives the type of kurtosis. A distribution with positive excess kurtosis ($K(X) - 3 > 0$), known as *leptokurtic*, has a more acute peak around the mean and heavier tails, meaning that the distribution puts more mass on the tails than a normal distribution. This implies that a random sample from such distribution tends to contain more extreme values, as is often the case of financial returns. On the other hand, a distribution with negative excess kurtosis ($K(X) - 3 < 0$), known as *platykurtic*, presents a wider and lower peak around the mean and thinner tails. Returns seldom have platykurtic distributions; hence models based on this kind of distribution should not be considered.

The normal distribution has other properties of interest of which we mention two:

P1 A normal distribution is invariant under linear transformations: If $X \sim N(\mu, \sigma^2)$ then $Y = aX + b$ is also normally distributed and further $Y \sim N(a\mu + b, a^2\sigma^2)$.
P2 Linear combinations of normal variables are normal: If $X_1, \ldots, X_k$ are independent, $X_i \sim N(\mu_i, \sigma_i^2)$, and $a_1, \ldots, a_k$ are constants, then $Y = a_1X_1 + \cdots + a_kX_k$ is normally distributed with mean $\mu = \sum_{i=1}^{k} a_i\mu_i$ and variance $\sigma^2 = \sum_{i=1}^{k} a_i^2\sigma_i^2$.

It follows from P1 that if $X \sim N(\mu, \sigma^2)$, then

$$Z = \frac{X - \mu}{\sigma} \sim N(0, 1) \tag{2.29}$$

Equation (2.29) is a very useful tool for "normalizing" a random variable X for which we do not know its distribution. This follows from the *Central Limit Theorem*.

Theorem 2.1 (Central Limit Theorem) If $X_1, \ldots, X_n$ is a random sample from a distribution of expected value given by μ and finite variance given by σ^2, then the limiting distribution of

$$Z_n = \frac{\sqrt{n}}{\sigma}\left(\frac{1}{n}\sum_{i=1}^{n} X_i - \mu\right) \tag{2.30}$$

is the standard normal distribution.   □

Thus, if we have an observation x of random variable X then the quantity $z = (x - \mu)/\sigma$, known as *z-score*, is a form of "normalizing" or "standardizing" x. The idea behind is that even if we don't know the distribution of X (but know its mean and variance), we compare a sample to the standard normal distribution by measuring how many standard deviations the observations is above or below the mean. The z-score is negative when the sample observation is below the mean, and positive when above. As an application of the z-score there is the computation of the LCL and UCL thresholds in R Example 2.2 (Eq. (2.26)): We wish to estimate L and U bounds such that random variable X lies within (L, U) with a 95 % probability; that is, $\mathbb{P}(L < X < U) = 0.95$. We use the standardization Z of X, and hence we want L and U such that

$$\mathbb{P}\left(\frac{L - \mu}{\sigma} < Z < \frac{U - \mu}{\sigma}\right) = 0.95$$

From this equation it follows that $L = \mu - z\sigma$ and $U = \mu + z\sigma$, where z is the quantile such that $\mathbb{P}(-z < Z < z) = 0.95$. By the Central Limit Theorem

$$0.95 = \mathbb{P}(|Z| < z) \approx \Phi(z) - \Phi(-z)$$

where $\Phi(z)$ is given by Eq. (2.28). Computing numerically the root of $\Phi(z) - \Phi(-z) = 2\Phi(z) - 1 = 0.95$ one gets $z = 1.96$. Below we show how to do this computation in R.

R Example 2.3 In R the function `qnorm(x)` computes $\Phi^{-1}(x)$. Hence, for $x = (0.95 + 1)/2 = 0.975$, run the command `qnorm(0.975)` to get the quantile 1.96.   □

For a proof of the Central Limit Theorem and more on its applications see Feller (1968).

2.2.4 Distributions of Financial Returns

Are returns normally distributed? Almost any histogram of an asset's return will present some bell-shaped curve, although not quite as smooth as the normal distribution. We show this empirical fact with a particular stock.

Fig. 2.4 Histogram of ALV
returns from 06/01/2009 -
30/12/2009, with an estimate
of its density from sample
data (*solid line*), and adjusted
normal distribution (*dashed
line*)

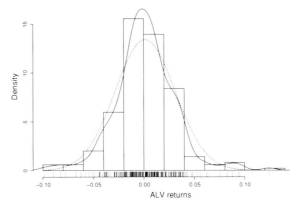

R Example 2.4 Consider the sequence of returns for Allianz (ALV) obtained in R
Example 2.1. We compute its histogram and on top we plot an estimate of its density
from the sample (with a solid line). Then plot the normal density function with the
sample mean and sample standard deviation of the given series (with a dashed line).

```
> alv <- na.omit(daxR$alvR); DS <- density(alv)
> yl=c(min(DS$y),max(DS$y)) #set y limits
> hist(alv,probability=T,xlab="ALV returns", main=NULL,ylim=yl)
> rug(alv); lines(DS); a=seq(min(alv),max(alv),0.001)
> points(a,dnorm(a,mean(alv),sd(alv)), type="l",lty=2)
> # if you rather have a red line for the normal distribution do:
> lines(a,dnorm(a,mean(alv), sd(alv)),col="red")
```

The output can be seen in Fig. 2.4.

The figure shows that a normal distribution does not fits well the sample estimate of the density of returns for the ALV stock. We can corroborate this empirical
observation with one of many statistical test for the null hypothesis that a sample
$R_1, \ldots, R_n$ of returns come from a normally distributed population. One popular such
test is Shapiro-Wilk. If the p-value that the test computes is less than a given confidence level (usually 0.05) then the null hypothesis should be rejected (i.e. the data
do not come from a normal distributed population). In R this test is implemented by
the function `shapiro.test()`. Running this function on `alv` we get the following
results:

```
> shapiro.test(alv)

Shapiro-Wilk normality test data:   alv W = 0.9701, p-value =
3.995e-05
```

The p-value is below 5×10^{-5}, hence the hypothesis of normality should be rejected.

In addition to these empirical observations and statistical tests, there are some
technical drawbacks to the assumption of normality for returns. One is that an asset
return has a lower bound in -1 and no upper bound, so it seems difficult to believe

in a symmetric distribution with tails going out to infinity in both directions (which are characteristics of normality). Another concern, more mathematically disturbing, is that multiperiod returns could not fit the normal distribution, since they are the product of simple returns, and the product of normal variables is not necessarily normal. This last mismatch between the properties of returns and normal variables steers our attention to log returns, since a multiperiod log return is the sum of simple log returns, just as the sum of normal variables is normal. Thus, a sensible alternative is to assume that the log returns are the ones normally distributed, which implies that the simple returns are *log-normally* distributed.

Example 2.3 (**The log-normal distribution**) A random variable X has the *log-normal distribution*, with parameters μ and σ^2, if $\ln X \sim N(\mu, \sigma^2)$. In this case we write $X \sim LogN(\mu, \sigma^2)$. The log-normal density function is given by

$$f_X(x) = \frac{1}{x\sigma\sqrt{2\pi}} \exp(-(\ln x - \mu)^2/2\sigma^2), \quad x > 0. \tag{2.31}$$

and the moments of the variable X are

$$E(X^n) = \exp\left(n\mu + \frac{1}{2}n^2\sigma^2\right), \quad n > 0$$

Therefore, if we assume that the simple return series $\{R_t\}$ is log-normally distributed with mean μ_R and variance σ_R^2, so that the log return series $\{r_t\}$ is such that

$$r_t = \ln(R_t + 1) \sim N(\mu_r, \sigma_r^2)$$

with mean μ_r and variance σ_r^2, we have that the respective moments for both series are related by the following equations

$$E(R_t) = \mu_R = e^{\mu_r + \sigma_r^2/2} - 1, \quad Var(R) = \sigma_R^2 = e^{2\mu_r + \sigma_r^2}(e^{\sigma_r^2} - 1) \tag{2.32}$$

As a first application of this hypothesis of normal distribution for the continuously compounded returns, we show how to compute bounds to the future price of the underlying asset with a 95% precision.

Example 2.4 Let $\{P_t\}$ be the price series of a stock (or any other financial asset) and $\{r_t\}$ its corresponding log return series, and assume r_t are independent and normally distributed. The τ-period log returns $r_t(\tau)$ are also normally distributed, because of Eq. (2.7) and the fact that a finite sum of iid normal random variables is normal. Therefore the mean of $r_t(\tau)$ is $\mu_r\tau$ and the variance is $\sigma_r\tau$. Now, consider an initial observed price P_0 at time $t = 0$, and we want to estimate the price P_T at a later time $t = T$. Then $\ln(P_T/P_0)$ is the continuously compounded return over the period $\tau = T$, and by the previous observations

$$\ln\left(\frac{P_T}{P_0}\right) = r_1 + \cdots + r_T \sim N(\mu_r T, \sigma_r^2 T) \tag{2.33}$$

where μ_r and σ_r^2 are the mean and variance of $\{r_t\}$. Let $Z_T = \dfrac{\ln(P_T/P_0) - \mu_r T}{\sigma_r \sqrt{T}}$.

Then $Z_T \sim N(0, 1)$, and we have seen as a consequence of the Central Limit Theorem that for the quantile $z = 1.96$ we have $\mathbb{P}(-z < Z_T < z) = 0.95$ (go back to Example 2.3). From this we have

$$\mu_r T - z\sigma_r \sqrt{T} < \ln\left(\frac{P_T}{P_0}\right) < \mu_r T + z\sigma_r \sqrt{T}$$

or, equivalently

$$P_0 \exp\left(\mu_r T - z\sigma_r \sqrt{T}\right) < P_T < P_0 \exp\left(\mu_r T + z\sigma_r \sqrt{T}\right) \tag{2.34}$$

These equations give bounds for the price at time $t = T$, with a 95 % probability if taking $z = 1.96$. On real data, one makes an estimate of μ_r and σ_r from a sample of the log returns $\{r_t\}$, and assumes these estimations of moments hold for the period $[0, T]$; that is, we must assume that the mean and variance of the log returns remain constant in time. Be aware then of all the considered hypotheses for these calculations to get the bounds in Eq. (2.34), so that estimations from real data should be taken as rough approximations to reality.   □

Although the log-normal distribution model is consistent with the linear invariance property of log returns and with their lower bound of zero, there is still some exper-imental evidence that makes the distribution of returns depart from log-normality. For example, it is often observed some degree of skewness, where in fact negative values are more frequent than positive; also some excessively high and low values present themselves (suggesting some kurtosis), and these extreme values can not be discarded as outliers, for here is where money is lost (or gained) in tremendous amounts.

Which is then an appropriate model for the distribution of stock returns? Given a collection of log returns $\{r_{it} : i = 1, \ldots, n; t = 1, \ldots, m\}$ corresponding to returns r_{it} of n different assets at times $t = 1, \ldots, m$, each considered as random variable, we should start with the most general model for this collection of returns which is its joint distribution function:

$$G(r_{11}, \ldots, r_{n1}, r_{12}, \ldots, r_{n2}, \ldots, r_{1m}, \ldots, r_{nm}; \theta), \tag{2.35}$$

where to simplify notation we use $G(X, Y; \theta)$ instead of the usual $F_{X,Y}(x, y; \theta)$, with the additional explicit mention of θ as the vector of fixed parameters that uniquely determines G (e.g. like μ and σ determine the normal distribution). More than a model, (2.35) is a general framework for building models of distribution of returns wherein we "*view financial econometrics as the statistical inference of θ, given G and realizations of $\{r_{it}\}$*".[3]

Therefore we should impose some restrictions to (2.35) to make it of practical use, but also to reflect our social or economic beliefs. For example, some financial models, such as the Capital Asset Pricing Model (to be studied in Chap. 8) consider the joint distribution of the n asset returns restricted to a single date t: $G(r_{1t}, \ldots, r_{nt}; \theta)$. This restriction implicitly assumes that returns are statistically independent through time and that the joint distribution of the cross-section of returns is identical across time. Other models focus on the dynamics of individual assets independently of any relations with other assets. In this case one is to consider the joint distribution of $\{r_{i1}, \ldots, r_{im}\}$, for a given asset i, as the product of its conditional distributions (cf. Eq. 2.14, and we drop θ for the sake of simplicity):

$$G(r_{i1}, \ldots, r_{im}) = G(r_{i1})G(r_{i2}|r_{i1})G(r_{i3}|r_{i2}, r_{i1}) \cdots G(r_{im}|r_{im-1}, \ldots, r_{i1})$$

$$= G(r_{i1}) \prod_{i=2}^{m} G(r_{im}|r_{im-1}, \ldots, r_{i1}) \qquad (2.36)$$

This product exhibits the possible temporal dependencies of the log return r_{it}. So, for example, if we believe in the weak form of the efficient market hypothesis, where returns are unpredictable from their past, then in this model the conditional distributions should be equal to the corresponding marginal distributions, and hence (2.36) turns into

$$G(r_{i1}, \ldots, r_{im}) = G(r_{i1})G(r_{i2}) \cdots G(r_{im}) = \prod_{i=1}^{m} G(r_{im}) \qquad (2.37)$$

We see then that under this framework issues of predictability of asset returns are related to their conditional distributions and how these evolve through time. By placing restrictions on the conditional distributions we shall be able to estimate the parameters θ implicit in (2.36). On the other hand, by focusing on the marginal or unconditional distribution we can approach the behavior of asset returns individually. This was the case of assuming a normal or log-normal distribution. From these basic distributions we can go on refining so as to capture, for example, the excess of kurtosis or non null skewness. Distributions such as *stable* distributions, or *Cauchy*, or a *mixture of distributions*, may capture these features found in returns better than log-normal distributions, but the downside is that the number of parameters in θ increases just as the computational difficulty to estimate them. In summary, fitting a model beyond the normal or log normal to the distribution of stock returns is a challenging albeit arduous problem.[4]

---

[3] Campbell et al. (1997 §1.4.2)

[4] For a more extensive discussion see Campbell et al. (1997 Chap. 1)

2.3 Stationarity and Autocovariance

Now, even if we do not believe that today's performance of returns is a reflection of their past behavior, we must believe that there exist some statistical properties of returns that remain stable through time; otherwise there is no meaningful statistical analysis of financial returns and no possible interesting models for their distribution at all. The invariance in time of the moments of the distribution of a random process is the stationarity hypothesis.

Definition 2.3 A random process $\{X_t\}$ is *strictly stationary* if for any finite set of time instants $\{t_1, \ldots, t_k\}$ and any time period τ the joint distribution of $\{X_{t_1}, \ldots, X_{t_k}\}$ is the same as the joint distribution of $\{X_{t_1+\tau}, \ldots, X_{t_k+\tau}\}$, i.e.,

$$F_{X_{t_1},\ldots,X_{t_k}}(x_1,\ldots,x_k) = F_{X_{t_1+\tau},\ldots,X_{t_k+\tau}}(x_1,\ldots,x_k). \tag{2.38}$$

One interesting property of strictly stationary processes is that once we have one we can produce many other strictly stationary processes by applying any "regular" operation on subsequences; e.g. moving averages, iterative products, and others. We state this important fact informally, and recommend the reading of Breiman (1992 Chap. 6, Prop. 6.6) for the precise mathematical details.[5]

Proposition 2.2 *Let $\{X_t\}$ be a strictly stationary process and Φ a function from $\mathbb{R}^{h+1}$ to $\mathbb{R}$. Then the process $\{Y_t\}$ defined by $Y_t = \Phi(X_t, X_{t-1}, \ldots, X_{t-h})$ is strictly stationary.* □

Stationary processes obtained as $Y_t = \Phi(X_t, X_{t-1}, \ldots, X_{t-h})$ can be classified by the scope of their variable dependency, in the following sense.

Definition 2.4 A random process $\{X_t\}$ is *m-dependent*, for an integer $m > 0$, if X_s and X_t are independent whenever $|t - s| > m$. □

For example, an iid sequence is 0-dependent; Y_t obtained as in Prop. 2.2 is *h*-dependent.

Example 2.5 A *white noise* $\{W_t\}$ is a sequence of iid random variables with finite mean and variance. This is a strictly stationary process: independence implies that $F_{W_1,\ldots,W_k}(w_1,\ldots,w_k) = \prod_{i=1}^{k} F_{W_i}(w_i)$ (cf. Eq. (2.15)), while being identically distributed implies that $F_{W_i}(w) = F_{W_{i+\tau}}(w) = F_{W_1}(w)$, for all i; hence, both hypotheses give Eq. (2.38). Next, consider the processes

---

[5] We remark that in many classical textbooks of probability and stochastic processes, such as Breiman (1992), strictly stationary is simply termed stationary, but in the modern literature of time series and financial applications, such as Brockwell and Davis (2002); Tsay (2010), it is more common to distinguish different categories of stationarity, in particular strict (as in Def. 2.3) and weak (to be defined later). We adhere to this latter specification of different levels of stationarity, and when using the term by itself –without adverb– is to refer to any, and all, of its possibilities.

- $Y_t = \alpha_0 W_t + \alpha_1 W_{t-1}$, with $\alpha_0, \alpha_1 \in \mathbb{R}$ (moving average);
- $Z_t = W_t W_{t-1}$

By Prop. 2.2 these are strictly stationary (and 1-dependent). □

It is not obvious that financial returns verify the strictly stationary hypothesis. However, it is a convenient assumption to ensure that one can estimate the moments of the returns by taking samples of data from any time intervals. Looking at some plots of returns (e.g., go back to Fig. 2.2), one often encounters that the mean is almost constant (and close to zero) and the variance bounded and describing a pattern that repeats through different periods of time. Therefore, an assumption perhaps more reasonable to the invariability in time of all moments of returns could be that the first moment is constant (hence invariant) and the second moment is in sync with its past (hence it can be well estimated from past data). To formalize this hypothesis we need as a first ingredient the notion of *covariance*.

Definition 2.5 The covariance of two random variables X and Y is

$$Cov\,(X, Y) = E((X - \mu_X)(Y - \mu_Y))$$

Note that $Cov\,(X, Y) = E(XY) - E(X)E(Y) = E(XY) - \mu_X\mu_Y$, and $Cov\,(X, X) = Var\,(X) = \sigma_X^2$. To be consistent with the σ notation, it is customary to denote $Cov\,(X, Y)$ by $\sigma_{X,Y}$. □

Remark 2.4 The expected value of the product of continuous random variables X and Y with joint density function $f_{X,Y}$ is

$$E(XY) = \int_{-\infty}^{\infty} \int_{-\infty}^{\infty} xy f_{X,Y}(x, y)dxdy \tag{2.39}$$

Now observe that if X and Y are independent then $Cov\,(X, Y) = 0$. This follows directly from the definition of covariance and Eqs. (2.39) and (2.15). The converse is not true, for we can have $Cov\,(X, Y) = 0$ and X and Y being functionally dependent. The popular example is to consider X *uniformly distributed* in $[-1, 1]$, and $Y = X^2$. Then X and Y are dependent, but $Cov\,(X, X^2) = E(X^3) - E(X)E(X^2) = 0 - 0 \cdot E(X^2) = 0$. Note that the dependency of X and $Y = X^2$ is non-linear.

We can estimate $Cov\,(X, Y)$ from a given sample $\{(x_i, y_i) : i = 1, \ldots, m\}$ by the sample covariance:

$$\widehat{Cov}\,(X, Y) = \frac{1}{m-1} \sum_{i=1}^{m} (x_i - \widehat{\mu}_X)(y_i - \widehat{\mu}_Y) \tag{2.40}$$

R Example 2.5 In R the function cov (x,y, ...) computes the sample covariance of vectors x and y, or if x is a matrix or table (and y=NULL) computes covariance between the columns. We use it to compute the covariance between pairs of the four stocks considered in R Example 2.2. Execute in your R console the command

```
> cov(daxRlog[,2:5],use="complete.obs")
```
The resulting table is

```
              alvR            bmwR            cbkR            tkaR
alvR 0.0008816302 0.0005196871 0.0008677487 0.0006721131
bmwR 0.0005196871 0.0008570846 0.0006724442 0.0006361113
cbkR 0.0008677487 0.0006724442 0.0023419267 0.0008754300
tkaR 0.0006721131 0.0006361113 0.0008754300 0.0011487287
```

We observe positive values in all entries of the sample covariance matrix in the example above. What can we deduce from these values? Observe that if X and Y are linearly dependent, say $Y = aX + b$, then $Cov(X, aX + b) = aVar(X)$. Since $Var(X) \geq 0$ always, we see that the sign of covariance depends on the sign of the slope of the linear function of Y on X; hence positive (resp. negative) covariance means that Y and X move in the same (resp. opposite) direction, *under the assumption of a linear model for the dependency relation*. How strong could this co-movement be? In $Cov(X, aX + b) = aVar(X)$ we could have large covariance due to a large $Var(X)$, while the factor of dependency a be so small, so as to be negligible. Hence, we need a way to measure the strength of the possible (linear) co-movement signaled by a non null covariance. The obvious solution suggested by the previous observation is to factor out the square root of both the variances of X and Y from their covariance: $\dfrac{Cov(X, Y)}{\sqrt{Var(X)Var(Y)}}$. This is the *correlation coefficient* of X and Y, which we will study in more detail in the next chapter.

We move on to the problem of determining the possible dependence of the variance of a random process with its past. For that matter one uses the autocovariance function.

Definition 2.6 For a random process $\{X_t\}$ with finite variance, its *autocovariance* function is defined by $\gamma_X(s, t) = Cov(X_s, X_t)$. □

The autocovariance is the concept we need to narrow strict stationarity of returns to just some of its key moments. This gives us weak stationarity.

Definition 2.7 A random process $\{X_t\}$ is *weakly stationary* (or *covariance stationary*) if it has finite variance ($Var(X_t) < \infty$), constant mean ($E(X_t) = \mu$) and its autocovariance is time invariant: $\gamma_X(s, t) = \gamma_X(s + h, t + h)$, for all $s, t, h \in \mathbb{Z}$. In other words, the autocovariance only depends on the time shift, or *lag*, $|t - s|$, and not on the times t or s. Hence, we can rewrite the autocovariance function of a weakly stationary process as

$$\gamma_X(h) = Cov(X_t, X_{t+h}) = Cov(X_{t+h}, X_t), \quad h = 0, \pm 1, \pm 2, \ldots \qquad (2.41)$$

$\gamma_X(h)$ is also called the lag-h autocovariance. □

Remark 2.5 A strictly stationary process $\{X_t\}$ with finite second moments is weakly stationary. The converse is not true: there are weakly stationary processes that are not strictly stationary.[6] However, if $\{X_t\}$ is a weakly stationary *Gaussian* process (i.e., the distribution of $\{X_t\}$ is multivariate normal), then $\{X_t\}$ is also strictly stationary. □

Remark 2.6 Some obvious but important properties of the autocovariance function of any stationary process (strictly or weakly) are: $\gamma(0) \geq 0$ (since $\gamma(0) = Var(X)$), $|\gamma(h)| \leq \gamma(0)$ and $\gamma(h) = \gamma(-h)$, for all h. □

Example 2.6 Let $\{W_t\}$ be a white noise with mean zero and variance σ^2, and consider the following sequence: $S_0 = 0$, and for $t > 0$, $S_t = W_1 + W_2 + \ldots + W_t$. Note that $S_t = S_{t-1} + W_t$. The process $S = \{S_t\}$ is a *random walk*. Compute the autocovariance for S. For $h > 0$,

$$\gamma_S(t, t+h) = Cov\,(S_t, S_{t+h}) = Cov\left(\sum_{i=1}^{t} W_i, \sum_{j=1}^{t+h} W_j\right)$$

$$= Var\left(\sum_{i=1}^{t} W_i\right) = t\sigma^2$$

The third equality follows from $Cov\,(W_i, W_j) = 0$, for $i \neq j$. The autocovariance depends on t; hence, the random walk S is not weakly stationary (and not strictly stationary). □

All is well in theory, but in practice if one doesn't have a model fitting the data, one can still determine the possibility of the underlying random process being weakly stationary by applying the following empirical method: look at a plot of the return time series, and if the values do not seem to fluctuate with a constant variation and around a constant level, then one can conclude with high confidence that the process is not weakly stationary; otherwise we can suspect of stationarity. It might be the case that the data needs first to be massaged to remove some observable *trend* (e.g. bull or bear market periods) or *seasonality* (e.g. values repeating periodically), to reveal the noisy component with the possible stationary behavior (we will comment on this decomposition of time series in Chap. 4). Do the appropriate transformation first and then try different models for the transformed data. For more on the theory and applications of stationary time series see Brockwell and Davis (1991).

---

[6] The following counterexample is from Brockwell and Davis (1991): let $\{X_t\}$ be a sequence of independent random variables such that X_t is exponentially distributed with mean 1 when t is odd and normally distributed with mean 1 and variance 1 when t is even, then $\{X_t\}$ is weakly stationary, but X_{2k+1} and X_{2k} have different distributions for each $k > 0$, hence $\{X_t\}$ cannot be strictly stationary.

2.4 Forecasting

In the analysis of financial time series we are interested in designing models for predicting future values. We are going to formalize the framework for forecasting in this chapter and postpone the model design for Chap. 4. Given a stationary stochastic process $\{X_t\}$ with zero mean and autocovariance function γ, we want to predict X_{t+h}, the value of the series h time periods into the future, based on some known set of information Z about X_t. The information set Z is comprised of random observations, and it is in practice a finite set of recent past values of the series, that is, $Z = \{X_t, X_{t-1}, \dots, X_{t-p}\}$. Given the set Z, we view an estimation $\widehat{X}_{t+h}$ of X_{t+h} as a function of Z, $\widehat{X}_{t+h} = F(Z)$, and we would like to know how good is $F(Z)$ as an estimator. The accepted criterion for measuring the performance of an estimator F of a random variable X, based on some information Z, is the following[7]

The *best estimator* or *predictor* of a random variable X, on the basis of random information Z about X, is the function $F(Z)$ that minimizes $E(X - F(Z))^2$.

This minimum mean square error predictor can be characterized in statistical terms as follows. Assuming the conditional density function of X given $Z = z$ exists (cf. Eq. (2.13)), we can consider the *conditional expectation of X given $Z = z$*, denoted $E(X|Z = z)$ (or, for the sake of simplicity, $E(X|Z)$), and defined as

$$E(X|Z = z) = \int_{-\infty}^{\infty} x f_{X|Z=z}(x)dx \tag{2.42}$$

Then the best estimator of X given the information Z is $E(X|Z = z)$. We show this fact in the following theorem.

Theorem 2.2 Let X and Z be two random variables with $E(X^2) < \infty$ and joint density function $f_{X,Z}$, and let $\widehat{X} = E(X|Z)$ the conditional expectation of X given Z. Then, for any function $F = F(Z)$, we have

$$E(X - \widehat{X})^2 \le E(X - F(Z))^2$$

Proof Let $F = F(Z)$ and consider the following equalities

$$E(X - F)^2 = E([(X - \widehat{X}) + (\widehat{X} - F)]^2)$$
$$= E(X - \widehat{X})^2 + 2E((X - \widehat{X})(\widehat{X} - F)) + E(\widehat{X} - F)^2 \tag{2.43}$$

Note that $\widehat{X} - F$ is a function of Z, so observed values of this random variable are denoted z' (i.e., given values $\widehat{X} - F = z'$). We decompose the second term in (2.43), using linearity of E, as follows

---

[7] see Box and Jenkins (1976); Breiman (1992) on properties of conditional expectation and best predictor discussed in this section.

$$E((X - \widehat{X})(\widehat{X} - F)) = E(X(\widehat{X} - F)) - E(\widehat{X}(\widehat{X} - F))$$

and now resolve, using (2.39), (2.15) and (2.42),

$$E(X(\widehat{X} - F)) = \int_{-\infty}^{\infty} \int_{-\infty}^{\infty} xz' f_{X,Z}(x, z')dxdz'$$

$$= \int_{-\infty}^{\infty} z' \left(\int_{-\infty}^{\infty} x f_{X|Z=z'}(x)dx \right) f_Z(z')dz' = \int_{-\infty}^{\infty} z' E(X|Z) f_Z(z')dz'$$

$$= E(\widehat{X}(\widehat{X} - F))$$

Therefore, $E((X - \widehat{X})(\widehat{X} - F)) = 0$, and
$$E(X - F)^2 = E(X - \widehat{X})^2 + E(\widehat{X} - F)^2 \ge E(X - \widehat{X})^2.$$

This theorem applies to the particular case of interest in which the information set is $Z = \{X_t, X_{t-1}, \ldots, X_{t-p}\}$, a recent history of X_t, and we want to make estimation of X_{t+h}. Then the best predictor in this case is the conditional expectation $E(X_{t+h}|X_t, X_{t-1}, \ldots, X_{t-p})$. We are thus interested in determining the form of this conditional expectation. For Gaussian processes this is well determined as stated in the following proposition whose proof we leave as exercise.

Proposition 2.3 *If the variables X_{t+h}, X_t, ..., X_{t-p} have a normal distribution, i.e., the process $\{X_t\}$ is Gaussian, then*

$$E(X_{t+h}|X_t, X_{t-1}, \ldots, X_{t-p}) = \alpha_0 X_t + \alpha_1 X_{t-1} + \cdots + \alpha_p X_{t-p}$$

where α_0, ..., α_p are real numbers. □

In view of this proposition, the problem of building a best predictor for a Gaussian process is solved by forming a linear regression. For any other process, not necessarily normally distributed, it is still desirable to design a predictor as a linear combination of its past history, even if this does not coincides with the conditional expectation of the process given its past history. In this case we want a linear function L of $\{X_t, X_{t-1}, \ldots, X_{t-p}\}$ that minimizes the mean square error $E(X_{t+h} - L)^2$; that is, we want to find coefficients $\alpha_0, \alpha_1, \ldots, \alpha_p$ to form $L(X_{t+h}) = \sum_{j=0}^{p} \alpha_j X_{t-j}$ and such that their values minimize $F(\alpha_0, \ldots, \alpha_p) = E\left(X_{t+h} - \sum_{j=0}^{p} \alpha_j X_{t-j}\right)^2$. This F is a quadratic function bounded below by 0, and hence there exists values of $(\alpha_0, \ldots, \alpha_p)$ that minimizes F, and this minimum satisfies $\dfrac{\partial F(\alpha_0, \ldots, \alpha_p)}{\partial \alpha_j} = 0, j = 0, \ldots, p$. Evaluating these derivatives we get:

$$E\left((X_{t+h} - \sum_{k=0}^{p} \alpha_k X_{t-k})X_{t-j}\right) = 0, \quad j = 0, \ldots, p. \tag{2.44}$$

Since $E(X_t) = 0$, $\gamma(h+j) = E(X_{t+h}X_{t-j})$ for $j = 0, 1, \ldots, p$, and consequently we obtain from Eq. (2.44) the following system of equations:

$$\gamma(h) = \alpha_0\gamma(0) + \alpha_1\gamma(1) + \cdots + \alpha_p\gamma(p)$$
$$\gamma(h+1) = \alpha_0\gamma(1) + \alpha_1\gamma(0) + \cdots + \alpha_p\gamma(p-1)$$

$$\vdots \qquad \vdots \qquad\qquad\qquad\qquad\qquad (2.45)$$

$$\gamma(h+p) = \alpha_0\gamma(p) + \alpha_1\gamma(p-1) + \cdots + \alpha_p\gamma(0)$$

Therefore, in terms of conditional expectation, the best linear h-step forecaster for the process $\{X_t\}$ is

$$L(X_{t+h}) = \sum_{j=0}^{p} \alpha_j X_{t-j} \qquad (2.46)$$

where $(\alpha_0, \alpha_1, \ldots, \alpha_p)$ satisfies Eq. (2.45). On the other hand, the mean square forecast error is given by the equation

$$E\left(X_{t+h} - L(X_{t+h})\right)^2 = \gamma(0) - 2\sum_{j=0}^{p}\alpha_j\gamma(h+j) + \sum_{i=0}^{p}\sum_{j=0}^{p}\alpha_i\alpha_j\gamma(i-j) \quad (2.47)$$

Note that if we don't know the autocovariance function γ we can make estimations using the sample covariance (Eq. (2.40)).

Thus, for a stationary process its best linear forecaster in terms of its past is completely determined by the covariances of the process. However, note that even though the system of linear Eq. (2.45) is computationally solvable, it can be time-consuming for large p. The solution of this system can be alleviated for processes which are themselves expressed by linear combinations of white noise, as it is the case of the ARMA processes to be studied in Chap. 4. For other arbitrary processes there are some very effective recursive procedures to compute the coefficients for a 1-step forecaster.[8]

2.5 Maximum Likelihood Methods

Once we have settled on a model for the distribution of returns, or for their second moments, an important next step is to estimate the parameters of the model from sample observations of the returns. The method of maximum likelihood is a much used technique in statistics for estimating the "most likely" values for the parameters of a model from observable data. It is an estimation method which is largely due, in its modern formulation, to R. A. Fischer.[9]

---

[8] One that we like very much is the Innovations Algorithm, see Brockwell and Davis (2002).

[9] see Hald (1999) for a historical account

Let X be a random variable for which we have sample observations $x_1, \ldots, x_m$, and let $\theta = (\theta_1, \ldots, \theta_p)$ be a vector of parameters, pertaining to some statistical model for X, and which we want to estimate. Assuming that the observations are made randomly and independently of each other, we can consider their joint distribution as the product of their conditional distributions (Eq. (2.36)). Then, if the distribution has density function $f(x; \theta)$, we define the likelihood of observing the m values of X, given θ, as

$$L(x_1, \ldots, x_m; \theta) = f(x_1; \theta) \cdot \prod_{i=2}^{m} f(x_i | x_{i-1}, \ldots, x_1; \theta) \qquad (2.48)$$

The value of θ that maximizes this likelihood function is the *maximum–likelihood estimate* (MLE) of θ. This MLE might not be unique, and might even not exists. In the case that it exists, but is not unique, any of the points where the maximum is attained is an MLE of θ. A useful mathematical trick for simplifying the computation of an MLE is to find instead a value of θ that maximizes the logarithm of L, called the *log likelihood function*:

$$l(x_1, \ldots, x_m; \theta) = \ln(L(x_1, \ldots, x_m; \theta)) = \ln f(x_1; \theta) + \sum_{i=2}^{m} \ln(f(x_i | x_{i-1}, \ldots, x_1; \theta))$$

$$(2.49)$$

Maximizing $\ln L$ is the same as maximizing L, because $\ln$ is a monotone function. The optimization of $l(x_1, \ldots, x_m; \theta)$ is often obtain only by numerical methods. In R this can be done with the function `mle()`, or by direct use of the optimization method `optim()`.

A further simplification to the likelihood computation is to assume that all random observations are made independently. In this case Eq. (2.48) and Eq. (2.49) turn into

$$L(x_1, \ldots, x_m; \theta) = \prod_{i=1}^{m} f(x_i; \theta) \quad \text{and} \quad l(x_1, \ldots, x_m; \theta) = \sum_{i=1}^{m} \ln(f(x_i; \theta)) \quad (2.50)$$

Although the independence assumption is quite strong for time series, it simplifies likelihood computation to analytical form. As an example of the MLE method we show how to obtain the best estimate of a constant variance from sample returns.

Example 2.7 Let $r_1, \ldots, r_m$ be a sequence of m independently observed log returns. We assume these returns have some constant variance σ^2, which we want to estimate under the normal distribution model, and a constant mean μ which we know. By the independence of the observations the log likelihood function is as in Eq. (2.50), with $\theta = (\sigma^2, \mu)$ and μ fixed, and the probability density function $f(r_i; \sigma^2, \mu)$ is given by the Gaussian density function (Eq. (2.27)). Therefore, the log likelihood of observing $r_1, \ldots, r_m$ is

$$l(r_1, \ldots, r_m; \sigma^2) = -\frac{1}{2} \sum_{i=1}^{m} \left[\ln(2\pi) + \ln(\sigma^2) + \frac{(r_i - \mu)^2}{\sigma^2} \right] \qquad (2.51)$$

By the maximum likelihood method the value of σ^2 that maximizes (2.51) is the best estimate (or more likely) value of σ^2. Differentiating $l(r_1, \ldots, r_m; \sigma^2)$ with respect to σ^2 and setting the resulting derivative to zero, we obtain that the maximum likelihood estimate of σ^2 is given by

$$\sigma_{mle}^2 = \frac{1}{m} \sum_{i=1}^{m} (r_i - \mu)^2$$

Note that this estimated value σ_{mle}^2 is biased (cf. Remark 2.2), since

$$E(\sigma_{mle}^2) = \frac{m-1}{m} E(\hat{\sigma}^2) = \frac{m-1}{m} \sigma^2$$

However, the bias of $-\sigma^2/m$ tends to 0 as $m \to \infty$, so the maximum likelihood estimate σ_{mle}^2 is *asymptotically unbiased* for σ^2. Thus, taking m sufficiently large we have σ_{mle}^2 a good estimator of the variance of log returns, sampled in a given interval of time, and which in fact, by construction, it is the most likely value that can be assigned to that variance. □

2.6 Volatility

One of the most important features of financial assets, and possibly the most relevant for professional investors, is the asset volatility. In practice volatility refers to a degree of fluctuation of the asset returns. However it is not something that can be directly observed. One can observe the return of a stock every day, by comparing the change of price from the previous to the current day, but one can not observe how the return fluctuates in a specific day. We need to make further observations of returns (and of the price) at different times on the same day to make an estimate of the way returns vary daily (so that we can talk about *daily volatility*), but these might not be sufficient to know precisely how returns will fluctuate. Therefore volatility can not be observed but estimated from some model of the asset returns. A general perspective, useful as a framework for volatility models, is to consider volatility as the conditional standard deviation of the asset returns.

Definition 2.8 Let r_t be the log-return of an asset at time t, and denote by F_{t-1} the information set available at time $t-1$. The conditional mean and variance of r_t given F_{t-1} are

$$\mu_t = E(r_t|F_{t-1}) \quad \text{and} \quad \sigma_t^2 = Var(r_t|F_{t-1}) = E((r_t - \mu_t)^2|F_{t-1}) \qquad (2.52)$$

The volatility of the asset at time t is $\sigma_t = \sqrt{Var(r_t|F_{t-1})}$. □

Now, to compute $Var(r_t|F_{t-1})$ we need to make some assumptions on the nature of F_{t-1}, and according to these assumptions we obtain different categories of volatilities. A common assumption is to consider $F_{t-1} = \{r_{t-1}, \ldots, r_{t-m}\}$; that is, F_{t-1} consists of finitely many past returns. If we assume further that the returns $\{r_t\}$ are iid and normal, then we can take for estimation sample returns in any time interval, which implies that the variance is considered constant in time, and in this case the best estimate is given by the maximum likelihood estimate of the sample variance (cf. Example 2.7):

$$\widehat{\sigma}^2 = \frac{1}{m} \sum_{k=1}^{m} (r_{t-k} - \widehat{\mu})^2,$$

where $\widehat{\mu}$ is the sample mean of the series. Because we are interested in quantifying the variability of returns, we might as well consider a non-centered variance ($\mu = 0$), and thus take as the (constant) volatility the non-centered sample standard deviation of returns[10]:

$$\sigma_o = \sqrt{\frac{1}{m} \sum_{k=1}^{m} r_{t-k}^2} \tag{2.53}$$

This is a form of volatility known as *historical volatility*, since the standard deviation, estimated from m past observations of a time series of continuously compounded returns, measures the historical degree of fluctuation of financial returns.

Scaling the volatility. Because we have assumed $r_t = \ln P_t - \ln P_{t-1} \sim N(\mu, \sigma^2)$, then for $h > 0$ successive equally spaced observations of the price, $r_{t+h} \sim N(\mu h, \sigma^2 h)$. Therefore, to annualize the historical volatility estimate we must multiply it by the scaling factor $\sqrt{h}$, for h being the number of time periods in a year. Thus the annualized historical volatility with respect to h periods per year is given by

$$\sigma_{ann} = (\sqrt{h})\sigma_o \tag{2.54}$$

In practice, if daily data is used then we take $h = 252$; if weekly data is used (i.e. prices are sample every week), then $h = 52$; and for monthly data, $h = 12$.

Range-based volatility estimates. Note that in practice the return r_t is computed by taking the closing prices of successive days, i.e. $r_t = \ln(C_t/C_{t-1})$, where C_t is the Close price of day t. A sharper estimate of daily volatility would be obtained if one considers variation of prices within the same day. The best one can do with publicly available data is to consider the daily range of prices, which is defined as the difference between the High and Low prices of the day, i.e., $H_t - L_t$. One such range-based estimator was defined by Parkinson (1980) as follows:

---

[10] There other practical reasons to discard the mean: on real financial data the mean is often close to 0; also, a non-centered standard deviation gives a more accurate estimated volatility, even compared to other type of models (see Figlewski (1994)).

$$\sigma_p = \sqrt{\frac{1}{4\ln 2} \cdot \frac{1}{m} \sum_{k=1}^{m} \ln\left(\frac{H_k}{L_k}\right)^2} \qquad (2.55)$$

Another range-based estimator was given by Garman and Klass (1980), defined as

$$\sigma_{gk} = \sqrt{\frac{1}{m} \sum_{k=1}^{m} \left[\frac{1}{2}\left(\ln\frac{H_k}{L_k}\right)^2 - (2\ln 2 - 1)\left(\ln\frac{C_k}{O_k}\right)^2\right]} \qquad (2.56)$$

where O_k is the Open price (and C_k, H_k and L_k the Close, High and Low price).

You can easily program these estimators of historical volatility, or use the function volatility in the R package TTR as we show in the next R Example.

R Example 2.6 Load into your R console the packages TTR and quantmod and with the function getSymbols retrieve data for Apple Inc. (AAPL) from *yahoo finance*. We will estimate the historical volatility to the end of May 2009 for this stock, considering 40 past observations. We use the TTR function volatility(ohlc, n=10, calc="close", N=260, ...), where ohlc is an object that is coercible to xts or matrix and contains Open-High-Low-Close prices; n is the length of the sample for the volatility estimate; calc is the type of estimator to use, where choices are: "close" for the Close-to-Close volatility (Eq. (2.53) with $r_t = \ln(C_t/C_{t-1})$); "parkinson" for the Parkinson estimator (Eq. (2.55)); "garman.klass" for the Garman and Klass estimator (Eq. (2.56)); and there are others (see Help documentation). Finally, N is the number of periods per year, as determined by the sample data. This is use to give the result in annualized form by multiplying the estimate by $\sqrt{N}$. Run the following commands:

```
> aapl=AAPL['2009-04/2009-05']; m=length(aapl\$AAPL.Close);
> ohlc <-aapl[,c("AAPL.Open","AAPL.High","AAPL.Low","AAPL.Close")]
> vClose <- volatility(ohlc, n= m,calc="close",N=252)
> vParkinson <- volatility(ohlc, n= m,calc="parkinson",N=252)
> vGK <- volatility(ohlc, n= m,calc="garman",N=252)
> vClose[m]; vParkinson[m]; vGK[m];
```

The resulting annualized volatility according to each estimator, as of 2009-05-29, is 0.356 for the Close-to-Close; 0.291 for the Parkinson; 0.288 for the Garman and Klass. □

For more on range-based volatility estimates see Tsay (2010), and for their effectiveness when continuous time models are considered see Alizadeh et al. (2002).

Time dependent weighted volatility. Now, volatility is not constant. We will argue on this theme in Chap. 4, Sect. 4.3, but right now we will give a first simple and intuitive approach to an estimation of volatility, in a scenario where returns are not necessarily iid, and their influence to the current variance is in some form diluted by how far away in the past they are. This idea can be modeled with the following weighted estimate of the variance

$$\sigma_w^2(t) = \sum_{k=1}^{m} \alpha_k r_{t-k}^2 \tag{2.57}$$

with $\sum_{k=1}^{m} \alpha_k = 1$ and so that $\alpha_1 > \alpha_2 > \ldots > \alpha_m$. In this way the values of returns that are further away in the past contribute less to the sum in Eq. (2.57). A standard assignment of weights that verify the strict decreasing property is to take $\alpha_k = (1 - \lambda)\lambda^{k-1}$, for $k = 1, \ldots, m$, where $0 < \lambda < 1$ and is known as the decay factor. This is the *Exponential Weighted Moving Average* (EWMA) model for the variance

$$\sigma_{ewma}^2(t) = (1 - \lambda) \sum_{k=1}^{m} \lambda^{k-1} r_{t-k}^2 \tag{2.58}$$

and it is this form which is usually considered at financial institutions. A good decay factor for developed markets has been estimated by J. P. Morgan's *RiskMetrics* methodology to be $\lambda = 0.94$. Observe that σ_{ewma}^2 can be obtained also from the following recursion, which is easily derived from Eq. (2.58)

$$\sigma_{ewma}^2(t) = \lambda \sigma_{ewma}^2(t-1) + (1 - \lambda) r_{t-1}^2 \tag{2.59}$$

This recurrence has the computational advantage of only needing to keep in memory the previous day return r_{t-1} and the previous estimate of σ_{ewma}^2. Therefore it is also suitable for on-line computation of volatility. In the R Lab 2.7.10 we propose to do a EWMA estimate of the volatility of AAPL and compare to the estimates done with previous methods.

2.7 Notes, Computer Lab and Problems

2.7.1 Bibliographic remarks: There are many textbooks on financial time series analysis and general statistics that can be recommended for complementing this chapter. Our main and preferred sources are Campbell et al. (1997) and Tsay (2010); in particular we recommend reading Chaps. 1 and 2 of Campbell et al. (1997) for a comprehensive and detailed exposition of the econometrics of financial time series. For mathematical statistics and probability in general we use and recommend Feller (1968), an all-time classic; also Breiman (1992) and Stirzaker (2005), which are nice complements. An encyclopedic account of 40 major probability distributions can be found in Forbes et al. (2011). A good source for the traditional core material of computational statistics through programmed examples in R is the book by Rizzo (2008).

2.7.2 R Lab: We show how to fit an exponential curve to the DJIA from 1978 to 2001 (see Fig. 2.1). If P_t is the price variable of the DJIA, we want to fit a model of the form $P_t = e^{a+bt}$. Taking logarithms the problem reduces to fit $\ln(P_t) = a + bt$, and we use regression analysis to estimate the coefficients a and b. R has the function `lm()` for fitting linear models using regression analysis, but it does not works well with `xts` objects (the data retrieved by `getSymbols` is encapsulated in that form), so we convert the data to numeric type. The R commands to do the `lm` fit and plot the results are the following:

```
> require(quantmod); getSymbols("DJIA",src="FRED")
> serie=DJIA["1978/2001"]
> price=as.numeric(serie) #extract numeric values of price
> time = index(serie) #extract the indices
> x=1:length(price)
> model=lm(log(price)~x)
> expo=exp(model$coef[1]+model$coef[2]*x)
> plot(x=time,y=price, main="Dow Jones",type="l")
> lines(time,expo,col=2,lwd=2)
```

2.7.3: In real life, every time you buy or sell an asset at a stock market you pay some fee for the transaction. Assume then, that the cost of a one-way transaction is some fixed percentage α of the asset's current price (for example, a reasonable transaction fee is in the range of 0.1 to 0.2 percent). Show that if we take into account this transaction fee then the k-period simple gross return is given by the equation

$$R_t[k] + 1 = \frac{P_t}{P_{t-k}} \cdot \frac{1 - \alpha}{1 + \alpha}.$$

2.7.4 Binomial distribution: An important discrete distribution is the binomial distribution, which can be obtained as the probability that n iid Bernoulli trials with probabilities p of success and $1 - p$ of failure result in k successes and $n - k$ failures. If S_n records the number of successes in the n Bernoulli trials, then S_n has the binomial distribution (written $S_n \sim Bin(n, p)$) with probability mass function

$$\mathbb{P}(S_n = k) = \left(\frac{n!}{(n - k)!k!} \right) p^k (1 - p)^{n-k}$$

Try proving that $E(S_n) = np$ and $Var(S_n) = np(1 - p)$. For further details on this particular distribution see Feller (1968). The binomial distribution plays an important role in the binomial tree option model to be studied in Chap. 4.

2.7.5: Let $\{\varepsilon_t\}$ be a time series with zero mean and covariance

$$Cov\,(\varepsilon_t, \varepsilon_s) = \begin{cases} \sigma^2 & \text{if } s = t \\ 0 & \text{if } s \neq t \end{cases}$$

ε_t is a weak form of white noise and is clearly weak stationary. Show that $X_t = a\varepsilon_t + b\varepsilon_{t-1}$, for constants a and b, defines a weak stationary time series.

2.7.6 R Lab: An R demonstration for doing some descriptive statistics of financial returns. We work with financial data from Allianz (ALV), Bayerische Motoren Werke (BMW), Commerzbank (CBK) and Thyssenkrupp (TKA), all German business trading in the Frankfurt Stock Exchange and listed in the main index DAX.

The data can be downloaded from the book's webpage, or if the reader wishes to build its own use the instructions in R Example 1.1. Each company's data is in a .csv file consisting of 5 columns labeled: Date, Open, High, Low, Close, Volume, AdjClose, and containing in each row a date, and for that date, the open price, highest price, lowest price, closing price, the total volume of transactions, and the adjusted close price. The data is ordered descending by date, beginning at 2009-12-30 down to 2003-01-02, and all files coincide by rows on the date.

```
> ### Part I: preprocessing the data ##########
> wdir="path-to-your-working-directory"; setwd(wdir)
>   # load the financial data from wdir
> ALV = read.csv(paste(wdir,"/ALV.csv", sep=""), header=T)
>   # extract 1 year of data from the AdjClose column
> alvAC= ALV$AdjClose[1:252]
>   ## repeat the previous instructs. with BMW, CBK, TKA.
> date= ALV$Date[1:252] # extract the column Date
> date <- as.Date(date) # and declare date as true date format
>   ##put all together into a data.frame
> dax =data.frame(date,alvAC,bmwAC,cbkAC,tkaAC)
>   # plot Adjusted prices vs date for ALV
> plot(dax$date,dax$alvAC, type="l",main="ALV.DE",
     +     xlab="dates",ylab="adj. close")
>   # Compute Returns. First define vectors of appropriate length
> alvR <- 1:252;  bmwR <- 1:252; cbkR <- 1:252; tkaR <- 1:252
> for (i in 1:252){alvR[i] <-(alvAC[i]/alvAC[i+1]) -1 }
>    #same with bmwR, cbkR, tkaR
>   # Remember dates are ordered descending. Make table Returns
> daxR =data.frame(dax$date,alvR,bmwR,cbkR,tkaR)
>   # Compute log returns (omit  column of dates)
> daxRlog <- log(daxR[2:5] +1)
> #plot returns and log returns (in red) and see coincidences:
> plot(dax$date,daxR$alvR, type="l",xlab="dates",ylab="returns")
> lines(dax$date,daxRlog$alvR, type="l",col="red")
> #### Part II: Basic statistics ############
> library(fBasics) ## load the library "fBasics"
> basicStats(daxRlog$alvR)
>   ## You can compute basic stats to a full data frame,
 + ## omitting non numeric data
> basicStats(na.omit(daxRlog[,2:5]))
>   ##Use a boxplot to help visualising and interpret results
> boxplot(daxRlog[,2:5])
> ##compute covariance matrix
> cov(daxRlog[,2:5],use="complete.obs")
> ####Extras: To save your table in your working directory
> write.table(dax,file="dax") ## or  as .csv use write.csv
>   ##To read the data saved in working directory
> dax = read.table("dax", header=T)
```

2.7.7: Prove Eq. (2.32). Use the definition of $E(X^n)$ for X log-normally distributed.

2.7.8 R Lab: With the data for one of the stocks analyzed in the previous R Lab, compute the bounds for the price of a stock one week ahead of some initial price P_0

chosen from the data, using the Eq. in (2.34). To estimate the mean and variance of the continuously compounded returns, use around 50 observations of prices previous to P_0. Compare your results with the real prices ahead of P_0.

2.7.9 R Lab: We explore in this lab the possibility of *aggregational normality* in stock returns. This refers to the fact that as one increases the time scale τ (daily, weekly and so on), the distribution of the τ-period returns looks more like a normal distribution. Execute in your R console the following commands (library(quantmod)):

```
> appl = getSymbols("AAPL",src="yahoo")
> apRd= periodReturn(appl,period="daily",type="log")
> dsd=density(apRd) #estimate density of daily log ret
> yl=c(min(dsd$y),max(dsd$y)) #set y limits
> plot(dsd,main=NULL,ylim=yl)
> ##plot the normal density with mean, stdv of apRd
> a=seq(min(apRd),max(apRd),0.001)
> points(a,dnorm(a,mean(apRd),sd(apRd)), type="l",lty=2)
```

Repeat the program with period $\in$ {weekly, monthly}. Comment your results.

2.7.10 R Lab: Implement in R the exponential weighted moving average (EWMA) model (Eq. (2.58)) to estimate the annualized volatility of AAPL by the end of May 2009, from its sample return. Use the data from R Example 2.6 and compare your estimate with the estimates done in that example.

Chapter 3
Correlations, Causalities and Similarities

It has been frequently observed that US markets leads other developed markets in Europe or Asia, and that at times the leader becomes the follower. Within a market, or even across different markets, some assets' returns are observed to behave like other assets' returns, or completely opposite, and thus may serve as pairs for a trading strategy or portfolio selection. The ability to assess the possible interdependence between two financial time series is what the study of different correlation measures will provide us. In this chapter we shall review and compare two widely used statistical measures of correlation: the ubiquitous in statistics Pearson's correlation coefficient ρ, and its non parametric counterpart Kendall's rank correlation coefficient τ. We shall see that in financial theory these two correlation measures serve for different purposes: while ρ is a measure of *linear* association between variables, τ can better capture monotone associations.

However, correlation is just a particular symmetric relation of dependency among stochastic variables, and so to know the degrees of asymmetry in the dependency of one financial time series with respect to another, whereby we may determine who leads and who follows, we must study measures of causality. Hence, we will study in this chapter the notion of causality introduced by Clive Granger in 1969, and review some of its application into testing for possible cause–effect links between different financial instruments.

For a more general and robust notion of dependency we shall adapt the data-mining concept of similarity to financial time series. The robustness is implicit in the conception of similarity as a distance measure, in the sense of being a metric. Thus the notion of two time series being close to each other or far apart has a precise mathematical meaning. This similarity metric will allow us to do unsupervised classification, or clustering, of financial time series, an important statistical technique for comparison and grouping according to predetermined properties. The chapter ends with a summary of the various statistical properties, or stylized facts, of asset returns.

A. Arratia, *Computational Finance*, Atlantis Studies in Computational Finance and Financial Engineering 1, DOI: 10.2991/978-94-6239-070-6_3, © Atlantis Press and the authors 2014

3.1 Correlation as a Measure of Association

As argue in Chap. 2 correlation is a key tool for measuring the strength of association between two random variables. There are different type of correlations. We have mentioned so far one motivated from exploiting the possible linear dependence among X and Y. We will see that in fact the linearity hypothesis is underlying the measure of correlation we arrived to in Chap. 2. We will then exhibit another measure of correlation which is not necessarily based on a possible linear relationship among the variables.

3.1.1 Linear Correlation

Definition 3.1 *(Pearson's ρ)* The Pearson's correlation coefficient ρ of two random variables X and Y is defined as

$$\rho(X, Y) = \frac{\text{Cov}(X, Y)}{\sqrt{\text{Var}(X)\text{Var}(Y)}} = \frac{E[(X - \mu_X)(Y - \mu_Y)]}{\sqrt{E[(X - \mu_X)^2]E[(Y - \mu_Y)^2]}} \tag{3.1}$$

Formally, ρ is a measure of the strength of *linear* dependence between X and Y. To see this, it is best to look at the following equivalent form of $\rho(X, Y)$:

$$\rho(X, Y)^2 = 1 - \frac{1}{\sigma_X^2} \min_{a,b} E[(Y - (aX + b))^2] \tag{3.2}$$

(Exercise: Show the above equation. In the notes at the end of the chapter we give a hint.)

Equation (3.2) shows that the correlation coefficient ρ of X and Y gives the relative reduction in the variance of Y by linear regression on X. It is clear from Eq. (3.2) that $\rho(X, Y) = \pm 1$ when there is perfect linear dependence between Y and X, that is, $\mathbb{P}(Y = aX + b) = 1$ for real numbers $a \neq 0$ and b. In the case of imperfect linear dependence we have $-1 < \rho(X, Y) < 1$; and if X and Y are independent, then $\rho(X, Y) = 0$ since $\text{Cov}(X, Y) = 0$.

In view of the above properties it is more precise to refer to ρ as *linear correlation*, which conveys the algebraic nature of dependence between the variables. It is partly because of these properties that makes linear correlation a very attractive measure of association. Other properties that add to its popularity are the following:

- It is easy to estimate from a sample $\{(x_i, y_i) : i = 1, \dots, m\}$ of (X, Y) by combining the formulas for sample variance and covariance, to get the sample linear correlation

$$\widehat{\rho}(X, Y) = \frac{\sum_{i=1}^m (x_i - \widehat{\mu}(X))(y_i - \widehat{\mu}(Y))}{\sqrt{\sum_{t=1}^m (x_t - \widehat{\mu}(X))^2 \sum_{t=1}^m (y_t - \widehat{\mu}(Y))^2}} \tag{3.3}$$

This $\widehat{\rho}$ is an unbiased estimator of ρ in the sense stated in Remark 2.2.

- It is invariant under changes of scale and location in X and Y:

$$\rho(aX + b, cY + d) = \text{sgn}(a \cdot b)\rho(X, Y),$$

where $\text{sgn}(\cdot)$ is the sign function.[1]
- It gives an elegant expression for the variance of a linear combination of random variables (see Problem 2 under Sect. 3.5).

We can now rewrite the covariance of X and Y as

$$\text{Cov}(X, Y) = \sigma_X \sigma_Y \rho(X, Y)$$

and this shows how the possible interdependency of X and Y is affected by their individual standard deviations. Thus, while covariance signals a possible dependence of one variable to the other, it is the correlation that determines the strength of this dependence.

R Example 3.1 In R the function `cor(x,y, ...)` computes the sample linear correlation $\hat{\rho}$. We first load to our workspace the data frame `daxRlog` build in the R Lab 2.7.6, containing 251 daily log-returns (for the period 06 Jan–30 Dec 2009) of ALV, BMW, CBK and TKA, stocks from the Frankfurt Stocks Exchange. Recall that `daxRlog`'s first column is `date` and we must omit this column for computing correlation. We do this by specifically setting to work with columns 2–5.

```
> cor(na.omit(daxRlog[,2:5]),method="pearson")
```

```
          alvR      bmwR      cbkR      tkaR
alvR 1.0000000 0.5978426 0.6038987 0.6678676
bmwR 0.5978426 1.0000000 0.4746328 0.6410805
cbkR 0.6038987 0.4746328 1.0000000 0.5337354
tkaR 0.6678676 0.6410805 0.5337354 1.0000000
```

The output is a correlation matrix giving at each entry the correlation value between pairs of stocks. The diagonal is obviously 1 since each stock is fully correlated with itself. We can see that in general there are significant values, so the dependency signaled by the covariance back in R Example 2.5 is now confirmed to be of relevance. We see that Allianz (`alvR`) and Thyssenkrupp (`tkaR`) have the highest correlation (0.66), while Commerzbank (`cbkR`) and BMW (`bmwR`) have the lowest correlation (0.47). □

Autocorrelation function (ACF). When dealing with time series of returns of financial assets, for which we assume are weakly stationary, it is of interest to assess the linear dependence of the series with its past; i.e., we look for *autocorrelation*. For a return time series $\{r_t : t = 1, \ldots, m\}$, the *lag-k autocorrelation* of r_t is the correlation coefficient between r_t and r_{t-k}, and is usually denoted by $\rho(k)$,

$$\rho(k) = \frac{\text{Cov}(r_t, r_{t-k})}{\sqrt{\text{Var}(r_t)\text{Var}(r_{t-k})}} = \frac{\text{Cov}(r_t, r_{t-k})}{\text{Var}(r_t)} = \frac{\gamma(k)}{\gamma(0)} \tag{3.4}$$

---

[1] $\text{sgn}(z)$ is either -1, 0 or 1, if $z < 0$, $z = 0$ or $z > 0$, respectively.

Note that the hypothesis of weakly stationarity is needed in order to have $\mathrm{Var}(r_t) = \mathrm{Var}(r_{t-k})$ and hence, $\rho(k)$ is only dependent on k and not of t, which also guarantees second moments are finite and $\rho(k)$ well-defined. The function $\rho(k)$ is referred to as the ACF of r_t.

For a given sample of returns $\{r_t : t = 1, \ldots, m\}$ and $0 \leq k < m$, the sample lag-k autocorrelation of r_t is

$$\widehat{\rho}(k) = \frac{\sum_{t=k+1}^{m} (r_t - \widehat{\mu}_m)(r_{t-k} - \widehat{\mu}_m)}{\sum_{t=1}^{m} (r_t - \widehat{\mu}_m)^2} = \frac{\widehat{\gamma}(k)}{\widehat{\gamma}(0)} \tag{3.5}$$

where $\widehat{\mu}_m = (1/m) \sum_{t=1}^{m} r_t$ is the sample mean of r_t.

Testing autocorrelation. There are various statistical tests to check whether the data are serially correlated. One such test is the *Ljung–Box* test, a so-called *portmanteau* test in the sense that takes into account all sample autocorrelations up to a given lag h. For a sample of m returns, the statistic is given by

$$Q(h) = m(m+2) \sum_{k=1}^{h} \frac{(\widehat{\rho}(k))^2}{m-k} \tag{3.6}$$

as a test for the null hypothesis $H_o : \rho(1) = \cdots = \rho(h) = 0$ against the alternative $H_a : \rho_i \neq 0$ for some $i \in \{1, \ldots, h\}$. H_o is rejected if $Q(h)$ is greater than the $100(1 - \alpha)$th percentile of a chi-squared distribution with h degrees of freedom. Alternatively, H_o is rejected if the p-value of $Q(h)$ is $\leq \alpha$, the significance level. For m time periods a good upper bound for the choice of lag is $h \approx \ln(m)$. Further details on this and other tests of autocorrelation can be learned from (Tsay 2010).

In R the function `acf(x, ...)` computes and plots estimates of the autocorrelation (by default) or autocovariance of a matrix of $N > 1$ time series or one time series object x, up to lag $5 \ln(m/N)$. This bound is justified by the Ljung–Box test and it is the default value; to give a specific number of lags use the parameter `lag.max`. The Ljung–Box test is implemented in R by the function `Box.test()`.

R Example 3.2 (Continue from R Example 3.1) From the data frame `daxRlog` extract the series of log returns for Allianz and compute all possible autocorrelations (ACF) with the instruction

```
> acf(na.omit(daxRlog$alvR),main="acf of ALV",ylim=c(-0.2,0.2))
```

We use `ylim` to scale the graph at the y-axis. The output is the plot shown in Fig. 3.1. The horizontal dashed lines mark the two standard error limits of the sample ACFs. Since all ACFs are within these limits, we can conclude that none of them is significantly different from zero with a 95 % of confidence; hence the log returns of Allianz in this time frame presents no autocorrelations (is true that at lag $= 17$ the ACF is above the limit, but not by a meaningful proportion, so we can discard it as a true autocorrelation).

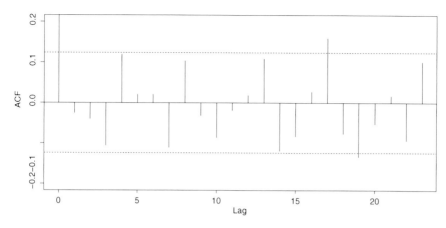

Fig. 3.1 ACF of ALV logreturns for the period 06 Jan–30 Dec 2009

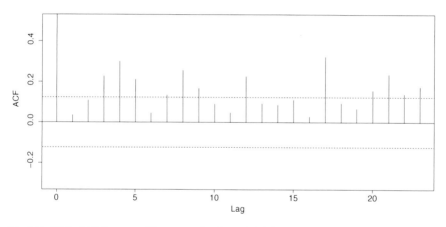

Fig. 3.2 ACF of ALV square of logreturns for the period 06 Jan–30 Dec 2009

Now compute the autocorrelations for the square of the log returns of Allianz in the same period 2009.

```
> acf(na.omit(daxRlog$alvR)^2,main=" ",ylim=c(-0.3,0.5))
```

The output is shown in Fig. 3.2. We observe significant correlations at lags 3, 4, 5, 8, 12, 17 and 21; that is, the autocorrelations at these lags are over the 5 % level. Since a good estimate of the variance of the return $\{r_t\}$ of a stock is the sum of the squares of $\{r_t\}$, this experimental result indicates that there is some linear dependence of the variance of AVL with its past values, at least at lag 3.

Deficiencies of linear correlation as dependency measure. A serious deficiency of the linear correlation ρ is that it is not invariant under all transformations of random variables X and Y for which order of magnitude is preserve. That is, for two real-valued random variables and arbitrary transformations $h, g : \mathbb{R} \to \mathbb{R}$, we have in general

$$\rho(h(X), g(Y)) \neq \rho(X, Y)$$

For example, for normally-distributed vectors (X, Y) and the standard normal distribution function $h(x) = \Phi(x)$, we have

$$\rho(h(X), h(Y)) = (6/\pi) \arcsin(\rho(X, Y)/2)$$

(see Joag-dev (1984, §1)). In general, under the normal distribution,

$$|\rho(h(X), g(Y))| \leq |\rho(X, Y)|$$

for arbitrary real-valued transformations h, g (cf. Embrechts et al. 2002). The quality of been invariant under arbitrary order-preserving transformations is desirable in order to obtain a distribution-free measure of dependency (and hence, non-parametric), since inferences can be made by relative magnitudes as opposed to absolute magnitudes of the variables. An additional deficiency of linear correlations is that the variances of X and Y must be finite for the correlation to be defined. This is a problem when working with heavy-tailed distributions, which is often the case for financial time series.

3.1.2 Properties of a Dependence Measure

Thus, what can be expected from a good dependency measure? Following the consensus of various researchers, summarized in Joag-dev (1984), Embrechts et al. (2002) and Gibbons and Chakraborti (2003, Chap. 11), among others, a "good" dependency measure $\delta : \mathbb{R}^2 \to \mathbb{R}$, which assigns a real number to any pair of real-valued random variables X and Y, should satisfy the following properties:

(1) $\delta(X, Y) = \delta(Y, X)$ (symmetry).
(2) $-1 \leq \delta(X, Y) \leq 1$ (normalization).
(3) $\delta(X, Y) = 1$ if and only if X, Y are co-monotonic; that is, for any two independent pairs of values (X_i, Y_i) and (X_j, Y_j) of (X, Y)

        $X_i < X_j$ whenever $Y_i < Y_j$ or $X_i > X_j$ whenever $Y_i > Y_j$.

This property of pairs is termed *perfect concordance* of (X_i, Y_i) and (X_j, Y_j).
(4) $\delta(X, Y) = -1$ if and only if X, Y are counter-monotonic; that is, for any two independent pairs of values (X_i, Y_i) and (X_j, Y_j) of (X, Y)

        $X_i < X_j$ whenever $Y_i > Y_j$ or $X_i > X_j$ whenever $Y_i < Y_j$.

In this case, it is said that the pairs (X_i, Y_i) and (X_j, Y_j) are in *perfect discordance*.
(5) δ is invariant under all transformations of X and Y for which order of magnitude is preserve. More precisely, for $h : \mathbb{R} \to \mathbb{R}$ strictly monotonic on the range of X,

$$\delta(h(X), Y) = \begin{cases} \delta(X, Y) & \text{if } h \text{ is increasing,} \\ -\delta(X, Y) & \text{if } h \text{ is decreasing.} \end{cases}$$

The linear correlation ρ satisfies properties (1) and (2) only. A measure of dependency which satisfies all five properties (1)–(5) is Kendall's τ. The trick is to based the definition on the probabilities of concordance and discordance of the random variables.

3.1.3 Rank Correlation

Definition 3.2 *(Kendall's τ)* Given random variables X and Y, the Kendall correlation coefficient τ (also known as rank correlation) between X and Y is defined as

$$\tau(X, Y) = p_c - p_d \tag{3.7}$$

where, for any two independent pairs of values (X_i, Y_i), (X_j, Y_j) from (X, Y),

$$p_c = \mathbb{P}((X_j - X_i)(Y_j - Y_i) > 0) \quad \text{and} \quad p_d = \mathbb{P}((X_j - X_i)(Y_j - Y_i) < 0)$$

p_c and p_d are the probabilities of concordance and discordance, respectively.

Thus, Kendall's measure of dependency reflects the agreement in monotonicity between two random variables. To estimate $\tau(X, Y)$ from n pairs of sample random values (x_1, y_1), (x_2, y_2), ..., (x_n, y_n), define

$$A_{ij} = \text{sgn}(x_j - x_i)\text{sgn}(y_j - y_i)$$

Then $A_{ij} = 1$ if these pairs are concordant; $A_{ij} = -1$ if the pairs are discordant; or 0 if the pairs are neither concordant nor discordant. An unbiased estimation of $\tau(X, Y)$ is given by

$$\widehat{\tau}_n(X, Y) = 2 \sum_{1 \le i < j \le n} \frac{A_{ij}}{n(n-1)} \tag{3.8}$$

The distribution of $\widehat{\tau}_n$ under the null hypothesis of no association ($\tau(X, Y) = 0$) is known. Tables for values of $\widehat{\tau}_n$ for $n \le 30$ can be found in Gibbons and Chakraborti (2003). Moreover, $\widehat{\tau}_n$ is asymptotically normal with zero mean and variance $2(2n + 5)/9n(n - 1)$.

R Example 3.3 The function `cor(x,y, ...)` can also compute correlation with the sample estimator of Kendall's tau by setting `method="kendall"`. We use the same data as in R Example 3.1 to compare with Pearson's linear correlation.

```
> cor(na.omit(daxRlog[,2:5]),method="kendall")
          alvR        bmwR        cbkR        tkaR
```

```
alvR 1.0000000 0.4132271 0.4484194 0.4714263
bmwR 0.4132271 1.0000000 0.3514914 0.4263586
cbkR 0.4484194 0.3514914 1.0000000 0.4032713
tkaR 0.4714263 0.4263586 0.4032713 1.0000000
```

These numbers give a measure of the strength of monotonic agreement, regardless of any possible linear dependency among the variables. It complements (and in this case, corroborates) some of the associations observed in R Example 3.1: Allianz (alvR) and Thyssenkrupp (tkaR) present the highest monotonic correlation, while Commerzbank (cbkR) and BMW (bmwR) present the lowest. It might be instructive to plot the four series together and look for graphical confirmation of both Pearson and Kendall correlations. Use the plot and lines functions (see usage in the R on-line manual).   □

3.2 Causality

There is a popular dictum that *"correlation is not causation"*, which should prevent researchers from inferring a cause–effect relationship between variables satisfying a strong correlation. Examples abound in the literature. The following plot has become a popular example in statistic courses given at universities in Spain.

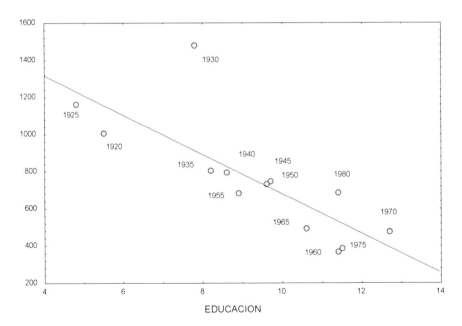

Fig. 3.3 Population of donkeys versus budget of education in Spain

Figure 3.3 shows a fairly good linear regression (hence, correlation) between the population of donkeys and the budget in education in Spain, from the years 1920 to 1975. One sees that the higher the budget in education, the fewer the living donkeys. Can we talk about causality here?

3.2.1 Granger Causality

The kind of cause–effect relation that we are interested in should be one suited for time series; one that considers the random nature of the variables and the flow of time. It should be a concept based purely on observations of past and present values, and where there is no possibility of external intervention to alter these values, as it is the case in experimental sciences where the estimation of causal effect from observational data is measured by changes in the distribution of the variables. The measure of causality for time dependent random variables that we need is the one given by Clive Granger (1969).

The basic idea of Granger causality, in the case of two random variables X and Y, is the following:

> X is said to (Granger) cause Y, if Y can be better predicted using the histories of both X and Y than it can by using the history of Y alone.

This idea can be mathematically realized by computing a regression of variable Y on the past values of itself and the past values of X, and testing the significance of coefficient estimates associated with X. There are various statistical tests of significance for this regression-based Granger causality. Geweke et al. (1983) analyze eight of the most popular such tests to conclude that the Wald variants are the more accurate. Toda and Yamamoto (1995) introduced a very simple Wald test statistic that asymptotically has a chi-square distribution, regardless of the variables being stationary or not (originally Granger assumed variables were strictly stationary for his test), and this version of causality testing has become one of the most widely applied (and as we shall see it is implemented in R). This form of Granger causality is the one we define below.

To mathematically test whether X (Granger) causes Y, one proceeds as follows:

(1) Consider a bivariate linear autoregressive model on X and Y, making Y dependent on the history of X and Y. This is the full model M_f. Additionally, consider a linear autoregressive model on Y, the restricted model M_r:

$$M_f : Y_t = a_0 + a_1 Y_{t-1} + \cdots + a_h Y_{t-h} + b_1 X_{t-1} + \cdots + b_h X_{t-h} + \varepsilon_t$$
$$M_r : Y_t = a_0 + a_1 Y_{t-1} + \cdots + a_h Y_{t-h} + \varepsilon_t$$

where h is the maximum number of lagged observations (for both X and Y). The coefficients a_i, b_i are the contributions of each lagged observation to the predicted value of Y_t, and ε_t is the residual or prediction error.

(2) Consider the null hypothesis

$$H_0 : X \text{ does not cause } Y$$

which is equivalent to $H_0 : b_1 = b_2 = \cdots = b_h = 0.$

Assuming a normal distribution of the data we can evaluate the null hypothesis through an F-test. This consists in doing the following:

(3) Calculate both Residual Sums of Squares,[2] RSS_f of the full model M_f and RSS_r of the restricted model M_r (Presumably RSS_f should fit the data better than RSS_r, this means that it should have lower error).

(4) Compute $F = \dfrac{RSS_r - RSS_f}{RSS_f} \cdot \dfrac{n - 2h - 1}{h}$. Under the null hypothesis H_0 (i.e. that model M_f does not provide a significantly better fit than model M_r), F will have an F *distribution*, with $(h, n - 2h - 1)$ degrees of freedom. The null hypothesis H_0 is rejected, meaning that X causes Y, if the F calculated from the data is greater than the critical value of the F distribution for some desired false-rejection probability (usually 0.05).

(5) As an alternative (and equivalent) significance test, statisticians compute the p-value for the F test, which is the probability, under the given null hypothesis, of observing an F at least as extreme as the one observed (i.e. the F_{obs} being the one calculated by the formula above). Formally, $p = \mathbb{P}_{H_0}(F > F_{obs})$. If the p-value is less than 0.05, one rejects the null hypothesis, and hence X causes Y. It is the p-value (and not the critical values of F distribution) what's usually computed numerically by statistics packages.

A first application. Christopher Sims (1972) made one of the earliest econometric application of the Granger causality measure to corroborate a thesis by Milton Friedman and Anna Schwartz asserting that money growth cause changes in the GNP of the US.[3] In the following R Example we reproduce Sims causality test of money to gross national product.

R Example 3.4 (Granger test in R) R has a function `grangertest` for performing a test for Granger causality, included in the package `lmtest`, a package for testing linear regression models. One of the key functions in this package is `lm` which builds regression models. The `grangertest` function just takes `lm` and builds for you the models M_r and M_f, with the lags given by the user; computes the residual sum of squares RSS; computes de F value and p-value for the null hypothesis. The usage is `lm(x ~ y)` which outputs a linear regression of x as function of y ($(x \sim y)$ denotes that $x = f(y)$).

To explore the possible causality direction between money and the GNP in the US we use the data `USMoney` in the package `AER` which contains a quarterly multiple time series from 1950 to 1983 with three variables: `gnp` (nominal GNP), `m1` (M1 measure of money stock), `deflator` (implicit price deflator for GNP). By executing the following commands at your R console, you will see the listed output

---

[2] For $y_t = f(x_t)$ and n observations, the residual sum of squares is $RSS = \sum_{t=1}^{n}(y_t - f(x_t))^2$.

[3] Friedman, M., & Schwartz, A. J. (1963). Money and business cycles. *The Review of Economics and Statistics*, 45, 32–64.

```
> library("lmtest")
> library("AER") ##Econometric library data
> data("USMoney")
> ##To know if m1 causes gnp
> grangertest(gnp~m1, order=3,data=USMoney)

Granger causality test
Model 1: gnp ~ Lags(gnp, 1:3) + Lags(m1, 1:3)
Model 2: gnp ~ Lags(gnp, 1:3)
  Res.Df Df     F    Pr(>F)
1    126
2    129 -3 6.0479 0.0007011 ***

> ##To know if gnp causes m1
> grangertest(m1~gnp, order=3,data=USMoney)

Granger causality test
Model 1: m1 ~ Lags(m1, 1:3) + Lags(gnp, 1:3)
Model 2: m1 ~ Lags(m1, 1:3)
  Res.Df Df    F   Pr(>F)
1    126
2    129 -3 2.734 0.04648 *
---
Signif. codes:  0 *** 0.001 ** 0.01 * 0.05 . 0.1  1
```

In this output Model 1 is M_f and Model 2 is M_r. The observed F_o value for the first test (gnp as a function of m1) is 6.0479 and its p-value, the $Pr(F > F_o)$ is 0.0007011. Thus we can surely reject the null hypothesis, and conclude that money causes GNP. The second (reverse) test gives p-value 0.04648, which is too close to the critical value and we can not reject the null hypothesis, thus cannot say for sure that GNP causes money.

```
## alternative way of specifying the last test
> grangertest(USMoney[,1],USMoney[,2],order=3)
```

Check the documentation on `grangertest` for further variations on its use. □

3.2.2 Non Parametric Granger Causality

There are two major drawbacks on applying the (parametric) Granger test just explained to financial time series. These are:

(1) it assumes a linear dependence on variables X and Y;
(2) it assumes data are normally distributed (the F-test works under this distribution).

These are seldom the cases for time series of financial returns. Thus a non-linear and nonparametric test for Granger causality hypothesis is desirable. (The crux of nonparametric testing is that does not rely on any assumption about the distribution of the data.)

The most frequently used nonparametric test of Granger causality is the Hiemstra and Jones (1994) test (HJ). However, a more recent work of Diks and Panchenko (2006) presents an improvement of the HJ test after having shown that the HJ test can severely over-reject if the null hypothesis of noncausality is not true. Hence the nonparametric causality test that we shall adopt is the one by Diks and Pachenko (DP), and it is this test which is explained below.

Let us begin with a more general formulation of the Granger causality in terms of distributions of the random variables. To say that $\{X_t\}$ Granger causes $\{Y_t\}$ is to say that the distribution of future values $(Y_{t+1}, \ldots, Y_{t+k})$ conditioned to past and current observations X_s and Y_s, $s \leq t$, is not equivalent to the distribution of $(Y_{t+1}, \ldots, Y_{t+k})$ conditioned to past and current Y-values alone. In practice it is often assume that $k = 1$; that is, to test for Granger causality comes down to comparing the one-step ahead conditional distribution of $\{Y_t\}$ with and without past and current values of $\{X_t\}$. Also, in practice, tests are confined to finite orders in $\{X_t\}$ and $\{Y_t\}$. Thus, to mathematically formalize all this, consider some time lags $l_X, l_Y \geq 1$ and define delay vectors $X_t^{l_X} = (X_{t-l_X+1}, \ldots, X_t)$ and $Y_t^{l_Y} = (Y_{t-l_Y+1}, \ldots, Y_t)$. Then it is said that $\{X_t\}$ Granger causes $\{Y_t\}$ if

$$Y_{t+1} | (X_t^{l_X}; Y_t^{l_Y}) \nsim Y_{t+1} | Y_t^{l_Y} \tag{3.9}$$

Observe that this definition of causality does not involve model assumptions, and hence, no particular distribution.

As before the null hypothesis of interest is that "$\{X_t\}$ is not Granger causing $\{Y_t\}$", which for the general formulation (3.9) translates to

$$H_0 : Y_{t+1} | (X_t^{l_X}; Y_t^{l_Y}) \sim Y_{t+1} | Y_t^{l_Y} \tag{3.10}$$

For a strictly stationary bivariate time series $\{(X_t, Y_t)\}$, Eq. (3.10) is a statement about the invariant distribution of the $(l_X + l_Y + 1)$-dimensional vector $W_t = (X_t^{l_X}, Y_t^{l_Y}, Z_t)$, where $Z_t = Y_{t+1}$. To simplify notation, and to bring about the fact that the null hypothesis is a statement about the invariant distribution of W_t, drop the time index t, and further assume that $l_X = l_Y = 1$. Then just write $W = (X, Y, Z)$, which is a 3-variate random variable with the invariant distribution of $W_t = (X_t, Y_t, Y_{t+1})$.

Now, restate the null hypothesis (3.10) in terms of ratios of joint distributions. Under the null the conditional distribution of Z given $(X, Y) = (x, y)$ is the same as that of Z given $Y = y$, so that the joint probability density function $f_{X,Y,Z}(x, y, z)$ and its marginals must satisfy

$$\frac{f_{X,Y,Z}(x, y, z)}{f_Y(y)} = \frac{f_{X,Y}(x, y)}{f_Y(y)} \cdot \frac{f_{Y,Z}(y, z)}{f_Y(y)} \tag{3.11}$$

for each vector (x, y, z) in the support of (X, Y, Z). Equation (3.11) states that X and Z are independent conditionally on $Y = y$, for each fixed value of y. Then Diks and Pachenko show that this reformulation of the null hypothesis H_0 implies

$$q \equiv E[f_{X,Y,Z}(X, Y, Z)f_Y(Y) - f_{X,Y}(X, Y)f_{Y,Z}(Y, Z)] = 0$$

and from this equation they obtained an estimator of q based on indicator functions: let $\widehat{f}_W(W_i)$ denote a local density estimator of a d_W-variate random vector W at W_i defined by

$$\widehat{f}_W(W_i) = \frac{(2\varepsilon_n)^{-d_W}}{n - 1} \sum_{j \neq i} I_{ij}^W$$

where $I_{ij}^W = I(\|W_i - W_j\| < \varepsilon_n)$ with $I(\cdot)$ the indicator function and ε_n the bandwidth, which depends on the sample size n. Given this estimator, the test statistic for estimating q is

$$T_n(\varepsilon_n) = \frac{n - 1}{n(n - 2)} \sum_i \left(\widehat{f}_{X,Y,Z}(X_i, Y_i, Z_i)\widehat{f}_Y(Y_i) - \widehat{f}_{X,Y}(X_i, Y_i)\widehat{f}_{Y,Z}(Y_i, Z_i) \right)$$

$$\tag{3.12}$$

It is then shown that for $d_X = d_Y = d_Z = 1$ and letting the bandwidth depend on sample size as $\varepsilon_n = Cn^{-\beta}$, for $C > 0$ and $1/4 < \beta < 1/3$, the test statistic T_n satisfies

$$\sqrt{n}\frac{(T_n(\varepsilon_n) - q)}{S_n} \xrightarrow{d} N(0, 1)$$

where $\xrightarrow{d}$ denotes convergence in distribution and S_n is an estimator of the asymptotic variance σ^2 of T_n.

Practical issues. To apply the T_n test use $C = 7.5$, and $\beta = 2/7$, to get ε_n. However, for small values of n (i.e. <900) we can get unrealistically large bandwidth; hence Panchenko (2006) recommend truncating the bandwidth as follows:

$$\varepsilon_n = \max(Cn^{-2/7}, 1.5) \tag{3.13}$$

See Table 1 of Diks and Panchenko (2006) for suggested values of ε_n for various values of n. Lags for both time series should be taken equal: $l_X = l_Y$ (although asymptotic convergence has only been proven for lag 1). See Sect. 3.5 for a description of a C program publicly available for computing T_n.

Example 3.1 (**A list of applications of Granger causality for your considera-tion**) There is a wealth of literature on uncovering causality relationships between different financial instruments, and a large amount of these works have been done employing the original parametric causality test of Granger, or at best with the nonlinear nonparametric HJ test, which, as has been pointed out, can over reject. Therefore, it is worth to re-explore many of these causality tests with the improved nonparametric causality test DP just explained, and further investigate improvements (for as we mentioned, the test statistic (3.12) has been shown to converge in distribution only for lag 1). Here is a short list of causality relations of general interest, and that you can explore with the tools exposed in this section.

- Begin with the classic money and GNP relation, considered by Sims and contemporary econometrists. Further explore the relation between money, inflation and growth through different epochs and for different countries.
- In stock markets, explore the price–volume relationship. It has been observed a positive correlation between volume and price. The question is, does volume causes price? The paper of Hiemstra and Jones (1994) deals with this question.
- The transmission of volatility or price movements between international stock markets. The question is, does the volatility of the index of a stock market (e.g. SP500) causes the volatility of the index of another stock market (e.g. Nikkei). This observed phenomenon is sometimes termed *volatility spillover*, and has been explored many times and with several markets. More recently it has been tested for a set of developed and emerging markets, using the DP test, in Gooijer and Sivarajasingham (2008).
- Between exchange rates.
- Between monetary fundamentals and exchange rates.
- Between high-frequent SP500 futures and cash indexes.[4]   □

3.3 Grouping by Similarities

In any given stock exchange market all business with shares outstanding are classified by *industrial sectors* (or *industry*) such as banking, technology, automotive, energy, and so on, and according to the business primary activities. In this, and other forms of classifying business and their stocks (by the nature of their share holders, by income statements, by revenues, etc.), there is an underlying idea of similarity for forming groups, or dissimilarity for separating elements, that goes beyond the straightforward structural dependencies among prices, or returns, or other statistics, although it is expected that being similar under any of these more general (and often subjective) criteria implies some form of structural dependency. For example, investors generally expect that the prices of stocks from the same market and industry should eventually move together. To investigate these broader forms of similarity classification of financial time series, the appropriate statistical technique to call for is *cluster analysis*.

---

[4] To learn the need of this causality relation see Dwyer, G. P., Locke, P., & Yu, W. (1996). Index arbitrage and nonlinear dynamics between the S&P 500 futures and cash. *Review of Financial Studies, 9*, 301–332.

Clustering is an unsupervised classification method which groups elements from a dataset under some similarity relation, in such a way that elements in the same group are closely similar while any two elements in different groups are far apart according to the stated similarity criterion. The choice of similarity relation depends on the nature of the data and the statistical properties we would like to deduce from them. A way to mathematically realized the notions of "closely" and "far apart" is to endow the similarity relation with the characteristics of a distance measure. Hence it is the dissimilarity between two observations that is determined by computing the distance separating the observations, and the clustering procedures are designed so as to maximized inter-clusters distances (dissimilarity) and minimized the within-a-cluster distances (similarity as minimal dissimilarity).

In the following section we review the fundamentals of data clustering and some of the most commonly used methods. The intention is to give the reader the minimal background so that following afterwards more specialized literature on the subject can carry on research on clustering, in particular related to financial time series.

3.3.1 Basics of Data Clustering

Clustering algorithms divide the data into groups or *clusters*. According to how this group division is performed a clustering method belongs to one of two major categories: hierarchical or partitional. Hierarchical methods begin with a gross partition of the data, which can be either to consider all the data as one single cluster, or each data element as a separate cluster, and successively produce other partitions using the previously established clusters and applying a similarity criterion, until certain convergence conditions are met. The case where the method begins with each element being a cluster by itself, and then in successive steps these are join together to form larger clusters, is called *agglomerative*. The other case, which begins with all data in one cluster and makes successive divisions to get smaller clusters, is called *divisive*. Partitional methods begin with a predetermined number of clusters (chosen arbitrarily or with some heuristic), and applying a similarity criterion elements are moved from one cluster to another, until convergence conditions are met. As commented before, it is more sound to work with a distance measure to quantify the dissimilarity between pairs of data elements.

Formally, given data as a set of vectors $x = (x_1, \ldots, x_s)$ in $\mathbb{R}^s$, where each component x_i is a feature or attribute, a distance metric on this space of real vectors is a bivariate function $d : \mathbb{R}^s \times \mathbb{R}^s \to \mathbb{R}$, which satisfies the following properties:

(1) $d(x, y) \geq 0$, for all x, y (positivity).
(2) $d(x, y) = 0$ if, and only if, $x = y$ (reflexivity).
(3) $d(x, y) = d(y, x)$ (symmetry).
(4) $d(x, y) \leq d(x, z) + d(z, y)$, for all x, y, z (triangle inequality).

On distance measures. The most commonly used distance measures for classifying data whose features are continuous belong to the family of L_p norms:

$$\|x - y\|_p = \left(\sum_{1 \le i \le s} |x_i - y_i|^p \right)^{1/p} \tag{3.14}$$

for $p \ge 1$. Among these, special attention is given to L_2, the Euclidean distance (which is more common to denote it by $\|x - y\|$, i.e. without the subscript 2), and L_1, known as the "Manhattan" distance (denoted $|x - y|$). A reason for the preference of L_p norm for $p = 1, 2$, and not higher is that for larger p it is more severe the tendency of the largest-scale feature to dominate the others, a problem that might be alleviated by normalization of the features (cf. Jain et al. 1999).

Example 3.2 To compare two stocks, A and B, it is usual to take as their vectors of features their respective time series of returns observed through some period of time, $r_A = (r_{A1}, \ldots, r_{As})$ and $r_B = (r_{B1}, \ldots, r_{Bs})$. Each observed return value r_i is a feature of the stock, and we can apply some L_p norm to quantify the dissimilarity among these stocks' features; for example, use the Euclidean distance

$$\|r_A - r_B\| = \left(\sum_{1 \le i \le s} |r_{Ai} - r_{Bi}|^2 \right)^{1/2}$$

In this case one is measuring the spread among the return history of the pair of stocks. □

In Aggarwal et al. (2001) it is shown that as p increases the contrast between the L_p-distances to different data points in high dimensional space is almost null. Hence, for classifying data in high dimensional space with L_p-norms, and in particular classifying financial time series over long periods of time, seems preferable to use lower values of p. This is a form of the *curse of dimensionality* suffered by the L_p metric. Thus, to cope with this problem, the Manhattan distance should be preferable over Euclidean; Euclidean preferable over L_3-norm, and so on. As a step further Aggarwal et al. (2001) propose to use a fractional L_p-norm, where p is taken as a fraction smaller than 1, and show that this metric is indeed better at displaying different degrees of proximity among objects of high dimension. Also, the L_1 and L_2 norms are sensitive to noise and outliers present in the time series, so here too some form of normalization or filtering of the data should be needed.

Another family of distance measures of interest are those based on linear correlation. Note that the correlation coefficient $\rho(x, y)$ is not by itself a metric, so it must be composed with other functions to turn it into a metric. The following two compositions give correlation-based distance metrics:

$$d_\rho^1(x, y) = 2(1 - \rho(x, y)) \tag{3.15}$$

and for $m > 0$,

$$d_\rho^2(x, y) = \left(\frac{1 - \rho(x, y)}{1 + \rho(x, y)}\right)^m \tag{3.16}$$

However, as it has been noted in Hastie et al. (2009, §14.3), if observations are normalized then $2(1 - \rho(x, y)) \approx \|x - y\|^2$; so that clustering based on linear correlation is equivalent to that based on squared Euclidean distance. As a consequence these distances based on correlation suffer from the same curse of dimension for L_2 norm mentioned above.

There are many others distance measures designed for quantifying similarity among objects of various nature: the Mahalanobis distance which corrects data with linearly correlated features, the maximum norm and the cosine distance are other measures for analyzing similarity of numerical data; for comparing sets and strings there are the Hamming distance, the Jaccard dissimilarity, Lempel-Ziv and information distances, and many others. For an encyclopedic account of distances see Deza and Deza (2009), but to keep within the context of statistical classification compare with Hastie et al. (2009).

3.3.2 Clustering Methods

The general goal of clustering can be formally stated as the following decision problem:

Given a finite set X of data points, a dissimilarity or distance function $d : X \times X \to \mathbb{R}^+$, an integer k and an objective function f defined for all k-partitions $X_1, \ldots, X_k$ of X, $f(X_1, \ldots, X_k) \in \mathbb{R}^+$, to find a k-partition with minimum objective function value, i.e., a partition $X_1^*, \ldots, X_k^*$ of X such that

$$f(X_1^*, \ldots, X_k^*) = \min[f(X_1, \ldots, X_k) : X_1, \ldots, X_k \text{ is a } k\text{-partition of } X].$$

Depending on the dimension of the data elements $x \in X$ (i.e., the number of features we want to compare), the number k of clusters, and for various objective functions f, this problem is, more often than not, NP-hard (see, e.g., Brucker 1977; Gonzalez 1985; Garey and Johnson 1979). Hence, clustering algorithms are build with the goal of optimizing the underlying objective function which is usually a measure of dissimilarity between clusters, and solutions are often approximations, not the best possible.

The computational complexity of the clustering problem has motivated a lot of research on clustering techniques. We shall limit our presentation to two of the most popular methods, the agglomerative hierarchical clustering algorithm and the

k-means partitional algorithm, and then show the equivalence of the clustering problem to the graph partition problem. This important link allows to export a broad box of graph combinatorial tools to aid for the solution of data clustering problems.

3.3.2.1 Hierarchical Clustering

In hierarchical clustering algorithms the dissimilarity between clusters is quantified by defining a distance between pairs of clusters as a positive real-valued function of the distance between elements in the clusters. Then, besides being agglomerative or divisive, we get further variants of the hierarchical clustering algorithms by changing the definition of the inter-clusters distance (which by being the objective function to optimize, it has an influence on the quality and precision of the resulting clusters).

The most popular variants of inter-cluster distances are the *single-link, complete-link, average-link* and *Ward* distance (also known as minimum variance), defined as follows. Given $X = \{x_1, \ldots, x_n\}$ a set of data points in $\mathbb{R}^s$ and $d : \mathbb{R}^s \times \mathbb{R}^s \to \mathbb{R}^+$ a distance metric, the distance between two clusters, C_i and C_j (subsets of X), is denoted $\widehat{d}(C_i, C_j)$. Then $\widehat{d}$ is

- single-link if $\widehat{d}(C_i, C_j) = \min[d(x_i, x_j) : x_i \in C_i, x_j \in C_j]$;
- complete-link if $\widehat{d}(C_i, C_j) = \max[d(x_i, x_j) : x_i \in C_i, x_j \in C_j]$;
- average-link if $\widehat{d}(C_i, C_j) = \dfrac{1}{n_i n_j} \sum_{x_i \in C_i} \sum_{x_j \in C_j} d(x_i, x_j)$, where n_i is the cardinality of C_i.
- Ward (or minimum variance) if $\widehat{d}(C_i, C_j) = \dfrac{n_i n_j}{n_i + n_j} \|c_i - c_j\|^2$, where $n_i = |C_i|$

as before, c_i is the centroid of C_i, and $\| \cdot \|^2$ is the squared Euclidean norm (Eq. (3.14)), which is the underlying distance between elements.[5] The centroid of C_i is defined as the point c_i in C_i such that for all x_j, x_k in C_i, $\|x_k - c_i\|^2 \leq \|x_k - x_j\|^2$.

The agglomerative hierarchical clustering algorithm is a bottom-up procedure beginning at level 0 with each element being its own cluster, and recursively merging a pair of clusters from previous level into a single cluster subject to minimizing the inter-cluster distance, as given by some form of $\widehat{d}$ as above, or other. The algorithm stops when all data elements are in one cluster, and this marks the top level. The details of this clustering heuristic are presented as Algorithm 3.1.

---

[5] Joe Ward (1963) considered cluster analysis as an analysis of variance problem, and consequently he considers as point-distance the squared Euclidean norm, and his objective function to minimize (the inter-cluster distance) when merging two clusters is the the sum of squared deviations or errors. This results in the formula given for $\widehat{d}$.

Algorithm 3.1 Agglomerative hierarchical clustering

0. **input**: $X = \{x_1, \ldots, x_n\} \subset \mathbb{R}^s$ data points;
   distance metric $d : X \times X \to \mathbb{R}^+$;
   inter-cluster distance: $\widehat{d} : \mathscr{P}(X) \times \mathscr{P}(X) \to \mathbb{R}^+$.
1. Define singleton clusters $C_1 = \{x_1\}, \ldots, C_n = \{x_n\}$;
   level $k = 0$; partition $S_k = \{C_1, \ldots, C_n\}$
2. **repeat:**
3.   Find the minimum inter-cluster distance at level k partition S_k:
   $$\widehat{d}_k := \min[\widehat{d}(C_i, C_j) : C_j, C_j \in S_k]$$
4.   **for each** pair of clusters C_i, C_j in S_k with $\widehat{d}(C_i, C_j) = \widehat{d}_k$
   merge them together to form new cluster $C_i \cup C_j$.
5.   Define next partition S_{k+1} consisting of all newly formed clusters
   (obtained by merging according to $\widehat{d}_k$) plus clusters in S_k
   that were not merged. Update level $k = k + 1$.
6. **until** all data points are in one cluster (i.e. top level partition $S_f = \{X\}$).

The algorithm generates a nested hierarchy of partitions $S_0, S_1, \ldots, S_f$, which is best visualized as a tree. This tree is obtained by representing every merge done in step 4 by an edge drawn between the pair of clusters. In fact, many implementations of the agglomerative hierarchical clustering algorithm does exactly that and outputs such a tree, called *dendrogram*. Then, a specific partition is chosen by the user by cutting the tree at a desired level k, which corresponds to choosing the level k partition S_k. A standard rule of thumb to compute this cut is to take the mean of the inter-clusters distances associated with the clustering method, and computed throughout the algorithm. Finally, note that the time complexity of this algorithm is of the order $O(n^2 \log n)$, where n is the number of data elements to classify.

R Example 3.5 In this R Lab we show how to do an agglomerative hierarchical clustering of 34 stocks belonging to the main Spanish market index IBEX.[6] For each stock we consider its series of daily returns, observed from 1 Dec 2008 to 1 Feb 2009, which amounts to 40 observations; each series of returns is a column in a table labelled `Ibex0809`. Agglomerative hierarchical clustering analysis can be done in R with the function `hclust(d, method, ...)` from the package `stats`. The parameter `d` is a dissimilarity structure produced by function `dist` (a distance measure), and `method` can be any of the four agglomerative methods. We experiment with all four methods and use as our dissimilarity measure the correlation-based distance from Eq. (3.15), namely $2(1 - \rho(x, y))$. Enter in your R console the following instructions

---

[6] The tickers for these stocks are: ABE, ABG, ACS, ACX, ANA, BBVA, BKT, BME, BTO, CRI, ELE, ENG, FCC, FER, GAM, GAS, GRF, IBE, IBLA, IBR, IDR, ITX, MAP, MTS, OHL, POP, REP, SAB, SAN, SYV, TEF, TL5, TRE, REE. The table `Ibex0809` can be downloaded from the web page of the book, http://computationalfinance.lsi.upc.edu.

```
> IBEX<-read.table("Ibex0809",sep="",header=T)
> dd <-as.dist(2*(1-cor(IBEX)))
> hc <-hclust(dd,method="ward") ##other: complete,single,average
> plot(hc,main=paste("ward"," method"),axes=TRUE,xlab="",sub="")
> #compute the cut  at mean height K:
> l<-length(hc$height); hh<-sort(hc$height); K<-mean(hh[1:l])
> abline(h=K,lty=2,lwd=2) ##draw the cut
> #branches below K make clusters, above go to singletons
> groups <- cutree(hc, h = K) ##obtain  clusters
> numgp <- max(groups) #number of clusters.
> #extract the names of each group and convert to list
> W <- list(names(groups[groups==1]))
> for(i in 2:numgp){W <- c(W,list(names(groups[groups==i])))}
> W ##see partition at level <=K
```

Repeating these instructions with each of the four values for method we get the four dendrograms shown in Fig. 3.4. The broken horizontal line is the cut which has been computed as the mean value of heights. Each height value corresponds to inter-clusters distance associated with the clustering method. Then we consider the partition at the level immediately below this cut: only the subtrees below the cut form a cluster. This partition is saved in W. For example, the clustering obtained with the Ward method consists of the 12 clusters: {ABE, ACS, FCC, GAM, GAS, IBE, TRE}, {ABG, ANA, IDR }, {ACX, BBVA, POP, REP, SAB, SAN, MTS, TEF, OHL}, {BKT, MAP}, {BME, TL5}, {BTO, ENG, IBLA}, {ELE, REE}, {IBR, ITX}, {FER}, {CRI}, {SYV}, {GRF}.   □

One observes that the single-link method produces clusters that are elongated, with elements linked in succession forming a chain, and as consequence tend to have large *diameter* (i.e. the maximum dissimilarity among its members), and hence highly unequal sizes. The complete-link method produces more compact clusters, with roughly the same small diameter, and hence roughly equal sizes. The average-link method give solutions somewhere in between single and complete linkage. The clusters are relatively compact, far apart and of reasonable balanced sizes. Ward method is based on optimizing the sum of squared errors, and so its performance is closer to the complete method based on L_2-norm. These observations on the performance of these methods are common to many experiments of hierarchical clustering of diverse data, not only related to financial time series. Therefore, the general consensus is that the complete and Ward methods are the ones that produce more meaningful partitions (Jain et al. 1999; Jain and Dubes 1988; Hastie et al. 2009, §14.3.12).

3.3.2.2 Partitional Clustering

Partitional clustering algorithms work with a single partition, exchanging elements among clusters so as to minimized certain objective function. The most natural, and statistically sound, objective function to use when classifying numerical data is the

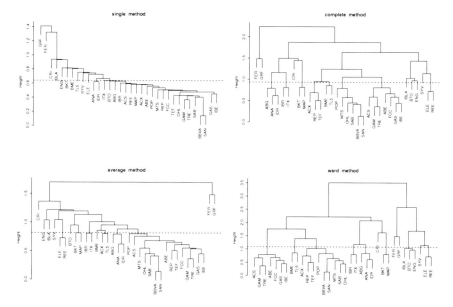

Fig. 3.4 Dendrograms obtained using four different inter-cluster linkages for IBEX stock's returns with dissimilarity measure $2(1 - \rho(x, y))$

sum of squared errors,[7] and under this criterion the partitional clustering problem amounts to:

> For a given set of n data points $X \subset \mathbb{R}^s$ and a given integer $k > 1$, find a set of k centers $\mathscr{C} = \{c_1, \dots, c_k\}$ so as to minimized the sum of squared errors
> $$f(C_1, \dots, C_k) = \sum_{i=1}^{k} \sum_{x \in C_i} ||x - c_i||^2,$$ where C_i is the cluster whose center
> is c_i.

Finding an optimal global solution to this problem is NP-hard (Drineas et al. 2004); hence we have to content ourselves with heuristics giving approximate solutions. The k-means algorithm (Algorithm 3.2) is the simplest and most popular method for finding solutions to the partitional clustering problem based on squared error. It begins with a random selection of centers, and builds neighbors (clusters) around these centers by assigning points that minimizes the sum of squared distances. It updates the centers and repeats the clustering process until the squared error ceases to decrease significantly, or a preassigned number of iterations is reached.

The popularity of k-means is due to its simplicity and its fast performance. Strictly speaking, the time complexity of k-means is $O(nkl)$, where n is the number of data elements, k is the number of clusters and l is the number of iterations. However, k and

---

[7] recall Ward's method that treats clustering as a problem of variance.

Algorithm 3.2 k-means clustering

---

0. **input**: $X = \{x_1, \ldots, x_n\} \subset \mathbb{R}^s$ data points; k number of clusters;
1. Randomly choose k cluster centers $\mathscr{C} = \{c_1, \ldots, c_k\}$
2. **repeat:**
3.   **for each** $i \in \{1, \ldots, k\}$ define cluster C_i to be
$$C_i = \{x \in X : \forall j \neq i (||x - c_i||^2 \leq ||x - c_j||^2)\}$$
4.   **for each** $i \in \{1, \ldots, k\}$ update each cluster's center:
set $c_i = (1/|C_i|) \sum_{x \in C_i} x$
5. **until** $\mathscr{C}$ no longer changes.

---

l are usually fixed in advance, so the algorithm has in practice linear time complexity $O(n)$.

The major disadvantage of k-means is that it is sensible to the selection of the initial partition, and so often converge to a local minimum solution.

R Example 3.6 We use the same data from R Example 3.5, written in the 40×34 table IBEX containing 40 daily returns observations, taken through the same period of time, for 34 stocks from IBEX. In R the k-means algorithm is implemented by the function kmeans from the package stats. We have to indicate the desired number k of clusters. This we set to 12, the same number of clusters obtained in R Example 3.5. The table must be transposed since kmeans works on rows. The instructions to enter in your R console are:

```
> k=12; ##number of clusters
> kc=kmeans(t(IBEX),k) ##t() gives transpose table
>##obtain the clusters
> groups <- kc$cluster; numgp <- max(groups)
> W <- list(names(groups[groups==1]))
> for(i in 2:numgp){W <- c(W,list(names(groups[groups==i])))}
> W
```

The clustering obtained with k-means on the IBEX data is: {BTO, ENG, GRF, IBLA, REE}, {ACS, GAM, GAS, IBE, TRE}, {ABE, ABG, FCC, IDR, REP, SAB, TEF, OHL}, {BBVA, POP, SAN}, {ACX, BME, TL5}, {IBR, ITX}, {ELE, SYV}, {BKT, MAP}, {CRI}, {FER}, {MTS}, {ANA}.   □

Be aware that the result of a clustering method depends to a large extent on features intrinsic to the method, mainly the element's similarity measure and the objective function (which induces the cluster's dissimilarity criterion), and these features are determined by the classification goal that we had in mind. In R Example 3.5 we used a correlation-based distance to gauge the similarity between return series, and so it should be expected that the clusters grouped correlated returns. We can run the command cor(IBEX[,1:34]) to check for correlations, and verify the extent to which the clusters reflect the correlation structure. In R Example 3.6 the similarity is based on squared L_2 norm, so the clusters reflect other statistics of returns beyond correlation, such as the spread among returns. Thus, in principle, the clusterings obtained in both examples for the same data are not comparable. Nonetheless, the question remains on how to determined the best performing clustering method for a given data set. We give some criteria in Sect. 3.3.3. But keep in mind, from what has been said, that the assessment of a variety of methods must be done, at least, under the same similarity measure and objective function.

In the next section we give a graphical view of the problem of clustering. One important consequence obtained from this graph framework is that balancing the sizes of the clusters (an important feature characterizing good clusterings) is computationally hard to do.

3.3.2.3 Graph Based Clustering

The problem of clustering any data set can be seen as a problem of partitioning a weighted graph in some optimal way. Given data set $X = \{x_1, \ldots, x_n\}$ and a similarity relation $S = (s_{ij})_{1 \leq i,j \leq n}$ on X (i.e. $s_{ij} = S(x_i, x_j)$, and is not necessarily a distance metric), define the *similarity weighted graph* $G = (X, E)$, whose vertex set is X and edge set $E \subset X \times X$ is given by $(x_i, x_j) \in E \iff s_{ij} > 0$, and the edge (x_i, x_j) has weight s_{ij}. Then the problem of clustering X becomes the *graph partition problem*:

> To find a partition of X such that edges between two different parts have low weight (hence, elements in different clusters are dissimilar from each other), and edges within the same part have high weight (i.e., elements within the same cluster are similar to each other).

As an optimization problem this graph partition problem can be stated in terms of a very simple objective function, defined for all partitions of X. Given a partition $A_1, \ldots, A_k$ of X, the *Cut* of this partition is defined as

$$Cut(A_1, \ldots, A_k) = \sum_{i=1}^{k} cut(A_i, \overline{A_i}) \tag{3.17}$$

where $cut(A, \overline{A}) = \sum_{x_i \in A, x_j \notin A} s_{ij}$. Now, the graph partition problem can be formalized as the *Min-cut* problem:

Min-cut: Given a graph $G = (X, E)$, to find a partition $A_1, \ldots, A_k$ of X which minimizes $Cut(A_1, \ldots, A_k)$.

This problem can be solve efficiently (i.e., deterministically in polynomial time) for $k = 2$, since in this case it is equivalent to the maximum network flow problem by the max-flow min-cut theorem (Papadimitriou and Steiglitz 1998, Chap. 6). For $k > 2$ you may get by with cleverly iterating the algorithm for the $k = 2$ case in the different parts. However, in practice, these network flow solutions of Min-cut often give clusters with only one element and one (or very few) clusters containing the rest of the elements. To prevent from these imbalanced solutions one can sharpen the objective function with some form of normalization that forces the clusters to take in more than one element. One possible such normalization is to consider the objective function

$$BCut(A_1, \ldots, A_k) = \sum_{i=1}^{k} \frac{cut(A_i, \overline{A_i})}{|A_i|} \qquad (3.18)$$

Note that the minimum of the sum $\sum_{i=1}^{k} \frac{1}{|A_i|}$ is attained when all $|A_i|$ are equal. Thus, with $BCut$ as objective function, solutions to the Min-cut problem consist of balanced clusters in terms of number of elements. The downside of this approach is that it turns the Min-cut into an NP-hard problem. In general any attempt to balance the sizes of the clustering solutions will make the graph partition a harder problem (see Wagner and Wagner 1993).

3.3.3 Clustering Validation and a Summary of Clustering Analysis

The most common and used empirical form of evaluating a clustering method is to create some artificial data set and for which one knows or forces to have a clustering fitting our desired objectives, i.e. create a controlled experiment. Then test the clustering method on this artificial data and see if it produces a solution identical to the already known clustering. For time series data the artificial set is usually obtained by doing simulations with a time series model, as an ARMA or GARCH, to be studied in Chap. 4.

This is the type of evaluation criteria where we know before hand some ground truth clustering, and we just use some cluster similarity measure to quantify the coincidence of the solution resulting from our clustering method with this ground truth. This can be formally done as follows. Let $C = \{C_1, \ldots, C_k\}$ be the set of clusters obtained by some clustering algorithm whose performance we want to evaluate against a previously known true clustering $V = \{V_1, \ldots, V_k\}$. Then a natural measure of evaluation is to calculate the proportion of coincidence , or similarity measure defined as

$$\text{sim}(V, C) = \frac{1}{k} \sum_{i=1}^{k} \max_{1 \leq j \leq k} \text{sim}(V_i, C_j) \qquad (3.19)$$

where $\text{sim}(V_i, C_j) = \frac{2|V_i \cap C_j|}{|V_i| + |C_j|}$. Other measures for comparing clusters are refinements of Eq. (3.19).

On the other hand, if we do not know (or are unable to build) a true sample of our clustering objective, then the evaluation criteria is limited to comparing the solutions obtained by the method, with the modeling assumptions; that is, to assess somehow if the lowest intra-cluster and highest inter-cluster distances are attained. A typical way of doing this evaluation is by setting the intra- and inter-cluster distances as functions of the number of cluster, and so focus on determining the appropriate number k of clusters. The *gap statistic* is one of many goodness of clustering measures based on estimating k (see Hastie et al. 2009, §14.3.11 and references therein). The gap statistic is programmed in R by the function `clusGap` from the package `cluster`. Other model selection criteria for determining k are listed in the survey by Liao (2005).

To recap this section, let us summarize the four stages of clustering analysis:

(1) Decide on data representation, which include cleaning, normalization or weighting of the features, and a general preparation to accommodate for the algorithm to be used.
(2) Define a proximity measure, which reflects the relation of similarity use to compare the data, and preferably is a distance metric.
(3) Choose a clustering strategy.
(4) Validate the results: usually by comparison to benchmarks, known or simulated solutions; performing the results of the clustering method on the data for which we know before hand the correct partition.

For financial time series, data representation refers to considering returns, or some other standardization of the series, or some model fitting the data, among other forms of representation. Cleaning refers to removing noise, missing data or outliers. On defining appropriate proximity measures we have already commented some advantages and disadvantages for L_p norms, correlation based distances, and others. Clustering methods you will find that the most recurred ones are the agglomerative hierarchical clustering and k-means; other clustering methods and techniques for validating results are discussed in (Jain et al. 1999; Jain and Dubes 1988; Hastie et al. 2009; Liao 2005).

3.3.4 Time Series Evolving Clusters Graph

Financial returns evolve in time, and consequently when trying to cluster financial instruments by similarities on their return behavior (or any of its moments), it would be desirable to have a representation of this dynamic. In this section we study a graphical tool for monitoring the temporal evolution of clusters of financial time series; that is, a representation of the *clusters in movement*. Through this graph some useful links between graph combinatorics and financial applications arise, and this allow us to use some known combinatorial methods from graph theory to produce sound answers for problems about financial markets. The general construction of this time evolving clusters graph follows.

Temporal graph of clusters. Let $\mathscr{S}$ be a set of financial time series and let T be the time period selected for analysis. Make a partition of T into m successive sub-periods of time T_1, T_2, ..., T_m, where m and the length of each sub-period are arbitrarily chosen.[8]

We have fixed a clustering algorithm and, for each $i = 1, \ldots, m$, apply it to the fragments of time series in $\mathscr{S}$ observed in sub period of time T_i. Let $\mathscr{C}_i$ be the collection of clusters (a clustering) obtained in each temporal interval T_i, and let $n_i = |\mathscr{C}_i|$. Let $S_{i,j}$, $1 \leq i \leq m$, $1 \leq j \leq n_i$, be the clusters in $\mathscr{C}_i$ with at least two elements, and Q_i be the reunion of elements that were not placed in any cluster by the clustering algorithm acting on time segment T_i, for $1 \leq i \leq m$. We define a directed graph $\mathscr{G}$ with vertex set the collection of subsets

$$\{S_{i,j} : 1 \leq i \leq m, 1 \leq j \leq n_i\} \cup \{Q_i : 1 \leq i \leq m\} \qquad (3.20)$$

and weighted edge set

$$\{(S_{i,j}, S_{i+1,k}, |S_{i,j} \cap S_{i+1,k}|) : 1 \leq i \leq m-1, 1 \leq j \leq n_i, 1 \leq k \leq n_{i+1}\} \quad (3.21)$$

that is, we link a cluster $S_{i,j}$ in the time segment T_i with a cluster $S_{i+1,k}$ in the next time segment T_{i+1} as long as their intersection is non empty, and the cardinality of their intersection is the weight assigned to the link. The sets Q_i remain as isolated vertices. We call $\mathscr{G}$ the *Temporal Graph of Clusters* (TGC) for the set of time series $\mathscr{S}$ in the time segmentation $T = \bigcup_{1 \leq i \leq m} T_i$.

Example 3.3 As an example of interest of a TGC consider the 34 stocks of IBEX treated in R Example 3.5, represented by daily returns series observed from 1 June 2008 to 1 Aug 2009. We divide this time period in seven successive bimonthly sub-periods $T_1 = [1$ June 2008, 1 Aug 2008$]$, $T_2 = [1$ Aug 2008, 1 Oct 2008$]$, ..., $T_7 = [1$ June 2009, 1 Aug 2009$]$. Consider as clustering goal to find among these stocks those whose returns series have similar monotone behavior, so as to obtain potential candidates for a *pairs trading strategy*.[9] Therefore, for the dissimilarity metric it is reasonable to work with some measure based on Kendall's correlation coefficient τ, since this reflects best the monotone relationship between random variables, and then apply the agglomerative hierarchical clustering with the complete or Ward method. However, since the order in which the data elements are selected has an influence in the resulting clustering from the hierarchical method, we shall consider the following proximity measure (based on τ) that avoids this "order of selection problem". For each pair of stocks A and B define

$$d(A, B) = 1 - \frac{|\mathscr{G}_A \cap \mathscr{G}_B|}{|\mathscr{G}_A \cup \mathscr{G}_B|} \qquad (3.22)$$

---

[8] An interesting subject of research is to determined this partition.

[9] This strategy attempts to capture the spread between two correlated stocks as they reverse to the mean price. It consists on taking long and short positions on the pair of stocks as they move together, changing positions at the mean (Vidyamurthy 2004).

where $\mathcal{G}_A$ is a set comprised of all stocks X whose series of returns $\mathbf{x}$, on the same time span of the returns $\mathbf{a}$ of A, have (sample) correlation with $\mathbf{a}$ higher than a positive δ; that is,

$$X \in \mathcal{G}_A \iff \widehat{\tau}_n(\mathbf{a}, \mathbf{x}) > \delta.$$

The δ is determined from the sample estimation given by Eq. (3.8) and depends on the sample size. For example, for 40 days period one should take $\delta = 0.26$, the 0.01 asymptotic quantile of the distribution of $\widehat{\tau}_n$, to ensure a significant correlation. The measure $d(A, B)$ is the Jaccard distance, and the reader should verify that is a metric. One can see that if A and B are similar, in the sense that their respective groups $\mathcal{G}_A$ and $\mathcal{G}_B$ have almost the same members, then $d(A, B)$ is close to 0; whilst if $\mathcal{G}_A$ and $\mathcal{G}_B$ greatly differ then $d(A, B)$ is close to 1. Note that d is a stronger measure of similarity than taking direct point-wise distance based on correlation (as given by Eqs. (3.15) and (3.16)), since it says that *two stocks are similar, not only if they are correlated but also correlated to almost all the same stocks.*

Next, on each sub-period T_i ($i = 1, \ldots, 7$), we apply agglomerative hierarchical clustering with distance d given by Eq. (3.22), where the decision of the partition is made by the standard procedure of cutting at mid level the dendrogram representation of the hierarchical clustering. After the cut, the non-clustered elements go into Q_i. The full TGC is shown in Fig. 3.5: the green boxes (boxes in first row from bottom) show the time periods for each return series; the pink boxes (boxes in second row from bottom) collect companies with correlations below our established threshold (hence, no conclusion can be drawn for these); and the blue boxes (all remaining boxes above second row from bottom) represent non singleton clusters, which also contain a label ij to indicate its position in the adjacency matrix representation of the graph; for example, cluster 12 is {ELE, IBLA}. The edges are weighted by the intersection measure.   □

Graph combinatorics applied to financial analytics. The TGC provides a framework for solving problems relevant to financial analytics with techniques proper to graph combinatorics. For example, by considering the cardinality of the intersection of the clusters as the weight for the edges in the TGC, we can seek to identify the most persistent or stable clusters of asset returns through time: these would be located on the paths of heaviest weight (considering the weight of a path as the sum of the weights of the edges that comprise it). We formalized this problem as follows.

Stable clusters: consist on finding the clusters (or sub-clusters) of asset returns that appear more frequently in consecutive intervals of time. On the TGC $\mathcal{G}$ this amounts to finding the heaviest paths from the first time interval to the last. The problem of deciding the heaviest weighted path in a graph is in general NP-hard; however, on acyclic directed graphs polynomial time solutions are known using dynamic programming (Garey and Johnson 1979, problem ND29). We shall make use of such algorithmic methodology for producing a solution to this problem on $\mathcal{G}$. The relevance of this graph problem to financial analytics is of a large extent. For example, it helps detect those stocks that correlate through different but con-

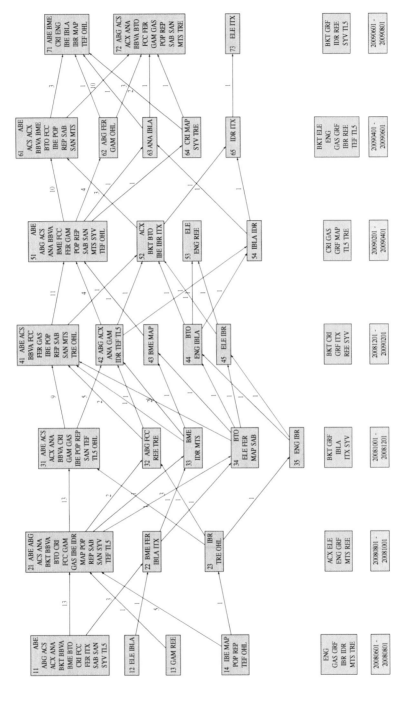

Fig. 3.5 Temporal Graph of Clusters for IBEX from 1 June 2008 to 1 Aug 2009

tinuous time intervals, and hence become candidates for a pairs trading investment strategy.

Another interesting graph combinatorial problem we can formulate on the TGC is

Stock cover: consist on finding the smallest possible set of stocks returns that intersects every (non singleton) cluster in the TGC $\mathcal{G}$. This is the Hitting Set problem (Garey and Johnson 1979, problem SP8), which is a form of the Vertex Cover problem on graphs, both NP-hard. Hence, the best we can aim for is an approximate solution. The interest of this computational query is to discover those stocks that essentially represent the market behavior through a sequence of time periods; thus, conforming a portfolio that *covers* the market.

We present efficient algorithms to solve these problems.

Algorithm for the Stable Clusters problem. Given a positive integer $k \geq 1$ and TGC $\mathcal{G} = \langle V, E \rangle$, with $V = \{S_{i,j} : 1 \leq i \leq m, 1 \leq j \leq n_i\}$ (excluding the isolated vertices Q_i), we want to find the k first heaviest paths starting at any cluster $S_{i,j}$ with no ingoing edge.

Let $\{T_1, \ldots, T_m\}$ be the sequence of time segments which partition the total time period T. We view the edges in $\mathcal{G}$ as directed going from time segment T_i to time segment T_{i+1}. Then we add an extra (dummy) cluster $S_{m+1,1}$ after time period T_m, and which serves as a sink in the sense that every cluster $S \in V$ that do not have an outgoing edge (i.e. is terminal) is connected by an edge of weight 1 to $S_{m+1,1}$. Now, a brute force solution is to find all paths that end in $S_{m+1,1}$ (say by Breadth First Search), order them by weight and keep the k heaviest. This, however, would incur in exponentially many comparisons, as we can have a worst case containing $O(n^m)$ many paths, where n is the maximum number of clusters per time segment and m is the number of time segments. Instead we apply a bottom-up approach (aka dynamic programming) where at step i, $1 \leq i \leq m$, we look at time segment T_i and define, for each cluster $S_{i,j}$, a heap $h_{i,j}$ (i.e. a tree with nodes having a positive value or *key*) where the root, labelled $S_{i,j}$, has key 0, its descendent will be all heaps $h_{i',j'}$ defined in previous time segments $T_{i'}$, $i' < i$, for which there is an edge from $S_{i',j'}$ to $S_{i,j}$, and the keys of all *leaves* of these descendent heaps are updated by adding the weight of the edge $(S_{i',j'}, S_{i,j})$, denoted $w(S_{i',j'}, S_{i,j})$.[10] Observe that the total weight of a path in the heap of root $S_{i,j}$ is the key of the leaf or initial node of such path in the heap. We order these leaves by key value and keep the paths for the k leaves with highest key, removing the paths of the remaining leaves before proceeding to step $i + 1$ of algorithm. By adding a sink to the graph, that is an extra dummy set $S_{m+1,1}$, the algorithm will collect at the last and extra step $i = m + 1$ all k-first heaviest paths in the heap $h_{m+1,1}$, as the paths from the leaves to the root. Algorithm 3.3 summarizes this strategy. We use the following abbreviations for functions whose

---

[10] By a leaf of a heap we understand a node with no ingoing edges. By updating the keys of the leaves only we significantly reduce computation time since we do not need to traverse all the heap. Also the information about the key value of inner nodes is irrelevant for our purposes. For further details on heaps see Cormen et al. (1990).

implementation we do not specify but are obvious: for a vertex S, $key[S]$ keeps the key value of S in a heap, which should be clear from context; for a heap h, $leaves(h)$ is the set of leaves of heap h; for two heaps h and h', $append(h', h)$, returns a heap made by drawing an edge from the root of heap h' to the root of heap h.

Algorithm 3.3 k first heaviest paths

1. **input:** $k > 0$, $V = \{S_{i,j} : 1 \leq i \leq m, 1 \leq j \leq n_i\}$, E
2. Add a sink $S_{m+1,1}$: {*preprocessing*}
3.   **for each** $S \in V$ with no outgoing edge **do**
4.   define new edge $(S, S_{m+1,1})$ of weight 1;
5.   **end for;**
6. **for each** $i \in \{1, \ldots, m+1\}$ **do**
7.   **for each** $j \in \{1, \ldots, n_i\}$ **do**
8.   define heap $h_{i,j}$ with unique element $S_{i,j}$ and $key[S_{i,j}] = 0$;
9.   **for each** $S_{i',j'}$ such that $(S_{i',j'}, S_{i,j}) \in E$ **do**
10. $h_{i,j} = append(h_{i',j'}, h_{i,j})$;
11.   **for each** $S \in leaves(h_{i',j'})$ **do**
12. $key[S] = key[S] + w(S_{i',j'}, S_{i,j})$;
13.   **end for;**
14.   **end for;**
15.   remove from $h_{i,j}$ all but k paths to $S_{i,j}$ of heaviest weight;
16.   {*The weight of each path can be read-off from the key of each leaf*}
17.   **end for;**
18. **end for;**
19. **output:** $h_{m+1,1}$;

One can see that for m time segments and a maximum of n many clusters per time segment, the algorithm has a worst-case running time bound of $O(mn^3)$, which is much better than the naive solution explained at the beginning.

Algorithm for the Stock Cover problem. Given a TGC $\mathcal{G} = \langle V, E \rangle$, with $V = \{S_{i,j} : 1 \leq i \leq m, 1 \leq j \leq n_i\}$, we want to find a (minimum) set of stocks that intersects every (non singleton) cluster in $\mathcal{G}$. We apply the following greedy strategy: pick the stock that shows up more often in the collection of clusters V; then remove the clusters containing this stock (i.e. covered by the stock). Repeat with remaining clusters.

In the pseudo-code implementation of the above strategy (Algorithm 3.4) we define an *incidence* matrix $\mathcal{M}$ for the set of clusters V, where rows are labelled by the companies ticker and columns labelled by the clusters. Π represents the set of tickers (which is given as input). Then there is a 1 in entry $\mathcal{M}(\pi, S)$, $\pi \in \Pi$ and $S \in V$, if stock represented by ticker π is in cluster S, or a 0 otherwise. By $dim(\mathcal{M})$ we denote the dimension of matrix $\mathcal{M}$, which is the number of rows multiplied by the number of columns.

Algorithm 3.4 Stock cover

1. **input:** $V = \{S_{i,j} : 1 \le i \le m, 1 \le j \le n_i\}$, Π
2. **initialize:** $\Gamma = \emptyset$;
3. define a matrix $\mathcal{M}$ with rows labelled by each ticker $\pi \in \Pi$ and columns labelled by each cluster $S \in V$, and there is a 1 in entry $\mathcal{M}(\pi, S)$, if stock represented by π is in cluster S, or a 0 otherwise;
4. **while** $dim(\mathcal{M}) \neq 0$ **do** $\{dim(\mathcal{M}) = (num.\,rows) \times (num.\,columns)\}$
5.   order the rows decreasingly with respect to the number of 1 appearing in it;
        {*The first row corresponds to the stock appearing more often in the clusters*}
6.   let π_1 be the ticker of the stock labeling the first row;
7. $\Gamma = \Gamma \cup \{\pi_1\}$;
8.   delete every column with label $S \in V$ for which $\mathcal{M}(\pi_1, S) = 1$;
9.   delete row labelled π_1;
10.   rename $\mathcal{M}$ the reduced matrix;
11. **end while**;
12. **output:** Γ;

The running time of this algorithm is bounded by $dim(\mathcal{M})$; that is by $|\Pi| \cdot |V|$. More importantly, the size of the solution is not too big with respect to the size of the full portfolio of companies involved; in fact, it is half the size of the original portfolio, as shown below.

Proposition 3.1 $|\Gamma| \le \frac{1}{2}|\Pi|$.

Proof Each cluster contains at least two stocks. The worst situation is that any stock appears in at most one cluster. The algorithm selects a stock π_1 (of highest weight) and removes the cluster where it belongs together with at least another stock that will not enter the solution set (otherwise its weight is higher than π_1's and would have been selected before). Thus, at most one half of the stocks enters Γ.   □

Thus, our algorithm guarantees a solution of size at most one half of the size of the set of companies Π. On the other hand, it is important to note that we, and nobody up to today, can not guarantee a solution of minimum size with current computer technology, since the problem of finding the minimum size vertex cover (of which our stock cover problem is a replica) is in general NP-hard (Garey and Johnson 1979). The hard way to find this minimum cover, following our greedy strategy, would be to check all possible combinations of assets that appear more often in the TGC in the first round of the while loop in the algorithm (lines 4–11), call this maximum number of appearances n_1, together with those that appear more often in the second round, say n_2 number of times, and so on until all sets are covered. This amounts to $\prod_{i=1}^{k} n_i$ combinations, where k is most half of the number of assets, by Proposition 3.1. Finally, a given cover of a given size might not necessarily be unique in its components. The composition of a cover depends on the order of selection of the elements in Π done by the algorithm.

Example 3.4 Back to Example 3.3. On the TGC computed for the set of stocks from the principal index IBEX of the Madrid Stock Exchange (Fig. 3.5) applying Algorithm 3.3 with $k = 3$, we found the following three heaviest paths (the number on top of the arrow is the weight of the corresponding edge):

$$11 \overset{13}{\mapsto} 21 \overset{13}{\mapsto} 31 \overset{9}{\mapsto} 41 \overset{11}{\mapsto} 51 \overset{10}{\mapsto} 61 \overset{10}{\mapsto} 72$$

$$11 \overset{13}{\mapsto} 21 \overset{13}{\mapsto} 31 \overset{9}{\mapsto} 41 \overset{11}{\mapsto} 51 \overset{10}{\mapsto} 61 \overset{3}{\mapsto} \mathbf{71}$$

$$11 \overset{13}{\mapsto} 21 \overset{13}{\mapsto} 31 \overset{5}{\mapsto} \mathbf{42} \overset{4}{\mapsto} 51 \overset{10}{\mapsto} 61 \overset{10}{\mapsto} 72$$

The first path weights 66 and the second heaviest weights 59 and the third weights 55. We have highlighted in boldface the indices of the sets where the paths differ. On these heaviest paths we can find the subsets of companies with higher tendency to correlate through time (correlation in movement). Once we computed a heaviest path we can take the intersection of all the sets belonging to the path in order to detect those assets that tend to behave similarly, in terms of their series of returns, most of the time. Doing this on the first heaviest path we found that ACS, BBVA and SAN remain consistently together through the clustering; hence any two of these three assets make a good candidate for a pairs trading strategy. However, the two banks SAN and BBVA tend to have a stronger correlation at all sub-periods.[11]

Applying the Stock Cover algorithm (Algorithm 3.4) we obtain the following cover:

{ABE, ELE, ABG, IBLA, IBR, MAP, IDR, GAM}

These eight companies represent a set of stocks that intersects all non isolated clusters in the TGC for IBEX through the period considered. It constitute a portfolio that would have replicated the behavior of the Spanish Exchange index throughout those fourteen months. Due to the computational hardness of the vertex cover problem, we can not guarantee in any feasible known way that this set of stocks is of minimum size. The best we can do is to repeatedly permute the order of the stocks ticker in the list Π and run the algorithm again, hoping to find in some of the possible exponentially many trials a better solution. We have been lucky with this brute force method and found another cover of smaller size (seven elements) by permuting FCC to the first place in the list.

{FCC, IBLA, ELE, GAM, IBR, MAP, IDR}

For the sake of completeness, one may add to the cover found by the algorithm those stocks that are not in any cluster, such is the case of GRF.   □

---

[11] The finding of the heaviest paths in this example confirmed a popular observation, known by many local brokers (at least as of 2010), that the banks BBVA and SAN conform the most stable cluster through any period of time, and further we see that this duo tends to participate in the largest clusters of IBEX components at different periods of time; hence being both together a driving force of the Spanish market.

There are many other graph combinatorial problems that can resemble questions on managing financial portfolios, and their algorithmic solutions be of used in answering the financial problems. One more example is given in Note 3.5.10. We encourage the reader to explore this subject further.

3.4 Stylized Empirical Facts of Asset Returns

We have by now learned several empirical properties of asset returns using a variety of tools for statistical analysis. Those statistical properties that have been observed independently in various studies, and common to many financial securities, across markets and time periods, are called *stylized facts*. Stylized facts describe a general picture of the distribution of asset returns and the behavior of their first and second moments, and emerge from the empirical study of asset returns, as we have seen in this and previous chapter. justifying some models. Asset returns, in general, show the following stylized properties:

(1) **Absence of autocorrelations**: log returns over daily or longer periods do not show significant autocorrelations; for very small intraday time periods (e.g. 20 min) the microstructure effects come into play (i.e. the bid-ask bounce, where transaction prices may take place either close to the bid or closer to the ask, and bounce between those limits, causing some negative autocorrelation).

(2) **Heavy tails**: the (unconditional) distribution of log returns seems to display a power-law or Pareto like tail, with finite tail index. The tail index κ of a distribution can be defined as the order of the highest absolute moment which is finite. The higher the κ the thinner the tail; for Gaussian distribution $\kappa = +\infty$ since all moments are finite. Thus, this empirical observation implies that log returns are not Gaussian.

(3) **Gain/loss asymmetry**: large drawdowns in stock prices and stock index values but not equally large upward movements. In other words, the distribution is skewed (This is not true for exchange rates).

(4) **Aggregational normality**: as one increases the time scale over which returns are calculated, their distribution looks more like a normal (or gaussian) distribution.

(5) **Volatility clustering**: different measures of sample volatility display a positive autocorrelation over several periods of time. This indicates that high-volatility events tend to cluster in time.

(6) **Taylor effect**: autocorrelations of powers of absolute returns are highest at power one.

(7) **Slow decay of autocorrelation in absolute returns**: the autocorrelation function of absolute returns decays slowly as a function of the time lag. This can be interpreted as evidence of long-range dependence.

(8) **Leverage effect**: most measures of volatility of an asset are negatively correlated with the returns of the asset.

(9) **Volume/volatility correlation**: trading volume is correlated with all measures of volatility.
(10) **Intermittency**: returns display, at any time scale, a high degree of variability.
(11) **Asymmetry in time scales**: coarse-grained measures of volatility predict fine-scale volatility better than the other way round.

We have observed some of these stylized properties of asset returns: the absence of autocorrelations of returns and volatility clustering have been experienced in R Example 3.2, and properties (2)–(4) in Chap. 2. We propose to explore the Taylor effect phenomenon in Note 3.5.6, and encourage the reader to run experiments on his preferred stock to verify the rest of the properties. These stylized properties, and some others, have been observed and documented by a long list of illustrous researchers. To name a few, Mandelbrot (1963) observed the heavy tail distribution of the returns of cotton prices, and studied alternative to the Gaussian model; Fama (1965) documented the low autocorrelation and non-normality of log returns; the Taylor effect is from Taylor (1986). For extensive pointers to the related literature and throughly discussion on this catalog of stylized properties see Cont (2001).

3.5 Notes, Computer Lab and Problems

3.5.1 Bibliographic remarks: Section 3.1 owes much to Gibbons and Chakraborti (2003) and Embrechts et al. (2002) for general discussion on properties of dependence measures. For a concise historical account of Pearson's correlation plus thirteen different ways of interpreting this measure of association read Rodgers and Nicewander (1988). Granger causality (Sect. 3.2) is due to Granger (1969). The nonparametric approach to causality is from Diks and Panchenko (2006), based on the work by Hiemstra and Jones (1994). Section 3.3 on clustering uses material from Hastie et al. (2009, Sect. 14.3), Jain et al. (1999), and Liao (2005). Jain and Dubes (1988). is a nice and easy to read compendium of clustering algorithms, data analysis and applications. Ward method is from (Ward, 1963), there one finds a clear exposition on hierarchical clustering. The k-means algorithm is due to Hartigan and Wong (1979). There are several research papers proposing improvements to the k-means, e.g., with respect to selecting the initial partition see Arthur and Vassilvitskii (2007), or improving efficiency for large data sets see Drineas et al. (2004). The graphical representation of time evolving clusters and graph combinatorics exported to financial analytics (Sect. 3.3.4) is from Arratia and Cabaña (2013). The summary of stylized facts presented here is from Cont (2001).

3.5.2 (The variance of a linear combination of random variables): Given n random variables $X_1, \ldots, X_n$, and n positive numbers $a_1, \ldots, a_n$, show that

$$\text{Var}\left(\sum_{i=1}^{n} a_i X_i\right) = \sum_{i=1}^{n}\sum_{j=1}^{n} a_i a_j \sigma_i \sigma_j \rho_{ij}$$

where σ_i stands for the standard deviation of variable X_i, and ρ_{ij} is the correlation coefficient of X_i and X_j (Hint: Compute $\mathrm{Var}(aX + bY)$ and then generalize to n variables by induction). This property of variance will be needed in Chap. 8.

3.5.3: Prove Eq. (3.2). Use that the $\min_c E[(Z - c)^2]$ is obtained at $c = E[Z]$, for random variable Z and real c.

3.5.4: Another desirable property of a measure of dependence δ is

$$\delta(X, Y) = 0 \iff X, Y \text{ are independent} \tag{3.23}$$

Both Pearson's and Kendall's correlation satisfy the right to left direction:

If X and Y are independent then $\rho(X, Y) = 0$ and $\tau(X, Y) = 0$. (*)

We have already shown (*) for ρ. Now prove it for τ (Hint: For τ prove the following equivalent expressions for p_c and p_d, the probabilities of concordance and discordance:

$$p_c = \mathbb{P}((X_i < X_j) \cap (Y_i < Y_j)) + \mathbb{P}((X_i > X_j) \cap (Y_i > Y_j))$$
$$p_d = \mathbb{P}((X_i < X_j) \cap (Y_i > Y_j)) + \mathbb{P}((X_i > X_j) \cap (Y_i < Y_j))$$

Then use that if X and Y are independent (and continuous random variables), $\mathbb{P}(X_i < X_j) = \mathbb{P}(X_i > X_j)$ and the joint probabilities in p_c or p_d are the product of the individual probabilities, to conclude that $p_c = p_d$).

The converse of (*) is in general not true (e.g. it is true in the case where X, Y have the bivariate normal distribution, see Gibbons and Chakraborti (2003, Chap. 11)).

3.5.5: However, the extra property (3.23) is incompatible with property (5) of Sect. 3.1.2:

Proposition 3.2 *There is no measure of dependence satisfying (5) of* Sect. 3.1.2 *and* (3.23). □

For the proof see Embrechts et al. (2002), or try proving it yourself! (it is a bit tricky).

3.5.6 R Lab: We shall explore the phenomenon known as *Taylor effect*. Taylor (1986) analyzes various financial return series and observes that the sample autocorrelations of absolute returns seem to be larger than the sample autocorrelations of the squared returns. It was later in Granger and Ding (1995) where this empirical property of absolute returns was termed Taylor effect, and used as grounds for a model of returns.

Go back to R Example 3.2, and now apply the function `acf` to the absolute value of the return series, i.e., `abs(na.omit(daxRlog$alvR))`. Do the same for higher powers of returns, and for other return series

3.5.7 R Lab: A curious application of Granger causality is the by now classic experiment of Thruman and Fisher (1988) to assess which came first, if the chicken or the egg. The necessary data is in the package `lmtest`, and it is the data set `ChickEgg` that contains the yearly production of chicken and eggs in the US from 1930 to 1983. To assess if chicken causes egg, execute `grangertest(egg chicken, order=3, data=ChickEgg)`. Do also the test for egg causing chicken, and write your conclusions.

3.5.8 C Lab: There is a C code (GCTtest.c) for the DP-test, available from the authors Diks and Panchenko (2006). The usage is as follows (from the package description):

GCTtest.exe is an executable program written in C that computes T_n test statistics (and p-values) for the nonparametric Granger causality test. Parameteres of the program:

- input file with series 1 (`file1`)
- input file with series 2 (`file2`)
- embedding dimension (`embdim`)
- bandwidth (`bandwidth`)
- output file (`output_file`)

Input file1 and input file2 are single column files of data and are to be of the same length. The parameters could be entered in two ways:

(1) From the command line in the following order (the parameters are optional and if some are not entered, they will be asked in the dialog regime later):

```
GCTtest.exe   file1   file2   embdim   bandwidth   output_file
```

(2) In the dialog regime after running the program

The program produces output stating the null hypothesis (input file1 does not cause input file2, and input file2 does not cause input file1) and the corresponding T statistics and the p-values.

   We have additional comments to these instructions: input file1 and input file2 are, for example, a column of log returns of stock prices of same length and coinciding on dates (you should enter a single column in each file, and the data in each file should coincide in date); the embedding dimension (variable embdim) refers to the dimension d_X, d_Y of vectors X and Y, and should be taken as 1 (so $embdim = 1$). The bandwidth should be taken as 1.5 for sample value $n < 1000$ and for greater n use the values given in Table 1 of (Diks and Panchenko 2006), or use the Eq. (3.13) with $C = 7.5$.

   As an application, consider the four return series in R Example 3.1 and test for non parametric causality among pairs of these series using the program GCTtest.exe.

3.5.9: Show that the correlation based similarity measures

$$d_\rho^1(x, y) = 2(1 - \rho(x, y)) \quad \text{and} \quad d_\rho^2(x, y) = \left(\frac{1 - \rho(x, y)}{1 + \rho(x, y)}\right)^m$$

verify the properties of a distance metric. Show that if $x = (x_1, \ldots, x_m)$ and $y = (y_1, \ldots, y_m)$ are normalized then $d_\rho^1(x, y) = ||x - y||^2$.

Show that the Jaccard distance $d(A, B) = 1 - \dfrac{|A \cap B|}{|A \cup B|}$, for sets A and B, is a metric.

3.5.10: Given a stock and a TGC as in Fig. 3.5, write an algorithm to determine the *trace of the stock*; that is, the stock's market associations through time. This is a simple path finding problem, where each node (cluster) of the path must contain the given stock.

Chapter 4
Time Series Models in Finance

This chapter presents some basic discrete-time models for financial time series. The general paradigm for modeling a time series $\{X_t\}$ is to consider each term composed of a deterministic component H_t and a random noise component Y_t, so that

$$X_t = H_t + Y_t. \tag{4.1}$$

Then one can readily estimate the deterministic component H_t, by some algebraic manipulations that we briefly indicate in Sect. 4.1, and we are left with the difficult task of approximating the values of Y_t, the random component. The basic structure of a time series model for Y_t has the form

$$Y_t = E(Y_t|F_{t-1}) + a_t \tag{4.2}$$

where F_{t-1} represents the information set available at time $t - 1$, $E(Y_t|F_{t-1})$ is the conditional mean, a_t is the stochastic shock (or innovation) and assumed to have zero conditional mean, and hence the conditional variance $Var(Y_t|F_{t-1}) = E(a_t^2|F_{t-1}) = \sigma_t^2$.

According to the form of $E(Y_t|F_{t-1})$ and $Var(Y_t|F_{t-1})$ as functions of F_{t-1}, we have models of different nature for Y_t. For example, if we fix the conditional variance to a constant value, $Var(Y_t|F_{t-1}) = \sigma^2$, and F_{t-1} is a finitely long recent past of Y_t, that is, $F_{t-1} = \{Y_{t-1}, \ldots, Y_{t-p}\}$, we obtain a linear model of the *Autoregressive Moving Average* (ARMA) type. We study the ARMA class of time series models in Sect. 4.2. If we leave both the conditional mean and conditional variance to vary as function of F_{t-1}, we obtain a nonlinear model of the *Autoregressive Conditional Heteroscedastic* (ARCH) type. This important class of nonlinear models is reviewed in Sect. 4.3. Other nonlinear models can be obtained by composition of the right-hand expression in Eq. (4.2) with some nonlinear function $g(\cdot)$, i.e., $Y_t = g(E(Y_t|F_{t-1}) + a_t)$, and assuming conditional mean or conditional variance constant or not. We look

A. Arratia, *Computational Finance*, Atlantis Studies in Computational Finance and Financial Engineering 1, DOI: 10.2991/978-94-6239-070-6_4, © Atlantis Press and the authors 2014

in Sect. 4.4 at two important machine learning models obtained this way: neural networks and support vector machines. We end the chapter with a case study of augmenting the set of information F_{t-1} with a sentiment measure drawn from a social networking platform, and an analysis of its relevance to forecasting financial markets.

4.1 On Trend and Seasonality

In the beginning, we treat an observed time series $\{X_t\}$ in the form given by Eq. (4.1). The deterministic component H_t comprises the *trend* and the *seasonality* that might be present individually or jointly in the series. From a plot of the series one can check for the existence of a trend (observing if several mean values of the series go up or down in general), and check for a possible seasonal component (observing if some values repeat periodically). If one perceives a trend or a seasonal component in the data then fits a model with trend or seasonality. We have then the *classical decomposition model*

$$X_t = m_t + s_t + Y_t$$

where m_t is the trend component, s_t is the seasonal component, and Y_t is the random component. The trend component is a slowly changing function (i.e., a polynomial in t: $m_t = a_0 + a_1 t + a_2 t^2 + \cdots + a_k t^k$, for some k), that can be approximated with least square regression; the seasonal component is a periodic function, with a certain period d (i.e., $s_{t-d} = s_t$), that can be approximated with harmonic regression. Once these deterministic components are estimated, their estimations $\widehat{H}_t = \widehat{m}_t + \widehat{s}_t$ are removed from the data to leave us only with the (sample) noise: $\widehat{Y}_t = X_t - \widehat{H}_t$. Alternatively, the trend or seasonality can be removed directly (without estimating them) by applying appropriate differencing operations to the original series X_t. For example, a linear trend is removed by taking first differences, $X_t - X_{t-1}$; a quadratic trend is removed taking second order differences, $X_t - 2X_{t-1} + X_{t-2}$, which is the same as taking first differences to $X_t - X_{t-1}$; and so on. This can be systematized by defining a difference operator (Box and Jenkins 1976). For further detail on estimation of trend and seasonality or removing them by difference operators see Brockwell and Davis (1991, Chap. 1). In either case (estimating and removing or removing directly), we are left with the random component Y_t which is assumed to be stationary, or weakly stationary, in order for some statistical modeling to be possible.[1] This is our departing point for the models we study in the following sections.

---

[1] cf. the discussion on stationary processes in Chap. 2.

4.2 Linear Processes and Autoregressive Moving Averages Models

A time series $\{X_t\}$ is a *linear process* if it has the form

$$X_t = \sum_{k=-\infty}^{\infty} \psi_k W_{t-k} \tag{4.3}$$

for all t, where $\{\psi_k\}$ is a sequence of constants with $\sum_{k=-\infty}^{\infty} |\psi_k| < \infty$, and $\{W_t\}$ is a (weak) white noise (uncorrelated random variables) with zero mean and variance σ^2; in symbols $\{W_t\} \sim WN(0, \sigma^2)$. The condition $\sum_{k=-\infty}^{\infty} |\psi_k| < \infty$ is to ensure that the series in Eq. (4.3) converges. A particular case of *Wold's decomposition theorem* asserts that any weakly stationary process $\{X_t\}$ has a linear representation of the form

$$X_t = \sum_{k=0}^{\infty} \psi_k W_{t-k} + V_t \tag{4.4}$$

where $\psi_0 = 1$, $\sum_{k=0}^{\infty} \psi_k^2 < \infty$, $\{W_t\} \sim WN(0, \sigma^2)$, $Cov(W_s, V_t) = 0$ for all s and t, and $\{V_t\}$ is *deterministic*.[2] Note that the linear process in the previous equation has $\psi_k = 0$, for all $k < 0$. This particular linear process is a (infinite order) moving average. From Wold's decomposition of $\{X_t\}$ in Eq. (4.4), one can obtain the structure of autocovariances for X_t in terms of the coefficients ψ_k (the moving average weights), as follows:

$$\gamma(0) = Var(X_t) = \sigma^2 \sum_{k=0}^{\infty} \psi_k^2$$

$$\gamma(h) = Cov(X_{t+h}, X_t) = \sigma^2 \sum_{k=0}^{\infty} \psi_k \psi_{k+h} \tag{4.5}$$

From these equations one obtains the autocorrelations $\rho(h) = \frac{\gamma(h)}{\gamma(0)} = \frac{\sum_{k=0}^{\infty} \psi_k \psi_{k+h}}{\sum_{k=0}^{\infty} \psi_k^2}$, and we will see that for particular cases of linear processes (such as the ARMA below), the deterministic component V_t equals the mean $E(X_t)$ times some constant.

Thus, Wold's representation is convenient for deducing properties of a stationary process. However, it is not of much use for modeling the behavior of the process, since estimation of $\{\psi_k\}$ from the infinite sums in Eq. (4.5) is not feasible. A class of models that can be seen as approximations to the Wold's form of a stationary process and for which estimation is possible are the autoregressive and moving averages (ARMA) processes.

---

[2] see Brockwell and Davis (2002, Chap. 2) for Wold's decomposition and the facts derived from it that we use in this section.

Definition 4.1 *(The class of ARMA processes)* Consider a weak white noise, $\{W_t\} \sim WN(0, \sigma^2)$, and let integers $p \geq 1$ and $q \geq 1$. A time series $\{X_t\}$ is

AR(p) (autoregressive of order p) if

$$X_t = W_t + \phi_1 X_{t-1} + \phi_2 X_{t-2} + \cdots + \phi_p X_{t-p} \qquad (4.6)$$

MA(q) (moving average of order q) if

$$X_t = W_t + \theta_1 W_{t-1} + \cdots + \theta_q W_{t-q} \qquad (4.7)$$

ARMA(p,q) (autoregressive and moving average of order p, q) if

$$X_t = \phi_1 X_{t-1} + \cdots + \phi_p X_{t-p} + W_t + \theta_1 W_{t-1} + \cdots + \theta_q W_{t-q} \qquad (4.8)$$

where $\phi_1, \ldots, \phi_p, \theta_1, \ldots, \theta_q$ are real numbers, in all three equations. $\square$

Remark 4.1 We are considering the series with zero mean to simplify notation, but there is no loss of generality in this assumption, since if a process Y_t has mean $E(Y_t) = \mu$, then fit the ARMA models to $X_t = Y_t - \mu$, which has zero mean. Note that this just adds a constant to the model. For example, suppose that Y_t follows an $AR(1)$ process, so that $Y_t - \mu = W_t + \phi_1(Y_{t-1} - \mu)$, and this is equivalent to $Y_t = W_t + \phi_1 Y_{t-1} + \phi_0$, where $\phi_0 = (1 - \phi_1)\mu$. If, on the other hand, Y_t follows a MA(1) process, then $Y_t = \mu + W_t + \theta_1 W_{t-1}$. $\square$

Any finite q-order moving average (MA(q)) process is easily seen to be weakly stationary for any values of θ_k, since their autocovariances are finite versions of the sums in Eq. (4.5), and independent of t:

$$\gamma(0) = \sigma^2(1 + \theta_1^2 + \cdots + \theta_q^2)$$
$$\gamma(h) = \begin{cases} \sigma^2(\theta_h + \theta_1\theta_{h+1} + \cdots + \theta_{q-h}\theta_q) & \text{for } h = 1, \ldots, q \\ 0 & \text{otherwise} \end{cases} \qquad (4.9)$$

Thus, a time series can be recognized as a MA(q) process if its ACF is non-zero up to lag q and is zero afterwards. In R use the function `acf` to estimate autocovariances and autocorrelations $\rho(h)$ (cf. Example 3.2). An infinite order moving average process (MA(∞)) is the special case of the general linear process in Eq. (4.3), with $\psi_k = 0$ for all $k < 0$. It is stationary provided $\sum_{k=0}^{\infty} \psi_k^2 < \infty$, and its autocovariance function is given by Eq. (4.5).

R Example 4.1 A natural finite order moving average process arises when one considers continuously compounded τ-period returns, for a given τ. Recall from Def. 2.2 that for a given series of continuously compounded returns $\{r_t\}$ and a given time period τ, the τ-period return $r_t(\tau)$ is expressed as a sum of log returns:

$$r_t(\tau) = r_t + r_{t-1} + \cdots + r_{t-\tau+1}.$$

Assuming that $\{r_t\}$ is white noise, more precisely, iid random variables with $r_t \sim WN(\mu, \sigma^2)$, then the second order moments of $r_t(\tau)$ satisfy

$$\gamma(0) = Var(r_t(\tau)) = \tau\sigma^2$$

$$\gamma(h) = Cov(r_t(\tau), r_{t-h}(\tau)) = \begin{cases} (\tau - h)\sigma^2 & \text{for } h < \tau \\ 0 & \text{for } h \geq \tau \end{cases}$$

and $E(r_t(\tau)) = \tau\mu$. This shows that the structure of finite moments of $\{r_t(\tau)\}$ is like a MA($\tau - 1$) process, and so it can be simulated as such a type of processes:

$$r_t(\tau) = \tau\mu + W_t + \theta_1 W_{t-1} + \cdots + \theta_{\tau-1} W_{t-\tau+1}$$

with $W_t \sim WN(0, \sigma^2)$. One can observe this empirically in R. Build the series of daily log returns from the price history of your favorite stock; check that there are no significant autocorrelations for any lag (using the `acf` function, or the Ljung-Box statistics—see Eq. (3.6)); build a series of τ-period returns, for say $\tau = 20$ (a monthly period), and compute the ACF of this series. An important lesson is learned from this example, and that is while $\{r_t\}$ might be white noise (iid random variables), the τ-period series $\{r_t(\tau)\}$ is not, and in fact has a smoother behavior. □

Now, to deal with the AR and ARMA models, it is better to rewrite their representation in a compact form using lag operators.

Lag operators and polynomials. The *lag operator*, or *backward shift*, acts on a time series by moving the time index back one time unit:

$$LX_t = X_{t-1}$$

Note that $L^2 X_t = L(LX_t) = LX_{t-1} = X_{t-2}$, and in general, the d-power of L moves X_t to its d-lag : $L^d X_t = X_{t-d}$. We can then define the polynomials in the lag operator,

$$\theta(L) = 1 + \theta_1 L + \cdots + \theta_q L^q \text{ and } \phi(L) = 1 - \phi_1 L - \cdots - \phi_p L^p.$$

With these polynomials in L, we can express the autoregressive and moving averages models in the following succinct forms:

$$\text{AR}(p): \quad \phi(L)X_t = W_t \qquad\qquad (4.10)$$
$$\text{MA}(q): \quad X_t = \theta(L)W_t \qquad\qquad (4.11)$$
$$\text{ARMA}(p, q): \quad \phi(L)X_t = \theta(L)W_t \qquad\qquad (4.12)$$

These representations allow to transform an AR(p) process into a MA(∞), and a MA(q) into an AR(∞), from where one can deduce conditions on the coefficients ϕ_k implying the stationarity of the AR (and ARMA) processes.

Going from AR to MA. The lag polynomials expressions suggest that to transform an AR(p) process into a MA we just have to multiply by the inverse polynomial $\psi(L)^{-1}$, so that, $X_t = \psi(L)^{-1}W_t$. But what is the inverse of $\psi(L)$ (if it exists)?

Consider and AR(1) process

$$X_t - \phi_1 X_{t-1} = (1 - \phi_1 L)X_t = W_t \tag{4.13}$$

To invert $(1 - \phi_1 L)$, recall the Taylor polynomial expansion of $1/(1 - z)$:

$$\frac{1}{1-z} = 1 + z + z^2 + \cdots = \sum_{k=0}^{\infty} z^k$$

defined for $|z| < 1$. This fact suggests that

$$X_t = (1 - \phi_1 L)^{-1} W_t = (1 + \phi_1 L + \phi_1^2 L^2 + \cdots)W_t = \sum_{k=0}^{\infty} \phi_1^k W_{t-k} \tag{4.14}$$

provided $|\phi_1| < 1$. Although this is an informal argument (we have not said much about the operator L, and are assuming that $|\phi_1| < 1$ implies $|\phi_1 L| < 1$), one can verify that the obtained MA(∞) process, $X_t = \sum_{k=0}^{\infty} \phi_1^k W_{t-k}$, is the (unique) stationary solution of (4.13). Thus, the zero mean AR(1) process (4.13) is stationary if $|\phi_1| < 1$, in which case it has the moving average (or Wold) representation given by Eq. (4.14).

Autocovariances and autocorrelations of an AR(1). From Eq. (4.15), one can compute

$$\gamma(0) = Var(X_t) = \sigma^2 \sum_{k=0}^{\infty} \phi_1^{2k} = \frac{\sigma^2}{1 - \phi_1^2}$$

$$\gamma(h) = \sigma^2 \sum_{k=0}^{\infty} \phi_1^k \phi_1^{k+h} = \sigma^2 \phi_1^h \sum_{k=0}^{\infty} \phi_1^{2k} = \frac{\sigma^2 \phi_1^h}{1 - \phi_1^2} \tag{4.15}$$

and the autocorrelations, $\rho(0) = 1$ and $\rho(h) = \gamma(h)/\gamma(0) = \phi_1^h$. This last equation says that the ACF of an AR(1) process decays exponentially with rate ϕ_1 and starting value $\rho(0) = 1$. Thus observing an inverse exponential curve in the plot of the (sample) ACF of a given time series is a sign of the existence of an underlying AR(1) process.

There are many financial time series that can be well modeled by an AR(1) process. Examples are interest rates, real exchange rates, and various valuation ratios, such as price-to-earning, dividend-to-price and others that will be studied in Chap. 6.

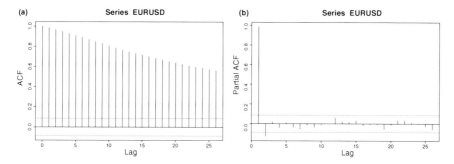

Fig. 4.1 **a** ACF and **b** PACF for the EUR/USD series

R Example 4.1 We explore the possible existence of autocorrelation in the EUR/USD exchange rates. Load the library of `quantmod` functions, and then execute the commands:

```
> getFX("EUR/USD") #download EUR/USD rates from oanda.com
> plot(EURUSD)
> acf(EURUSD)
> pacf(EURUSD)
```

The ACF is shown in Fig. 4.1a. One can see a slow exponential decay for successive lags, hence revealing that the series EUR/USD does behaves as an AR(1) process. The Partial ACF (PACF) function (to be explained later), and plotted in Fig. 4.1b, confirms that the order of the underlying autoregressive process is 1.

Now, in order to apply a similar idea for transforming an AR(p) process, given in the form of Eq. (4.10), into a MA(∞) process, factor the lag polynomial $\phi(L) = 1 - \phi_1 L - \phi_2 L^2 - \cdots - \phi_p L^p$ into a product of linear factors

$$\phi(L) = (1 - \lambda_1 L)(1 - \lambda_2 L) \cdots (1 - \lambda_p L) \tag{4.16}$$

where the λ_k can be complex numbers. Assuming that $\lambda_1, \cdots, \lambda_p$ are all different, we have

$$\phi(L)^{-1} = \frac{1}{(1 - \lambda_1 L)(1 - \lambda_2 L) \cdots (1 - \lambda_p L)} = \frac{A_1}{(1 - \lambda_1 L)} + \frac{A_2}{(1 - \lambda_2 L)} + \cdots + \frac{A_p}{(1 - \lambda_p L)}$$

where the coefficients $A_1, \ldots, A_p$ are obtained by a partial fractions decomposition, and depend on $\lambda_1, \ldots, \lambda_p$. If we assume that for $h = 1, \ldots, p, |\lambda_h| < 1$, we can write each factor $\dfrac{1}{(1 - \lambda_h L)} = \sum_{k=0}^{\infty} \lambda_h^k L^k$. Then, the AR($p$) process

$$X_t = \phi(L)^{-1} W_t = \left(A_1 \sum_{k=0}^{\infty} \lambda_1^k L^k + \cdots + A_p \sum_{k=0}^{\infty} \lambda_p^k L^k \right) W_t$$

$$= \sum_{k=0}^{\infty} \left(\sum_{h=0}^{p} A_h \lambda_h^k \right) W_{t-k} \tag{4.17}$$

is a MA(∞) stationary process, provided $|\lambda_1| < 1, \ldots, |\lambda_p| < 1$. This last condition can be shown to be equivalent to

$$|\phi_1 + \phi_2 + \cdots + \phi_p| < 1 \tag{4.18}$$

Thus, the AR(p) process (4.10) is stationary if Eq. (4.18) holds, and a solution has the general form (4.17). It will be very instructive for the reader to work out the details for the AR(2) process, and obtain the explicit values for λ_1, λ_2, A_1 and A_2 in terms of ϕ_1 and ϕ_2, see Problem 4.7.2.

Remark 4.2 If there are repeated factors $(1 - z)^k = (1 - \lambda L)^k$ of degree $k > 1$ in Eq. (4.16), then in the partial fraction decomposition appear terms $A_1/(1 - z) + A_2/(1 - z)^2 + \cdots + A_k/(1 - z)^k$. Each term $1/(1 - z)^s$, for $s > 1$, is the derivative of order s of $1/(1 - z)$ times some constant; hence taking derivatives to the Taylor expansion of $1/(1 - z)$ we get the desired series for the powers of this fraction; specifically $\dfrac{(s - 1)!}{(1 - z)^s} = \sum_{k>s} k(k - 1) \cdots (k - s) z^{k-s+1}$.   □

Autocovariances of an AR(p) and Yule-Walker equations. The structure of the autocovariance function for a process AR(p), $p > 1$, is not as easy to get directly from its MA(∞) (or Wold) expression (Eq. (4.17)) as in the AR(1) case. Here the trick is to consider the AR(p) process in its original form

$$X_t - \phi_1 X_{t-1} - \phi_2 X_{t-2} - \cdots - \phi_p X_{t-p} = W_t, \quad \text{with } \{W_t\} \sim WN(0, \sigma^2) \tag{4.19}$$

and note that since $E(X_t) = 0$, the autocovariances $\gamma(h) = E(X_t X_{t-h})$. Then multiplying each side of (4.19) by X_{t-h}, $h = 0, \ldots, p$, taking expectations and using that $X_t = \sum_{k=0}^{\infty} \psi_k W_{t-k}$, where $\psi_k = \sum_{h=0}^{p} A_h \lambda_h^k$ (Eq. (4.17)), to evaluate the right-hand sides (i.e., to compute $E(X_{t-h} W_t)$), we obtain the following system of equations for the moments of the AR(p) process

$$\gamma(0) = \phi_1 \gamma(1) + \phi_2 \gamma(2) + \cdots \phi_p \gamma(p) + \sigma^2$$
$$\gamma(h) = \phi_1 \gamma(h - 1) + \phi_2 \gamma(h - 2) + \cdots \phi_p \gamma(h - p), \quad \text{for } h = 1, 2, \ldots, p. \tag{4.20}$$

These are the *Yule-Walker* equations. In matrix form,

$$\gamma(0) = \Phi_p^t \boldsymbol{\gamma}_p + \sigma^2$$
$$\boldsymbol{\gamma}_p = \Gamma_p \Phi_p \tag{4.21}$$

where $\boldsymbol{\gamma}_p = (\gamma(1), \ldots, \gamma(p))'$, $\Phi_p = (\phi_1, \ldots, \phi_p)'$ and $\Gamma_p = [\gamma(i-j)]_{1 \le i, j \le p}$ is the autocovariance matrix (recall from Remark 2.36 that $\gamma(h) = \gamma(-h)$).

The Yule-Walker equations are a useful tool to estimate the coefficients $\phi_1, \ldots, \phi_p$ of an AR(p) model to fit to a given observed process $\{X_t\}_{t \ge 0}$. Consider the sample autocovariance,

$$\widehat{\gamma}(h) = \widehat{Cov}(X_t, X_{t-h}) = \frac{1}{m-1} \sum_{t=h}^m (X_t - \widehat{\mu}_X)(X_{t-h} - \widehat{\mu}_X)$$

for $h = 0, 1, \ldots, p$. Then, we have that the sample covariance matrix

$$\widehat{\Gamma}_p = \begin{pmatrix} \widehat{\gamma}(0) & \widehat{\gamma}(1) & \cdots & \widehat{\gamma}(p-1) \\ \widehat{\gamma}(1) & \widehat{\gamma}(0) & \cdots & \widehat{\gamma}(p-2) \\ \vdots & \vdots & \ddots & \vdots \\ \widehat{\gamma}(p-1) & \widehat{\gamma}(p-2) & \cdots & \widehat{\gamma}(0) \end{pmatrix}$$

is invertible,[3] and from Eq. (4.21) we have

$$\widehat{\Phi}_p = \widehat{\Gamma}_p^{-1} \widehat{\gamma}_p \tag{4.22}$$

The solution $\widehat{\Phi}_p$ is an unbiased estimation of Φ_p provided the sample size m is taken large enough, and $h < m$. A rough guide is to take $m \ge 50$ and $h \le m/4$ (Box and Jenkins 1976).

Forecasting with AR(p). The reader should have noticed the similarities of the Yule-Walker equations with the linear equations (2.45) in Sect. 2.4 verified by the coefficients of a one-step linear forecaster. The procedures to derive both systems of equations are exactly the same. Indeed, if we consider an AR($p+1$) process

$$X_{t+1} = \phi_0 X_t + \phi_1 X_{t-1} + \phi_2 X_{t-2} + \cdots + \phi_p X_{t-p} + W_t$$

the coefficients $\phi_0, \phi_1, \ldots, \phi_p$ satisfy the Yule-Walker equations (4.20) with $p+1$ variables, which can be written as

$$\gamma(h+1) = \phi_0 \gamma(h) + \phi_1 \gamma(h-1) + \cdots + \phi_p \gamma(h-p), \quad h = 0, 1, \ldots, p.$$

On the other hand, the best linear forecaster of X_{t+1} in terms of $\{X_t, X_{t-1}, \ldots, X_{t-p}\}$ is

$$L(X_{t+1}) = \alpha_0 X_t + \alpha_1 X_{t-1} + \cdots + \alpha_p X_{t-p}$$

where $\alpha_0, \ldots, \alpha_p$ satisfy the linear equations (2.45):

$$\gamma(h+1) = \alpha_0 \gamma(h) + \alpha_1 \gamma(h-1) + \cdots + \alpha_p \gamma(h-p), \quad h = 0, 1, \ldots, p.$$

---

[3] The reader can take our word on that or find this fact in Brockwell and Davis (1991).

By the unicity of the solutions to these equations, it must be

$$\alpha_0 = \phi_0, \alpha_1 = \phi_1, \ldots, \alpha_p = \phi_p$$

Therefore, for an AR(p) process, the best linear forecaster of X_{t+1} in terms of $\{X_t, X_{t-1}, \ldots, X_{t-p}\}$ is

$$L(X_{t+1}) = \phi_0 X_t + \phi_1 X_{t-1} + \cdots + \phi_p X_{t-p}$$

(i.e., it is determined by the same coefficients that define the autoregressive process). The mean square error of the prediction, $E (X_{t+1} - L(X_{t+1}))^2$, is given by Eq. (2.47).

The particular case of forecasting with an AR(1) process is quite simple. The best linear one-step forecaster of an AR(1) process $X_t = \phi X_{t-1} + W_t$ (where $|\phi| < 1$ and $W_t \sim WN(0, \sigma^2)$) in terms of its past is

$$L(X_{t+1}) = \phi X_t$$

and using Eq. (4.15) we can express the mean square error of the prediction as

$$E (X_{t+1} - L(X_{t+1}))^2 = \gamma(0) - 2\phi\gamma(1) + \phi^2\gamma(0) = \frac{\sigma^2}{1 - \phi^2} - \phi\gamma(1) = \sigma^2$$

A rule to determine the order p. What remains to know is a method to determine the order p of the AR(p) model. Just as in an AR(1) process, a plot of the ACF function will show a sinuous shape, a mixture of sines, cosines and exponential decay, and this gives an indication that an AR model of certain order is possible. But to actually determine the order p of the autoregressive model use the *partial autocorrelation function* (PACF), which is based on setting up the following AR models listed in increasing orders:

$$X_t = \phi_{11} X_{t-1} + W_{1t}$$
$$X_t = \phi_{21} X_{t-1} + \phi_{22} X_{t-2} + W_{2t}$$
$$\vdots$$
$$X_t = \phi_{p1} X_{t-1} + \phi_{p2} X_{t-2} + \cdots + \phi_{pp} X_{t-p} + W_{pt}$$

and solve this multiple linear regression by the least squares method. Then the rule for fitting an AR(p) model is that the first p coefficients $\phi_{ii}, i = 1, \ldots, p$ are non-zero and for the rest $i > p$, $\phi_{ii} = 0$. The PACF criterion is programmed in R with the function `pacf` (and we have already use it in R Example 4.1). The AR(p) model estimation based on the Yule-Walker equations (Eq. (4.22)) is done in R with the function `ar()`.

R Example 4.2 (Continue from R Example 4.1) The EUR/USD exchange rates behave as an AR(1) process: its ACF function showed an exponential decay and its PACF function indicated a lag of order 1. Therefore we fit here an autoregressive of order 1 to this series, using the `armaFit` from the package `fArma`, for this is a more robust implementation of `ar()` estimator. Run the commands

```
> library(fArma)
> ar1=armaFit(~ar(1),method="yule-walker",data=EURUSD)
> summary(ar1)
> predict(ar1,n.ahead=5,n.back=80)
```

The `summary` gives as output the coefficient $\phi_1 = 0.985$, and the intercept value of $\phi_0 = 1.308$. This last one corresponds to the more general centered form of AR(1) process, $X_t = \phi_0 + \phi_1 X_{t-1} + W_t$, as explained in Remark 4.1. Thus, the corresponding AR(1) process for the EUR/USD series is

$$X_t = 1.308 + 0.985 X_{t-1} + W_t, \quad \{W_t\} \sim N(0, \sigma^2)$$

Using the fitted AR(1) time series, `predict` plots `n.back` points of the process and forecasts and plots `n.ahead` values.   □

Going from MA to AR. Applying similar ideas as the transformation of a finite order AR process into an infinite order MA, we can go from a MA(q) process to a AR(∞) process by inverting the polynomial $\theta(L)$ in Eq. (4.11), so that $\theta(L)^{-1} X_t = W_t$ becomes an autoregressive process of infinite order. The polynomial $\theta(L)$ is invertible if $|\theta_i| < 1$ holds for all $i = 1, \ldots, q$. The MA process is invertible if the polynomial $\theta(L)$ is invertible. Let us illustrate this transformation explicitly for the case of moving averages of order 1.

R Example 4.2 We are given a MA(1) process in lag operator form,

$$X_t = W_t + \theta W_{t-1} = (1 + \theta L) W_t.$$

Using the Taylor expansion for the function $f(z) = 1/(1 + z)$, which is defined for $|z| < 1$ as follows

$$\frac{1}{1+z} = 1 - z + z^2 - z^3 + \cdots = \sum_{k=0}^{\infty} (-1)^k z^k,$$

we obtain

$$W_t = \frac{1}{1 + \theta L} X_t = \sum_{k=0}^{\infty} (-1)^k \theta^k X_{t-k} = X_t + \sum_{k=1}^{\infty} (-1)^k \theta^k X_{t-k}$$

or equivalently, $X_t = \sum_{k=1}^{\infty} (-1)^{k+1} \theta^k X_{t-k} + W_t$, which is clearly an AR(∞) process.   □

What we can observe from this representation is that the PACF for an invertible MA process has an exponential decay towards zero.

ARMA(p, q) autocovariances and model estimation. From the back-and-forth transformations between AR and MA process, what we can readily say about the autocorrelation functions of an ARMA(p, q) process is that for $p > q$ the ACF behaves similarly to the ACF of an AR(p) process, while for $p < q$ the PACF behaves like the PACF of a MA(q) process, and, consequently, both ACF and PACF of the ARMA(p, q) process decay exponentially for large p and q. To compute an instrumental expression for the ACF of an ARMA(p, q) process we proceed with an analysis similar to the AR case (i.e., a derivation like Yule-Walker's). Consider $\{X_t\}$ an ARMA(p, q) process with zero mean or mean–corrected (i.e., the mean has been subtracted from the series), written in the form

$$X_t - \phi_1 X_{t-1} - \cdots - \phi_p X_{t-p} = W_t + \theta_1 W_{t-1} + \cdots + \theta_q W_{t-q},$$

with $\{W_t\} \sim WN(0, \sigma^2)$. Multiply both sides of this equation by X_{t-h}, for $h = 0, 1, 2, \cdots$, and take expectations. Using the Wold representation, $X_t = \sum_{k=0}^{\infty} \psi_k W_{t-k}$, and that $\gamma(h) = E(X_t X_{t-h})$ since $E(X_t) = 0$, we obtain from the previous operations the following equations:

$$\gamma(h) - \phi_1 \gamma(h-1) - \cdots - \phi_p \gamma(h-p) = \sigma^2 \sum_{k=0}^{\infty} \theta_{h+k} \psi_k, \quad \text{for } 0 \leq h < m, \quad (4.23)$$

and

$$\gamma(h) - \phi_1 \gamma(h-1) - \cdots - \phi_p \gamma(h-p) = 0, \quad \text{for } h \geq m, \quad (4.24)$$

where $m = \max(p, q+1)$, $\psi_k = 0$ for $k < 0$, $\theta_0 = 1$, and $\theta_k = 0$ for $k \notin \{0, \ldots, q\}$. Assuming that the lag polynomial, $\phi(z) = 1 - \phi_1 z - \cdots - \phi_p z$, has p distinct roots, then the system of homogeneous linear difference Eq. (4.24) has a solution of the form

$$\gamma(h) = \alpha_1 \xi_1^{-h} + \alpha_2 \xi_2^{-h} + \cdots + \alpha_p \xi_p^{-h}, \quad \text{for } h \geq m - p, \quad (4.25)$$

where $\xi_1, \ldots, \xi_p$ are the distinct solutions of $\phi(z) = 0$, and $\alpha_1, \ldots, \alpha_p$ are arbitrary constants.[4] Since we want a solution of Eq. (4.24) that also satisfies Eq. (4.23), substitute the general solution (4.25) into Eq. (4.23) to obtain a set of m linear equations that uniquely determine the constants $\alpha_1, \cdots, \alpha_p$, the autocovariances $\gamma(h)$, for $h = 0, \ldots, m - p$, and consequently, the autocorrelations $\rho(h) = \gamma(h)/\gamma(0)$.

The procedure just described to compute the autocorrelations of an ARMA(p, q) process is implemented by the function ARMAacf of the stats package in R.

R Example 4.3 Execute in your R console the following instruction:

```
> rho = ARMAacf(ar=c(1.38, -5.854e-01),
```

---

[4] see Brockwell and Davis (1991, Sect. 3.6) for the mathematical details of solving Eq. (4.24) including the case where $\phi(z) = 0$ has repeated solutions.

Fig. 4.2 Exponential decay
for $\rho(h)$ of an ARMA(2,4)

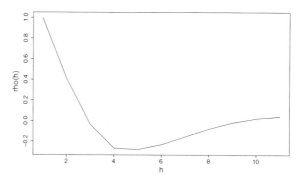

```
ma=c(-1.098,0,0,1.053e-01),lag.max=10)
```

ARMAacf returns a list of the theoretical values of the autocorrelations up to lag 10 of an ARMA process whose AR and MA coefficients are, respectively,

$$(\phi_1, \phi_2) = (1.38, -0.5854) \quad \text{and} \quad (\theta_1, \theta_2, \theta_3, \theta_4) = (-1.098, 0, 0, 0.1053)$$

(This ARMA will be obtained from modeling real data in R Example 4.4 below.)

If we now plot the 10 values in rho and draw the curve through them, we will see the exponential decay of $\rho(h)$ for this ARMA(2,4) process:

```
plot(rho,type="l",xlab="h",ylab="rho(h)")
```

The plot is shown in Fig. 4.2. Adding the parameter pacf=TRUE to the ARMAacf function above, gives the theoretical values of the PACF. □

Now, to fit an ARMA(p, q) model to a given zero-mean time series $\{X_t\}$ one first chooses a desired order pair (p, q), and assuming that the observed data is stationary and Gaussian one uses the maximum likelihood estimation (MLE) method with respect to parameters $\boldsymbol{\phi} = (\phi_1, \ldots, \phi_p)$,

σ^2, to find estimators $\widehat{\boldsymbol{\phi}}, \widehat{\boldsymbol{\theta}}$ and $\widehat{\sigma}^2$. These maximum likelihood estimators can be obtained as solutions to the following system of equations:

$$\widehat{\sigma}^2 = \frac{1}{n} S(\widehat{\boldsymbol{\phi}}, \widehat{\boldsymbol{\theta}}), \tag{4.26}$$

where

$$S(\widehat{\boldsymbol{\phi}}, \widehat{\boldsymbol{\theta}}) = \sum_{j=1}^{n} (X_j - \widehat{X}_j)^2 / r_{j-1}, \tag{4.27}$$

and $\widehat{\boldsymbol{\phi}}, \widehat{\boldsymbol{\theta}}$ are the values of $\boldsymbol{\phi}$ and $\boldsymbol{\theta}$ that minimize

$$l(\boldsymbol{\phi}, \boldsymbol{\theta}) = \ln \left(n^{-1} S(\boldsymbol{\phi}, \boldsymbol{\theta}) \right) + n^{-1} \sum_{j=1}^{n} \ln r_{j-1} \tag{4.28}$$

where n is the size of the sample observations, $\{X_1, \ldots, X_n\}$; $\widehat{X}_1 = 0$ and $\widehat{X}_j = E(X_j | X_1, \ldots, X_{j-1})$, for $j \geq 2$, are the one-step predictors; the r_j are obtained from the covariances of $\{X_t\}$ and are independent of σ^2. $l(\boldsymbol{\phi}, \boldsymbol{\theta})$ is a "reduced log-likelihood". The derivation of this system of equations is not difficult but involved, and can be studied in full details in Brockwell and Davis (1991, Chap. 8). The MLE for ARMA is a nonlinear optimization problem and so it requires a numerical procedure to iteratively find the values of the parameters that minimized the square prediction error. It needs to be given some initial values for the parameters $\phi_1, \ldots, \phi_p, \theta_1, \ldots, \theta_q$ to start, and convergence and correctness will depend to some extend on these initial values. Hence, some appropriate approximation methods should be employed to find good initial $p + q$ values. Two such initialization methods are the innovation algorithm and Hannan-Rissanen algorithm (again see Brockwell and Davis op. cit.).

Finally, to select appropriate values for the orders p and q one uses the Akaike information criterion (AIC), which consists on choosing $p, q, \boldsymbol{\phi}$ and $\boldsymbol{\theta}$ that minimize

$$AIC = -2 \ln L(\boldsymbol{\phi}, \boldsymbol{\theta}, S(\boldsymbol{\phi}, \boldsymbol{\theta})n^{-1}) + 2(p + q + 1)n/(n - p - q - 2) \quad (4.29)$$

where $L(\boldsymbol{\phi}, \boldsymbol{\theta}, \sigma^2)$ is the Gaussian likelihood for an ARMA(p, q) process. Thus, in practice the AIC criterion works as follows: choose some upper bounds, P and Q, for the AR and MA order respectively. Then fit all possible ARMA(p, q) models for $p \leq P$ and $q \leq Q$, on the same sample series of size T, and keep the model whose parameters $p, q, \boldsymbol{\phi}$ and $\boldsymbol{\theta}$ minimize AIC. Beware though that this criterion asymptotically overestimates the order, that is, as $T \to \infty$ larger values of p and q are chosen. It can also be a time-consuming optimization procedure for high orders p and q. However, the general wisdom asserts that in financial applications small values of p and q (say, less than 5) are sufficient. In the following R Example we program the AIC criterion and use it to fit an ARMA model to financial data.

R Example 4.4 We consider adjusting an ARMA model to Robert Shiller's Price-to-Earning (P/E) ratio[5] for the S&P 500. This series, also known as Shiller's PE 10, is based on the average inflation-adjusted earnings of the S&P 500 taken over the past 10 years, and computed every month since 1881. The data is publicly available from Shiller's webpage,[6] and we've saved it in our disk with the label SP500_shiller.csv. We implement the AIC criterion by setting up a double loop to apply for all pairs of (p, q), from $(0, 0)$ to upper bound in $(5, 5)$, the armaFit method on the data, and keep the model with minimum AIC. To extract the AIC value computed in each execution of armaFit one uses the expression tsARMA@fit$aic, where tsARMA is the name of the variable that saves the current computed ARMA. Because the method armaFit might fail to compute a model, in which case it returns an error or warning message producing an interruption of the loop, we execute armaFit within tryCatch to handle these error messages. The full R program follows.

---

[5] P/E ratio is defined in Chap. 6.

[6] http://www.econ.yale.edu/$\sim$shiller/data.htm, although we use a cvs version from http://data.okfn.org/data/.

```
> library(fArma)
> sp500=read.csv("SP500_shiller.csv")
> sp500PE = na.omit(sp500$P.E10)
> ts=diff(log(sp500PE))  ##compute returns
> ts=ts-mean(ts) ##subtract mean
> bestAIC=1e9
> for(p in 0:5) for (q in 0:5)
  { formula = as.formula(paste(sep="","ts~arma(",p,",",q,")"))
     tsARMA = tryCatch(armaFit(formula, data=ts),
                    error=function(e) FALSE,
                    warning=function(w) FALSE )
    if( !is.logical(tsARMA) ){
      AIC = tsARMA@fit$aic
      if(AIC < bestAIC){
        bestAIC=AIC
        bestARMA=tsARMA
        bestPQ= c(p,q) }
    }
    else print("FALSE")
  }
> bestPQ
> summary(bestARMA)
```

The program outputs 2 and 4 as the best pair (p, q); with summary we obtain
the coefficients together with some error analysis that includes standard errors,
t values (from a t-statistic to test the null hypothesis of each coefficient being zero)
and corresponding p values. This information is most useful to refine our model. In
this example it outputs the following table

```
Coefficient(s):
              Estimate  Std. Error  t value Pr(>|t|)
ar1          1.380e+00   1.369e-01   10.081  < 2e-16 ***
ar2         -5.854e-01   1.521e-01   -3.848 0.000119 ***
ma1         -1.098e+00   1.368e-01   -8.024 1.11e-15 ***
ma2          2.248e-01   1.244e-01    1.807 0.070818 .
ma3          7.947e-02   5.702e-02    1.394 0.163393
ma4          1.053e-01   2.767e-02    3.804 0.000142 ***
intercept  -9.574e-06   1.522e-03   -0.006 0.994982
```

From this table we see that the estimates of the MA coefficients θ_2 and θ_3 (ma2
and ma3 in the table) are not statistically significant at the 5 % level, and consequently
can be discarded. Therefore, the best ARMA for the zero mean returns of Shiller's
PE 10 series is the following ARMA(2,4) model

$$X_t = 1.38X_{t-1} - 0.585X_{t-2} + W_t - 1.098W_{t-1} + 0.105W_{t-4}$$

We can forecast 5 monthly values of PE 10 returns with
```
predict(bestARMA,n.ahead=5,n.back=80)
```
Back in R Example 4.3 we've computed the theoretical ACF for this ARMA process.□

Remark 4.3 Not all financial time series sampled from real data can be fitted with an
AIC–optimal ARMA(p, q), with $p > 0$ and $q > 0$. The P/E ratios and long interest

rates (also obtainable from Shiller's dataset) are among the exceptions. The reader can use the previous R implementation for the AIC criterion to test this hypothesis. The series of returns from your favorite stock, or market index (S&P 500, DJIA, etc.), returns of bond yields or crude prices, all yield through this AIC test an ARMA(p, 0) model, namely an AR(p), and more often than not with $p < 3$. Alas! in any case, this little program can help you find your optimal AR, MA or ARMA model for your time series at hand.   □

4.3 Nonlinear Models ARCH and GARCH

Financial time series often exhibit daily returns with similar high variance for a succession of time periods and similar low variance for other string of periods. We have empirically observed this dependance of the return's variance with its past by computing the ACF function of squared returns and obtained significant lag values. These observations motivates models of financial returns where their variance evolves over time, or what are known as *conditional heteroscedastic* models. In the general structure of stationary random time series model given by Eq. (4.2), we want $\sigma_t^2 = Var(Y_t|F_{t-1})$ to vary with F_{t-1}.

4.3.1 The ARCH Model

The *Autoregressive Conditional Heteroscedastic* model of order p, ARCH(p), was conceived by Robert Engle under the premise that the conditional variance is a linear function of the past p squared innovations.[7] Thus,

$$Y_t = \mu_t + a_t$$
$$\sigma_t^2 = \omega + \alpha_1 a_{t-1}^2 + \cdots + \alpha_p a_{t-p}^2 \qquad (4.30)$$

For this model to be well defined and the conditional variance to be positive, the parameters must satisfy $\omega > 0$ and $\alpha_1 \geq 0, \ldots, \alpha_p \geq 0$. Defining $W_t = a_t^2 - \sigma_t^2$, we can rewrite Eq. (4.30) as

$$a_t^2 = \omega + \sum_{k=1}^{p} \alpha_k a_{t-k}^2 + W_t \qquad (4.31)$$

and since $E(W_t|F_{t-1}) = 0$, the above equation says that the ARCH(p) models corresponds to an AR(p) model for the squared innovations, a_t^2. Therefore the process

---

[7] R. Engle, Autoregressive conditional heteroscedasticity with estimates of the variance of UK inflation, *Econometrica*, 50 (1982) 987–1008.

is weakly stationary if and only if $|\alpha_1 + \cdots + \alpha_p| < 1$ (cf. Eq. (4.18)), in which case the unconditional variance (hence, constant variance) is, taking expectation in (4.31),

$$\sigma^2 = Var(a_t) = \frac{\omega}{1 - \alpha_1 - \cdots - \alpha_p}$$

Remark 4.4 For making inference with ARCH type models it is often assumed that $a_t = \sigma_t \varepsilon_t$, where $\{\varepsilon_t\}$ is a sequence of iid random variables with $\varepsilon_t \sim N(0, 1)$. Observe that although this realization of a_t is serially uncorrelated, by Eq. (4.31) is clearly not independent through time. We have then that large (small) values of volatility are to be followed by other large (small) values of volatility. Thus, the ARCH model is in accordance with the stylized fact termed volatility clustering (cf. Sect. 3.4).   □

Steps to build an ARCH model. Given a sample return series $\{r_t : t = 1, \ldots, T\}$ that we want to fit to an ARCH model, the first step is to determine the order of the model. Consider the residuals $a_t = r_t - \mu_t$ of the mean Eq. (4.30) (with $Y_t = r_t$), and since the ARCH model is an autoregression on the squared residuals (Eq. (4.31)), we can compute the PACF for $\{a_t^2\}$ to determine the order p. The next step is to estimate the parameters $\omega, \alpha_1, \ldots, \alpha_p$ of the model. This is done with the maximum likelihood estimation method. Assuming that the probability distribution of a_t conditional on the variance is normal (i.e., $a_t \sim N(0, \sigma_t^2)$), then the likelihood function is[8]

$$L(\omega, \alpha_1, \ldots, \alpha_p) = \prod_{t=p+1}^{T} \left[\frac{1}{\sqrt{2\pi \sigma_t^2}} \exp\left(\frac{-a_t^2}{2\sigma_t^2}\right) \right] \qquad (4.32)$$

Maximizing this function is the same as maximizing the logarithm

$$-\frac{1}{2} \sum_{t=p+1}^{T} \left(\ln(2\pi) + \ln \sigma_t^2 + \frac{a_t^2}{\sigma_t^2} \right) \qquad (4.33)$$

The values that maximize the above formula are the same for the formula without additive constants. Thus, the log-likelihood function to consider is

$$l(\omega, \alpha_1, \ldots, \alpha_p) = -\frac{1}{2} \sum_{t=p+1}^{T} \left(\ln \sigma_t^2 + \frac{a_t^2}{\sigma_t^2} \right) \qquad (4.34)$$

One proceeds to search iteratively for the values of $\omega, \alpha_1, \ldots, \alpha_p$ that when substituted in the model (4.30) for σ_t^2 maximize the value of $l(\omega, \alpha_1, \ldots, \alpha_p)$, beginning with the variance at time $t = p + 1$ estimated with past p times residuals $a_1, \ldots, a_p$, that

---

[8] These equations are obtained by similar reasoning as in Example 2.7 for constant variance.

is, $\sigma_{p+1}^2 = \omega + \alpha_1 a_p^2 + \alpha_2 a_{p-1}^2 + \cdots + \alpha_p a_1^2$, where $a_i^2 = (r_i - \widehat{\mu})^2$ ($\widehat{\mu}$ being the sample mean).

Forecasting with ARCH(p) model. This is done similarly as with an AR(p). Considering the origin of the forecast at time t, the one-step ahead forecast of the variance σ_{t+1}^2, denoted $\sigma_{t+1|t}^2$ is given by

$$\sigma_{t+1|t}^2 = \omega + \alpha_1 a_t^2 + \cdots + \alpha_p a_{t+1-p}^2$$

By Remark 4.4, the square of future residuals a_{t+1} can be approximated by $\sigma_{t+1|t}^2$. Then the 2-step ahead forecast of the variance is

$$\sigma_{t+2|t}^2 = \omega + \alpha_1 \sigma_{t+1|t}^2 + \alpha_2 a_t^2 + \cdots + \alpha_p a_{t+2-p}^2$$

and continuing this way, the h-step ahead forecast is

$$\sigma_{t+h|t}^2 = \omega + \alpha_1 \sigma_{t+1|t}^2 + \alpha_2 \sigma_{t+2|t}^2 + \cdots + \alpha_{h-1} \sigma_{t+h-1|t}^2 + \alpha_h a_h^2 + \cdots + \alpha_p a_{t+h-p}^2$$

To fit an ARCH model to a sample return series and make predictions of future variance in R, we use `garchFit()` and the associated suite of functions from the package `fGarch`.

R Example 4.5 We fit an ARCH model to the daily log-return of Allianz (DAX), sample from January 6 to December 30, 2009 (cf. R Lab 2.7.6). The data is saved as a column labelled `alvR` in the table `daxRlog`. We use `pacf` to estimate appropriate order for the model. Run the following commands

```
> library(fGarch);
> daxRl = read.table("daxRlog", header=T)
> alvRl =na.omit(daxRl$alvR) ##extract log returns of alv
> alvRl2 <- alvRl^2   ##compute square log returns
> acf(alvRl2)
> pacf(alvRl2)
> arch5=garchFit(formula=~garch(5,0),data=alvRl,cond.dist="norm")
> summary(arch5)
```

The ACF and PACF for the square log returns shows the existence of possible autoregression and up to lag 5; hence suggesting to fit a ARCH(5) model to `alvRl`, see Fig. 4.3. The `garchFit` method works as follows: parameter `formula` takes a formula object describing the mean and the variance equations. In our experiment we only want an ARCH(5) for the variance, hence we set `formula` = ~garch(5,0) (in the next section we will see that GARCH($p, 0$) = ARCH(p)). If we wanted a combined model ARMA for the mean and ARCH for the variance, we set `formula` = ~arma(n,m) + garch(p,0) (we will do this in R Example 4.6). In any case for the type of model, note that the input data is always the return series, not the squares. The instruction `summary` gives, as always, a long report that includes the coefficients and an error analysis

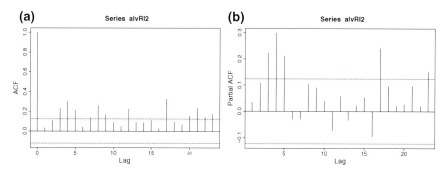

Fig. 4.3 ACF **a** and PACF **b** of ALV squared log returns

```
Coefficient(s):
          mu         omega        alpha1        alpha2
  0.00165830   0.00026441    0.00000001    0.03941390
      alpha3        alpha4        alpha5
  0.34698768   0.14304642    0.19924847
```

These are the estimated coefficients. From the error analysis only ω, α_3, α_4 and α_5 can be considered significantly different from zero (see an explanation of the error analysis in R Example 4.4). This gives us the following ARCH(5) model

$$r_t = a_t,$$
$$\sigma_t^2 = \omega + \alpha_3 a_{t-3} + \alpha_4 a_{t-4} + \alpha_5 a_{t-5}$$
$$= 2.64 \times 10^{-4} + (3.47 \times 10^{-1})a_{t-3} + (1.43 \times 10^{-1})a_{t-4} + (1.99 \times 10^{-1})a_{t-5}$$

We can make a graphical diagnosis of the fitted ARCH with the `plot` method (select 1 in the plot's menu):

```
> plot(arch5)
```

Compare this with the plot of the sample data:

```
> plot(alvRl,type='l')
```

To make up to 7-steps ahead variance predictions run the command

```
> predict(arch5,7)
```

To make predictions and plots (including a 95 % confidence band):

```
> predict(arch5,7,plot=TRUE)
```
                                                                          □

4.3.2 The GARCH Model

The *Generalized Autoregressive Conditional Heteroscedasticity* model of order p and q, GARCH(p,q), was proposed by Tim Bollerslev in 1986 to circumvent the estimation of ARCH(p) models for large values of p, which often arise in practice.[9]

---

[9] T. Bollerslev, Generalized Autoregressive Conditional Heteroscedasticity, *Journal of Econometrics*, 31 (1986) 307–327.

Bollerslev's idea was to make the variance also dependent on its q recent past values, hence obtaining the following recursive formula for σ_t^2:

$$Y_t = \mu_t + a_t$$

$$\sigma_t^2 = \omega + \sum_{i=1}^{p} \alpha_i a_{t-i}^2 + \sum_{j=1}^{q} \beta_j \sigma_{t-j}^2 \qquad (4.35)$$

For the conditional variance in this GARCH(p,q) model to be well defined and positive, and the process be weakly stationary, the coefficients must satisfy

$$\omega > 0, \ \alpha_1 \geq 0, \ldots, \alpha_p \geq 0, \ \beta_1 \geq 0, \ldots, \beta_q \geq 0 \ \text{ and } \ \sum_{i=1}^{\max(p,q)} (\alpha_i + \beta_i) < 1 \qquad (4.36)$$

These conditions on the coefficients of GARCH(p,q) model are deduced by reasoning similarly as for the ARCH(p), as follows. We can rewrite Eq. (4.35) as

$$a_t^2 = \omega + \sum_{k=1}^{\max(p,q)} (\alpha_k + \beta_k) a_{t-k}^2 - \sum_{k=1}^{q} \beta_k W_{t-k} + W_t \qquad (4.37)$$

with $W_{t-k} = a_{t-k}^2 - \sigma_{t-k}^2$, $k = 0, 1, \ldots, q$. This equation defines an ARMA(max $(p, q), q$) model for a_t^2. From this it follows that the process is weakly stationary if and only if $\sum_{k=1}^{\max(p,q)} (\alpha_k + \beta_k) < 1$. Moreover, the unconditional variance for the GARCH(p,q) model is obtained by taking expectation in (4.37) as

$$\sigma^2 = Var(a_t) = \frac{\omega}{1 - \sum_{k=1}^{\max(p,q)} (\alpha_k + \beta_k)} \qquad (4.38)$$

This unconditional variance can be interpreted as a long-run predicted variance. To see this consider a 1-step prediction with the simpler GARCH(1,1) model. This is just a straightforward extension of the ARCH(1) forecast, and it has the form

$$\sigma_{t+1|t}^2 = \omega + \alpha a_t^2 + \beta \sigma_t^2$$

For $h > 1$, the h-step ahead prediction with GARCH(1,1) has the form

$$\sigma_{t+h|t}^2 = \omega + \alpha \sigma_{t+h-1|t}^2 + \beta \sigma_{t+h-1|t}^2$$

Taking limit as $h \to \infty$ we get the long-run predicted variance

$$V_L = \lim_{h \to \infty} \sigma_{t+h|t}^2 = \frac{\omega}{1 - \alpha - \beta} \qquad (4.39)$$

This is the mean variance per day (or per time period considered) implied by the GARCH model. The implied daily volatility is $\sqrt{V_L}$. This also provides another interpretation for the GARCH model as an approximation to the conditional variance by a long-run variance rate, V_L, plus squared innovations and recent past variances. For the GARCH(1,1) model this is

$$Y_t = \mu_t + a_t$$
$$\sigma_t^2 = \gamma V_L + \alpha a_{t-1}^2 + \beta \sigma_{t-1}^2$$

with $\gamma = 1 - \alpha - \beta$ and $\alpha + \beta < 1$.

Building a GARCH model. The steps are similar to the building of an ARCH model for a given return series $\{r_t\}$. However, determining the order (p, q) of the model is not as easy. We have seen that the square of the residuals $a_t = r_t - \mu_t$ verify an ARMA$(\max(p, q), q)$ model, so we could apply a PACF test to estimate a number p' for the AR part of the model, but this $p' = \max(p, q)$, and we must then trial different combinations of p and q. Nonetheless, in practice, lower order models suffice and, in fact, GARCH(1,1) model with $\alpha_1 + \beta_1$ close to one have been found to approximate the data quite well.[10]

Next, the parameters ω, αs, βs and μ are estimated with MLE method, where the log-likelihood function has the same form as for the ARCH model and given by Eq. (4.34).

General criterion for building a volatility model. Now that we know how to fit an ARCH or GARCH model from observable data, it is important to have some criterion to decide whether such model is appropriate or not. We have seen that estimation of the ACF for the squared residuals (i.e., the squares of $a_t = r_t - \mu_t$) is used to decide on a possible dependance of the variance with its past. Then, to effectively test for conditional heteroscedasticity, or what is known as the *ARCH effect*, one possibility is to apply the Ljung-Box statistics $Q(h)$ to the $\{a_t^2\}$ series,[11] where the null hypothesis to test is that the first h lags of the ACF of $\{a_t^2\}$ are zero. If this null hypothesis is rejected (e.g., p-values of $Q(h)$ are less than the 5 % significance level), then we have evidence of ARCH effect, and can well proceed to fit a volatility model.

R Example 4.6 We consider building an ARMA+GARCH(1,1) model for the series of returns of Shiller's PE 10 series. We will first test for ARCH effect, employing the Ljung-Box criterion, in order to justify the need for such volatility model. In R Example 4.4 we had determined that the best ARMA for the mean is of the type ARMA(2,4), and in R Example 4.5 we explained how to set up such combined model. Bear in mind that now we need to make a joint likelihood estimate of both sets of parameters for the ARMA and the GARCH, so one can not expect these to have the same values as those from models estimated separately. Execute the following commands:

---

[10] Explanations for this phenomenon are discussed in Bollerslev et al. (1994)
[11] see Chap. 3, Eq. (3.6)

```
> library(fGarch)
> sp500=read.csv("SP500_shiller.csv")
> sp500PE = na.omit(sp500$P.E10)
> ts=diff(log(sp500PE))   ##compute returns
> ts=ts-mean(ts)
> Box.test(ts^2)
```

The p-value resulting from the Ljung-Box test to the squared residuals is 1.171×10^{-6}, which is way below the significance level; hence we do have ARCH effect in the series. We continue fitting a GARCH model.

```
> armaGarch = garchFit(formula=~arma(2,4)+garch(1,1),data=ts)
> summary(armaGarch)
```

The error analysis in summary shows the following results:

```
Error Analysis:
            Estimate   Std. Error   t value  Pr(>|t|)
mu         3.141e-18    1.002e-03     0.000   1.00000
ar1        6.551e-01    2.171e-01     3.018   0.00254 **
ar2       -4.954e-01    2.041e-01    -2.427   0.01522 *
ma1       -3.785e-01    2.177e-01    -1.738   0.08219 .
ma2        3.488e-01    1.872e-01     1.864   0.06235 .
ma3        1.064e-01    5.566e-02     1.912   0.05584 .
ma4        7.444e-02    2.693e-02     2.764   0.00571 **
omega      7.147e-05    1.712e-05     4.174  3.00e-05 ***
alpha1     1.205e-01    1.735e-02     6.945  3.78e-12 ***
beta1      8.362e-01    2.130e-02    39.256   < 2e-16 ***
```

Thus, only the coefficients ar1, ar2, ma4, omega, alpha1 and beta1 are significant at the 5 % level. We then have the following ARMA(2,4)+GARCH(1,1) model

$$r_t = 0.6551 r_{t-1} - 0.4954 r_{t-2} + 0.0744 a_{t-4} + a_t,$$
$$\sigma_t^2 = 7.147 \times 10^{-5} + 0.1205 a_{t-1}^2 + 0.8362 \sigma_{t-1}^2, \quad a_t = \sigma_t \varepsilon_t, \ \varepsilon_t \sim N(0,1) \quad \square$$

4.4 Nonlinear Semiparametric Models

The next two nonlinear models that we are going to study, *Neural Networks* (NNet) and *Support Vector Machines* (SVM), fall in the category of semiparametric models. What this means is that although the NNet and SVM, just as the other methods seen in this chapter, try to relate a set of input variables and weights to a set of one or more output variables, they differ from the ARMA, ARCH or GARCH methods in that their parameters are not limited in number, nor do they have to be all known or computed for the method to give an approximate solution.

The reader might be acquainted with other semiparametric estimation models. For example, Taylor polynomials are instances of this class of models. Here the degree of the polynomials is not fixed, although the higher the degree and the more terms computed, the better the approximation to the target function. However, the NNet and SVM differ from polynomial approximation schemes, such as Taylor's, in other fundamental aspect, and it is that they do not give closed form solutions. For Taylor polynomials we have a definitive formula for computing their coefficients in terms of derivatives of the function we want to approximate. But for NNet and SVM models their parameters have to be estimated from data and adjusted iteratively, much the same as it is done in maximum likelihood estimation for ARCH models. In this regard is that these are known as machine learning models, since they fit in the concept of being algorithms that improve their performance at some task through experience. The task of our interest is forecasting future values of a given time series, and the experience is constituted by the past values of such time series and other statistics. The models progressively tune their parameters by a two step process wherein the data is divided into a *training* set, where the learning from experience process takes place, and a *testing* set, where the performance of the model is evaluated with respect to some performance measure (e.g. minimizing residual sum of squares), and succeeding adjustments of the model are made. In the case of time series usually the training and the testing sets are taken as two consecutive time periods of the series of varying length. Details will be given in the following sections.

The quality of being semiparametric or parametric, and the existence of closed form solutions provides two important dimensions to characterize all the forecasting models seen so far and the two models of this section. This model typology suggested by McNelis (2005) is summarized in the following table

Closed-form solutions	Parametric	Semiparametric
YES	Linear regression	Taylor polynomial
NO	ARCH / GARCH	NNet / SVM

Thus, in one extreme of McNelis model spectrum are the linear regression models, which are parametric and have easy to compute closed form solutions (e.g. the parameters of AR models are obtained by Yule-Walker equations). The downside of these models is mainly that solutions can be very imprecise due to the linearity and other constraints. On the other extreme of the spectrum are the semiparametric NNet and SVM models that give approximate solutions and are harder to estimate, but can capture the non linearity of the process and can be more precise in their forecasts.

4.4.1 Neural Networks

The feed-forward neural network is one of the most popular methods to approximate a multivariate nonlinear function. Their distinctive feature is the "hidden layers", in which input variables activate some nonlinear transformation functions producing a continuous signal to output nodes. This hidden layer scheme represents a very efficient parallel distributed model of nonlinear statistical processes. In fact,

Fig. 4.4 A 3-4-1 feed forward
neural network with one
hidden layer

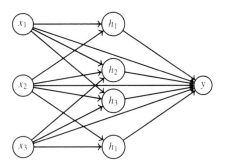

feed-forward neural networks have been shown to be capable of approximate any
continuous function uniformly on compact sets, for a suitable number of nodes in
the hidden layer and a sufficiently large training sample from data (see Hornik et al.
(1989), Ripley (1994) and references therein) .

Let us begin with an schematic presentation of a feed-forward neural network.
Figure 4.4 shows one such network with three input nodes x_1, x_2 and x_3, one output
node y, and four hidden nodes or neurons, h_1, h_2, h_3 and h_4, which constitute the
hidden layer.

Each neuron h_i receives the input information and goes active, transmitting a signal
to output, if a certain linear combination of the inputs, $\sum_{i \to j} \omega_{ij} x_i$, is greater than
some threshold value α_j; that is, h_j transmits a non-zero value if

$$z_j = \sum_{i \to j} \omega_{ij} x_i - \alpha_j > 0$$

where the sum is taken over all input nodes x_i feeding to neuron h_j, and ω_{ij} are
the weights. The signal is produced by an activation function applied to z_j, which is
usually taken as a logistic function,

$$\mathscr{L}(z) = \frac{\exp(z)}{1 + \exp(z)} = \frac{1}{1 + \exp(-z)}$$

If there is no activation ($z_j \leq 0$) the signal is 0, and when there is activation the value
of the signal is $0 < \mathscr{L}(z_j) < 1$. Composing a linear combination of the possible
values of signals transmitted by the hidden layer, plus some value a to account for the
bias in the connection, with an activation function f for the output node, we obtain a
mathematical expression for the basic feed-forward network with connections from
inputs to hidden nodes, and from these to output:

$$y = f\left(a + \sum_{j=1}^{q} \theta_j \mathscr{L}\left(\alpha_j + \sum_{i \to j} \omega_{ij} x_i\right)\right) \qquad (4.40)$$

For a more complex feed-forward network one allows direct connections from input
to output nodes, and thus we have a more general expression for these neural networks

$$y = f\left(a + \sum_{i=1}^{p} \phi_i x_i + \sum_{j=1}^{q} \theta_j \mathscr{L}\left(\alpha_j + \sum_{i \to j} \omega_{ij} x_i\right)\right) \tag{4.41}$$

where $x_1, \ldots, x_p$ are the input values, y is the output, $\mathscr{L}$ is the activation function for each of the q hidden nodes, α_j is the threshold and ω_{ij} are the weights along the connection for the jth hidden node, ϕ_i and θ_j are the weights for the linear combinations of order p and q respectively defined, a is the bias in the connection, and f is the activation function for the output node. This f can also be logistic, or a linear function, or a threshold function

$$f(z) = \begin{cases} 1 & \text{if } z > 0 \\ 0 & \text{otherwise,} \end{cases}$$

among others. Note that the case of f being linear gives Eq. (4.41) the form of an autoregressive moving average model for a given time series $\{y_t\}$. Indeed, taking $x_1 = y_{t-1}, \ldots, x_p = y_{t-p}$ (i.e., as p different lags of the series), and the output $y = y_t$ as the time series value to forecast, and f linear in Eq. (4.41), we have

$$y_t = a + \sum_{i=1}^{p} \phi_i y_{t-i} + \sum_{j=1}^{q} \theta_j \mathscr{L}\left(\alpha_j + \sum_{i \to j} \omega_{ij} y_{t-i}\right)$$

(here the moving average part is being modelled by a nonlinear function on the input lags). This shows that the feed forward neural network is a generalization of ARMA(p, q) model.

The case of f being a threshold function is also of interest in time series applications. If $y = r_t$ is a return to be computed, and $x_1, \ldots, x_p$ are different lags of this return series $\{r_t\}$, or some other statistics (e.g. its historical volatility), then Eq. (4.41) with f threshold can be interpreted as a model to forecast the direction of returns, where $f(z) = 1$ implies positive returns and $f(z) = 0$ negative returns.

Model estimation and forecasting. We are interested in fitting a feed-forward neural network to a time series of returns. Let $\{(x_t, r_t) : t = 1, \ldots, T\}$ be the available data where r_t is the return of some financial asset and x_t is vector of inputs (also called features) which can contain various lags of the series, or the volume, the variance or any fundamental indicator of the series. The model fitting for a neural network requires a division of the data in two subsamples, so that one subsample is used for training the model and the other subsample is used for testing. In the training step the parameters (i.e., the weights, the thresholds and connection bias) in Eq. (4.41) are chosen so that some forecasting error measure is minimized; usually it is required that the paremeters chosen minimized the mean squared error

$$R(w) = \frac{1}{T}\sum_{t=1}^{T}(r_t - y_t)^2$$

where w is the vector of required parameters and y_t is the output of Eq. (4.41) on inputs x_t, and r_t the actual return. This is a non-linear estimation problem that can be solved by some iterative method. One popular method is the back propagation algorithm which works backward starting with the output and by using a gradient rule adjusts the weights iteratively (Ripley 1994, Sect. 2.1).

In the testing step the network build in the training step is used on the second subsample data to predict some values and compare the estimations with the actual sampled values. The steps are repeated several times to tune the parameters, with the possible application of cross-validation to select the best performing network with respect to forecasting accuracy. In spite of its popularity the cross-validation method can fail to find the best parameters in the presence of local minima, and may required a lot of computer time in processing neural networks. Other statistical methods for selecting networks based on maximum likelihood methods or Bayes factor are appropriate alternatives (Ripley 1995).

NNet models in R. We show how to fit a NNet to a time series in the appendix of this chapter.

4.4.2 Support Vector Machines

The Support Vector Machine (SVM) is a learning algorithm for obtaining a model from training data, developed by Vladimir Vapnik and coworkers,[12] as an alternative to neural networks and other machine learning methods that are based on the *empirical risk minimization principle*. This principle refers to methods that seek to minimize the deviation from the correct solution of the training data (e.g., as measured by the mean squared error), whereas the SVM is an approximate implementation of the *structural risk minimization principle*, by trying to estimate a function that classifies the data by minimizing an upper bound of the generalization error. This is achieve by solving a linearly constraint quadratic programming problem, and as a consequence the SVM trained solution is unique and globally optimal, as opposed to neural networks training which require nonlinear optimization and solutions can thus get stuck at local minimum.

Support vector machines for regression. In the context of forecasting time series, the SVM is more suited as a regression technique, as opposed to a classifier. Therefore we restrict our exposition of the theory of support vector machines for regression.

Given a data set $G = \{(x_i, y_i) : i = 1, \ldots, N\}$ which is produced by some unknown function that maps each vector of real values $x_i \in \mathbb{R}^d$ to the real value y_i, we want to approximate this function by regression on the vectors x_i. The SVM considers regressions of the following form:

[12] Cortes, C. and Vapnik, V. (1995). Support vector networks. *Machine Learning*, 20: 273–297; Golowich, S.E., Smola, A. and Vapnik, V. (1996). Support vector method for function approximation, regression estimation, and signal processing. In *Advances in neural information processing systems 8*, San Mateo, CA; and the textbook by Vapnik (2000), originally published in 1995.

$$f(x, w) = \sum_{i=1}^{D} w_i \phi_i(x) + b \tag{4.42}$$

where the functions $\{\phi_i(x)\}_{i=1}^{D}$ are the features of inputs and $w = \{w_i\}_{i=1}^{D}$ and b are coefficients which are estimated from the data by minimizing the risk functional:

$$R(w) = \frac{1}{N} \sum_{i=1}^{N} |y_i - f(x_i, w)|_\varepsilon + \frac{1}{2} ||w||^2 \tag{4.43}$$

with respect to the linear ε-insensitive loss function

$$|y_i - f(x_i, w)|_\varepsilon = \begin{cases} 0 & \text{if } |y_i - f(x_i, w)| < \varepsilon \\ |y_i - f(x_i, w)| & \text{otherwise} \end{cases}$$

The approximation defined by Eq. (4.42) can be interpreted as a hyperplane in the D-dimensional feature space defined by the functions $\phi_i(x)$, and the goal is to find a hyperplane $f(x, w)$ that minimizes $R(w)$. It is shown in Vapnik (2000, Chap. 6) that such minimum is attained by functions having the following form:

$$f(x, \alpha, \alpha') = \sum_{i=1}^{N} (\alpha_i' - \alpha_i) K(x, x_i) + b \tag{4.44}$$

where $K(x, y)$ describes the inner product in the D-dimensional feature space, that is, $K(x, y) = \sum_{i=1}^{D} \phi_i(x) \phi_i(y)$ and is called the *kernel* function, and the coefficients α_i, α_i' are Lagrange multipliers that satisfy the equations $\alpha_i \alpha_i' = 0$, $\alpha_i \geq 0$, $\alpha_i' \geq 0$, for $i = 1, \ldots, N$, and are obtained by maximizing the following quadratic form:

$$Q(\alpha, \alpha') = \sum_{i=1}^{N} y_i(\alpha_i' - \alpha_i) - \varepsilon \sum_{i=1}^{N} (\alpha_i' + \alpha_i)$$
$$- \frac{1}{2} \sum_{i=1}^{N} \sum_{j=1}^{N} (\alpha_i' - \alpha_i)(\alpha_j' - \alpha_j) K(x_i, x_j)$$

subject to the constraints:

$$\sum_{i=1}^{N} (\alpha_i' - \alpha_i) = 0, \quad 0 \leq \alpha_i \leq C, \quad 0 \leq \alpha_i' \leq C, \; i = 1, \ldots, N. \tag{4.45}$$

The parameters C and ε are free and chosen by the user. The nature of this quadratic programming problem forces a certain number of the coefficients $\alpha'_i - \alpha_i$ to be different from zero, and the data points associated to them are called *support vectors*.

Now, one of the interesting facts about the SVM method is that the kernel $K(x, x_i)$ can be computed analytically, thus saving us the trouble of computing the features functions $\phi_i(x)$ and their inner product. Any function that satisfies Mercer's condition can be used as a kernel function, thus reducing the computation of an inner product in high dimensional feature space to a scalar computation in input space (see Vapnik (2000, Sect. 5.6.2)). There are many choices for the kernel function, although the most commonly used are the following:

- the polynomial kernel (with degree d): $K(x, y) = (x \cdot y + 1)^d$;
- the Gaussian radial basis function (RBF)[13]: $K(x, y) = \exp(-||x - y||^2/\sigma^2)$, with bandwidth σ^2, and
- the sigmoid kernel: $K(x, y) = \tanh(\kappa x \cdot y - \sigma)$.

Model estimation and forecasting. The estimation of parameters in SVN model is done much the same way as for neural networks. The data is divided into two subsamples, one for training the model and the other for testing. In the training stage the parameters are estimated under the structural risk minimization principle (Eq. (4.43)), and after measuring the accuracy of the predictions in the testing stage, one repeats the process to tune the parameters for a number of iterations. At this stage cross-validation is also useful. In the next section we give some criteria for measuring the accuracy of predictions, applicable to this and all the other forecasting models.

There are at least four R packages that implements SVM models: e1071, kernlab, klaR and svmpath. The first two are the most robust and complete implementations of SVM, based on the C++ award winning libsvm library of support vector machines classification, regression and kernel methods. The other two have less computing features and have simpler (and slower) implementations of SVM, so we do not consider their use. For an in-depth review of these four R packages for SVM and a comparison of their performance, see Karatzoglou et al. (2006).

SVM models in R. We fit a SVM to a time series in the appendix of this chapter.

4.5 Model Adequacy and Model Evaluation

Modeling time series involves a certain amount of subjective judgement, and what's desirable is to keep this subjectiveness as low as possible. With that in mind we can apply (and we should apply) various statistical tests to guide the selection of appropriate models. To fit an ARMA model we use the ACF and PACF functions to reveal a possible autoregressive relation on returns, and determine the order of this autoregression from the plot of those functions, or information criteria as the AIC

---

[13] $||x - y||^2$ is the Euclidean norm as defined in Chap. 3.

(Eq. (4.29)). We have a test for the ARCH effect which helps to decide for fitting a volatility model. To select input features for the machine learning models we can run causality tests (Sect. 3.2) and feed the models with those time series that present a significant cause–effect relation with the target series (e.g., lags of the target series, its volume, or any other feature from the series or other). In a more general, and possibly primer stage of selection of models, there is to determine the possibility of the time series having a nonlinear relation with its lagged values, or with another time series, so that we focus our model selection to nonlinear models. Let us review some commonly used tests for nonlinearity relation among time series.

4.5.1 Tests for Nonlinearity

There are various statistical test to assess if a group of time series are nonlinearly related. Two of these tests, based on neural networks, are Teraesvirta's test and White's test[14] for *neglected nonlinearity*. What this means is the following.

Given the general regression model, $Y_t = E(Y_t|F_{t-1}) + a_t$, where we view the information set F_{t-1} as a k-dimensional vector that may contain lagged values of Y_t, and $E(Y_t|F_{t-1})$ is the true unknown regression function. Y_t is *linear in mean* conditional on F_{t-1}, if $\mathbb{P}\left(E(Y_t|F_{t-1}) = F'_{t-1}\theta^*\right) = 1$ for some $\theta^* \in \mathbb{R}^k$, where θ^* is the parameter vector of the optimal linear least squares approximation to $E(Y_t|F_{t-1})$. The null hypothesis of linearity and the alternate hypothesis of interest are as follows:

$$H_0 : \mathbb{P}\left(E(Y_t|F_{t-1}) = Y'_{t-1}\theta^*\right) = 1 \quad \text{for some} \quad \theta^* \in \mathbb{R}^k$$
$$H_1 : \mathbb{P}\left(E(Y_t|F_{t-1}) = Y'_{t-1}\theta^*\right) < 1 \quad \text{for all} \quad \theta \in \mathbb{R}^k$$

When the null hypothesis is rejected the model is said to suffer from *neglected nonlinearity*, meaning that a nonlinear model may provide better forecasts than those obtained with the linear model. The neural network test for neglected nonlinearity, formulated by White, uses a single hidden layer feedforward neural network with nonlinear activation functions capable of approximating an arbitrary nonlinear mapping. Teraesvirta's test is based on, and improves, White's neural network test by using a Taylor series expansion of the activation function to obtain a suitable test statistic. Both tests are implemented in the R package `tseries` by the functions `terasvirta.test()` and `white.test()`, and returns *p*-values of a Chi-squared test (by default) or an F-test (if `type` is set to F).

R Example 4.7 We test for nonlinearity among the S&P 500 and Shiller's P/E ratio for this market index, namely the series PE 10, analyzed before. We will use Teraesvirta test. The data, as we know, is in a `.csv` table, and so we want to transform it

---

[14] Teraesvirta, T., Lin, C. F., Granger, C. W. J., 1993. Power of the neural network linearity test. *Journal of Time Series Analysis* 14, 209–220; and: Lee, T. H., White, H., Granger, C. W. J., 1993. Testing for neglected nonlinearity in time series models. *Journal of Econometrics* 56, 269–290.

to xts form for handling the data by dates, since we want to consider the series from 1900 to 2012 (just for illustration purposes). Note that we specify the format of the date variable, as it appears in the file. This is to prevent errors in loading the data.

```
> library(tseries); library(xts)
> inData = as.xts(read.zoo('SP500_shiller.csv',sep=',',
           header=T,format='%Y-%m-%d'))
> data=inData['1900/2012']
> sp500 = na.omit(data$SP500)
> sp500PE= na.omit(data$P.E10)
> terasvirta.test(x=sp500PE,y=sp500)
```

The result of the test that you should see is:

```
Teraesvirta Neural Network Test
data:   sp500PE and sp500
X-squared = 177.5502, df = 2, p-value < 2.2e-16
```

The small p value indicates that a nonlinear relation holds from PE 10 series to the S& P 500. Thus we can try fitting a nonlinear model to the S& P 500, for example a NNET or SVM, using the PE 10 series as part of the known information. We will do that in the last section of this chapter.   □

4.5.2 Tests of Model Performance

After fitting a time series model the next and final step is to assess the performance of the model; that is, its accuracy as a predictor. Time series prediction evaluation is usually done by holding out an independent subset of data (the testing set) from the data used for fitting the model (the training set), and measuring the accuracy of predicting the values in the testing set by the model. Let $\{y_t : t = 1, \ldots, m\}$ be the testing set from the time dependent data that's been modeled, and let $\{\widehat{y}_t : t = 1, \ldots, m\}$ be the values predicted by the model in the same time period of the testing set. Some common measures to determined how close are the predicted values to the actual values in the testing set are the *mean squared error* (MSE), the *mean absolute error* (MAE) and the *root mean squared error* (RMSE), defined as follows:

$$MSE = \frac{1}{m}\sum_{t=1}^{m}(y_t - \widehat{y}_t)^2 \qquad MAE = \frac{1}{m}\sum_{t=1}^{m}|y_t - \widehat{y}_t|$$

$$RMSE = \sqrt{\frac{1}{m}\sum_{t=1}^{m}(y_t - \widehat{y}_t)^2} = \sqrt{MSE}$$

Note that these are scale-dependent measures, and hence, are useful when comparing different methods applied to the same data, but should not be used when comparing across data sets that have different scales. The RSME is often preferred to the MSE since it is on the same scale as the data, but both are more sensitive to outliers than MAE. An experimental comparison of these and other scale dependent, and independent, measures of forecast accuracy is done in Hyndman and Koehler (2006). These prediction error measures, and variants, are all available in the R package of evaluation metrics for machine learning, Metrics.

Alternatively, to evaluate the models, instead of measuring the accuracy of the predicted values, one can look at how well the model predicts the direction of the series, that is, if it has gone up or down with respect to the preceding value. This is the same as treating the forecaster as a binary classifier, and consequently appropriate measures of its accuracy are obtained from the information recorded in a contingency table of hits and misses of predicted against actual upward or downward moves. The general form of a contingency table for recording predictions of direction is like the one shown below,

Actual	Predicted		
	up	down	+
up	m_{11}	m_{12}	m_{10}
down	m_{21}	m_{22}	m_{20}
+	m_{01}	m_{02}	m

where m is the size of the testing set (equal to the number of predicted ups and downs), m_{11} (resp. m_{22}) is the number of hits in predicting upward (resp. downward) moves, m_{12} (resp. m_{21}) is the number of misses in predicting upward (resp. downward) moves, etc. The performance of the tested model can then be measured with the information given by this table with three different metrics:

Accuracy: computed as a simple percentage of correct predictions: $(m_{11} + m_{22})/m$.

Cohen's Kappa (Cohen, 1960): this measure takes into account the probability of random agreement between the predicted and the actual observed values and it is computed as $\kappa = \dfrac{P(a) - P(e)}{1 - P(e)}$, where $P(a) = (m_{11} + m_{22})/m$ is the observed agreement and $P(e) = (m_{10}m_{01} + m_{20}m_{02})/m^2$ is the probability of random agreement, that is, the probability that the actual and the predicted coincide assuming independence between predictions and actual values. Therefore, $\kappa = 1$ when the predicted and the actual values completely agree.

Directional Measure: it is computed out from the contingency table as $\chi^2 = \displaystyle\sum_{i=1}^{2}\sum_{j=1}^{2} \dfrac{(m_{ij} - m_{i0}m_{0j}/m)^2}{m_{i0}m_{0j}/m}$. Similar to Cohen's Kappa, large values of χ^2 tell us that the model outperforms the chance of random agreement. χ^2 behaves like a chi-squared distribution with 1 degree of freedom, and we can use this information to compute the quantile with respect to a given significance level (Tsay 2010).

The models with best performance, according to one or a combination of these measures, are selected, and then their parameters can be re-calibrated and the new models tested again until some threshold in the measure valuations is reached or these values stabilize.

4.6 Appendix: NNet and SVM Modeling in R

John Campbell and Robert Shiller made several studies on the usefulness of price-to-earnings and dividend-price ratios as forecasting variables for the stock markets; a complete update of these studies is in Campbell and Shiller (2005). We are going to test Campbell and Shiller's thesis by using the PE 10 series, examined in previous examples, as an additional feature to forecast the monthly returns of the S&P 500 with a neural network and a support vector machine models. We have confirmed in R Example 4.7 the nonlinearity relationship between these two series; hence the use of NNet and SVM is thus justified.

R Example 4.8   We use the SP500_shiller.csv data set from where we can extract the monthly observations of the PE 10 series and the price of the S&P 500 from 1900 to 2012. Libraries nnet and kernlab are use for neural networks and support vector machines modeling. To train the models, and iteratively estimate and tune the parameters to obtain the best fitted models, we use the functions provided by the caret package for classification and regression training (see a user's guide written by Kuhn (2008)). For regression problems (e.g. predicting returns, as we do in this example) the best model is selected with respect to the RSME measure (by default), while for classification problems (e.g. trying to predict direction) the Accuracy and Kappa measures are used. In the case of SVM one chooses the kernel (we choose a radial kernel, see svmFit below), and the caret function train tunes over the scale parameter σ and the cost value C used to control the constraints boundary (see Eq. (4.45)); for the NNet the tuning is done over the size of the hidden layer(s), where the number of hidden layers can be selected but we leave the default value of 1, and the weight decay term used in the weight updating rule (a Gradient descent rule). The train function also facilitates the pre-processing of the data; thus, we scale and center the data. We add the following features to the models: lags 1, 2, and 3 of the returns of the S& P 500 (our target series); the PE 10 series and its lags 1 and 2.

```
> library(xts); library(quantmod);
> library(nnet); library(kernlab);
> library(caret) ##for some data handling functions
> library(Metrics)##Measures of prediction accuracy
> sp500 = as.xts(read.zoo('SP500_shiller.csv',sep=',',
+                  header=T, format='%Y-%m-%d'))
> data=sp500['1900/2012']
> sp500PE = na.omit(data$P.E10) ##feature:P/E MA(10)
> ret=diff(log(data$SP500))   ##compute returns
> ret=na.trim(ret-mean(na.omit(ret)))
```

```
##Compute some features: lags 1,2,3; PE, lags 1,2
> feat = merge(na.trim(lag(ret,1)),na.trim(lag(ret,2)),
+        na.trim(lag(ret,3)),sp500PE,na.trim(lag(sp500PE,1)),
+        na.trim(lag(sp500PE,2)),  all=FALSE)
##TARGET to predict: returns (ret)
> dataset = merge(feat,ret,all=FALSE)   ##(1)
##Label columns of dataset
> colnames(dataset) = c("lag.1", "lag.2","lag.3","PE10",
+                       "PE10.1","PE10.2", "TARGET")

##Divide data in training (75%) and testing (25%) with caret
> index = 1:nrow(dataset)
> trainindex= createDataPartition(index,p=0.75,list=FALSE)
> ##process class sets as data frames
> training = as.data.frame(dataset[trainindex,])
> rownames(training) = NULL
> testing = as.data.frame(dataset[-trainindex,])
> rownames(testing) = NULL

##Build and tune SVM & Nnet models, bootstrapping 200 reps.
> bootControl <- trainControl(number=200)
> prePro <- c("center", "scale") #data is centered and scaled
> set.seed(2)
> idxTra = ncol(training)
##SVM model:
> svmFit <- train(training[,-idxTra],training[,idxTra],
+     method="svmRadial",tuneLength=5,
+     trControl=bootControl,preProc=prePro)
> svmFit
> svmBest <- svmFit$finalModel ##extract best model
> svmBest

##Nnet model
> nnetFit<-train(training[,-idxTra],training[,idxTra)],
+        method="nnet",tuneLength=5,trace=FALSE,
+        trControl=bootControl,preProc=prePro)
> nnetFit
> nnetBest <- nnetFit$finalModel ##tune parameters size,decay
> summary(nnetBest)
```

The best SVM Radial model found has 721 support vectors, $\sigma = 0.393$ and cost $C = 4$; the best NNet is a 6-3-1 feedforward net with 25 non zero weights. The instruction summary(nnetBest) describes the network completely. Now proceed to compute the predictions on the testing set and evaluate the accuracy of the models. The output is shown.

```
> ###Build Predictors
> predsvm <- predict(svmBest,testing[,-ncol(testing)])
> prednet<-predict(nnetBest,testing[,-ncol(testing)],type="raw")

> ###EVALUATION (MSE, MAE)
> actualTS<-testing[,ncol(testing)] ##the true series
```

```
##For SVM
> predicTS<-predsvm
> mse(actual=actualTS,predicted=predicTS)
[1] 0.002044547
> mae(actual=actualTS,predicted=predicTS)
[1] 0.0325104
##For Nnet
> predicTS<-prednet
> mse(actual=actualTS,predicted=predicTS)
[1] 0.002359158
> mae(actual=actualTS,predicted=predicTS)
[1] 0.03923168
```

The results of MSE and MAE measures are very similar for both models (although we concede that SVM's prediction accuracy looks slightly better); hence for this particular experiment both models are promising.   □

4.7 Notes, Computer Lab and Problems

4.7.1 Bibliographic remarks: A complete treatment of autoregressive moving average models (ARMA) is given in Brockwell and Davis (1991), and from a more instrumental perspective by the same authors in Brockwell and Davis (2002). A classic in the subject of ARMA models and time series in general is Box and Jenkins (1976). For the variety of ARCH models and their applications read the survey by Bollerslev et al. (1994) and the book by Tsay (2010). Zivot and Wang (2006) gives a nice review of time series models and their implementation in S-plus, the commercial version of R. For a detailed statistical review of neural networks see Vapnik (2000), and from the perspective of econometrics and finance see the book by McNelis (2005). For details on support vector machines see Vapnik (2000) and the tutorial article by Burges (1998). A good summary of some of the best software packages available in R for time series analysis is in McLeod et al. (2012). We have used material from all of the previously cited works for writing this chapter.

4.7.2 AR(2) process: Show that the AR(2) process $(1 - \phi_1 L - \phi_2 L)X_t = W_t$, with $\{W_t\} \sim WN(0, \sigma^2)$, has as solution

$$X_t = \sum_{k=0}^{\infty} \left(\frac{\lambda_1}{\lambda_1 - \lambda_2} \lambda_1^k + \frac{\lambda_2}{\lambda_2 - \lambda_1} \lambda_2^k \right) W_{t-k}$$

where $\phi_1 = \lambda_1 + \lambda_2, \phi_2 = -\lambda_1\lambda_2$, and $|\lambda_1| < 1, |\lambda_2| < 1$. Show that the restrictions on λ_1, λ_2 are equivalent to $|\phi_1 + \phi_2| < 1$. (Hint: Find λ_1 and λ_2 such that $1 - \phi_1 L - \phi_2 L^2 = (1 - \lambda_1 L)(1 - \lambda_2 L)$, and find A_1, A_2, such that $\dfrac{1}{(1 - \lambda_1 L)(1 - \lambda_2 L)} = \dfrac{A_1}{(1 - \lambda_1 L)} + \dfrac{A_2}{(1 - \lambda_2 L)}$; see Eq. (4.17).)

4.7.3: A nice result from (Granger and Morris 1976) states that "if X_t is an ARMA(p_1, q_1) process and Y_t is an ARMA(p_2, q_2) process then $Z_t = X_t + Y_t$ is an ARMA($p_1 + p_2, \max(p_1 + q_2, p_2 + q_1)$) process". Try proving this result. As a warm up, show the following particular case: if X_t and Y_t are both AR(1) processes, then $Z_t = X_t + Y_t$ is a ARMA(2,1) process. (Hint: Use the lag operator expressions, $(1 - \phi_x L)X_t = W_t$ and $(1 - \phi_y L)Y_t = W_t'$. Then, what is $(1 - \phi_x L)(1 - \phi_y L)Z_t$?)

4.7.4: Show that the conditions for the GARCH model (Eq. (4.36)) guarantee that the unconditional variance of the residuals $a_t = r_t - \mu_t = \sigma_t \varepsilon_t$ is finite.

4.7.5: A random process is *mean reverting* if it always returns to its mean. More formally, X is mean reverting if for any observation X_t there exists another observation X_{t+s}, after a period of time $s > 0$, such that $X_{t+s} \approx E(X)$. Show that a stationary AR(1) process is mean reverting. (Hint: From Remark 4.1 an AR(1) process with mean μ has the form $X_t = \phi_0 + \phi_1 X_{t-1} + W_t$, with $\mu = \phi_0/(1 - \phi_1)$ and $\{W_t\} \sim WN(0, \sigma^2)$. What does Eq. (4.14) tell us with respect to the contribution of the shocks to $X_t - \mu$?)

4.7.6: Show that a stationary ARCH(1) processes is mean reverting in the volatility. Show that the same occurs for a stationary GARCH(1,1) process.

4.7.7 R Lab: To predict direction instead of returns in R Example 4.8, substitute the instruction marked `##(1)` by the following two instructions

```
dir = ifelse(ret >=0,1,-1) ##1=up, -1=down
dataset = merge(feat,dir,all=FALSE)
```

After the instructions defining the training and testing sets, include the following to define target as factor and change name of levels from -1, 1 to "down", "up".

```
training$TARGET = as.factor(training$TARGET)
testing$TARGET = as.factor(testing$TARGET)
levels(training$TARGET)= list(down="-1",up="1")
levels(testing$TARGET)= list(down="-1",up="1")
```

To evaluate the accuracy of predictors construct contingency table with

```
table(predict=predicTS, true=actualTS)
classAgreement(table(predict=predicTS, true=actualTS))
```

4.7.8 R Lab: Download historical data for your favorite stocks and fit an ARMA, a GARCH, NNet and SVM models, with different features. Compare their accuracy of predictions (either returns or direction), and plot the target series and their predictors.

Chapter 5
Brownian Motion, Binomial Trees and Monte Carlo Simulation

This chapter presents Brownian motion, also known as Wiener process. This is the most fundamental continuous-time model in finance. On top of the Brownian motion other more elaborated continuous processes are built, that explain better the behavior of stock prices. This is the case of the geometric Brownian motion, which is at the core of the Black-Scholes option pricing formula. We will study all these continuous-time processes and the tools to derive them as solutions of certain stochastic differential equations, including the important Itô's lemma.

We make a big detour from continuity to explain the discrete-time option valuation model of Cox, Ross and Rubinstein based on binomial trees. The importance of this model is that it allows to valuate a wider range of vanilla type of options (American and European), and it has the Black-Scholes model as a limiting case. We then look at Monte Carlo simulation, focusing on two basic approximation schemes, namely, Euler and Milstein, which hopefully will be enough to give a good idea of the power of these methods to valuate almost all kind of derivatives.

5.1 Continuous Time Processes

Brownian motion is formally represented by a continuous time Gaussian process, and has as a discrete time counterpart the random walk. By continuous time process it is meant that the process (e.g. the price, if we are modeling a stock) evolves continuously in time, although we generally observe its values at discrete times. The assumption of continuity has many advantages. An important one, is that the continuous model is independent of the timing intervals, and closed formulas for pricing might be possible (as opposed to recursive numerical procedures). Also the assumption of continuously evolving prices is fundamental for the derivation of the Black–Scholes option pricing formula, which we will study in Sect. 5.2.1, since it relies on the possibility of adjusting the positions in a portfolio continuously in time,

A. Arratia, *Computational Finance*, Atlantis Studies in Computational Finance and Financial Engineering 1, DOI: 10.2991/978-94-6239-070-6_5,
© Atlantis Press and the authors 2014

in order to replicate an options payoff exactly. In the Black–Scholes option pricing model the price of stocks are assumed to follow a geometric Brownian motion.

The mathematical formalization of Brownian motion is due to Norbert Wiener, and for that reason it is known as Wiener process. We present Brownian motion as a Wiener process, and extensions of this family of continuous Gaussian processes which have been fundamental in the development of stocks and options pricing formulas.

5.1.1 The Wiener Process

The *Wiener process* $w(t)$, $t \in \mathbb{R}$ is a Gaussian centered process, with $w(0) = 0$, characterized by the following properties:

(i) the increments $\{w(t) - w(s) : t < s\}$ are stationary and independent;
(ii) their variance $E(w(t) - w(s))^2 = |t - s|$, for $s, t \in \mathbb{R}$.

One consequence of these properties is that $w(t) - w(s) \sim N(0, t - s)$.

The Wiener process is a mathematical model for describing the erratic move of particles suspended in fluid, a physical phenomenon observed and documented first by the botanist Robert Brown from whom it takes the name of (standard) Brownian motion, and so often denoted $B(t)$ instead of $w(t)$ as above. The sample paths of a Brownian motion $B(t)$ can be simulated in an interval of time $[0, T]$ by partitioning the interval in finitely many time instants, $0 = t_0 < t_1 < \cdots < t_n = T$, and recursively compute

$$\Delta B(t_i) = B(t_i) - B(t_{i-1}) = \varepsilon_i \sqrt{\Delta t_i} \tag{5.1}$$

with $\varepsilon_i \sim N(0, 1)$ and iid random variables (a white noise), and $\Delta t_i = t_i - t_{i-1}$. Note that the discrete-time random process $\{B_t = B(t) : t = t_0, \ldots, t_n\}$ verifies all the properties of a Wiener process in the continuum.

One possible way to extend the Wiener process to other random processes (so called *stochastic processes*), is with the use of the Wiener integral. The *Wiener integral* of a function f with continuous derivative f' over the interval $[a, b]$ is defined by

$$\int_a^b f(s)dw(s) = [f(s)w(s)]_a^b - \int_a^b w(s)f'(s)ds \tag{5.2}$$

The *generalized Wiener process*, or *arithmetic Brownian motion*, is now defined for constants α and σ as the process $X(t)$ satisfying the following differential equation

$$dX(t) = \alpha dt + \sigma dw(t) \tag{5.3}$$

with $w(t)$ a Wiener process. This is a *stochastic differential equation* (SDE), very different from an ordinary differential equation. For example, note that the ratio of the increments of the Wiener with respect to time, namely $(w(t + h) - w(t))/h$ does not converge to a well-defined random variable as h goes to 0, since $Var\left(\dfrac{(w(t + h) - w(t))}{h}\right) = \dfrac{|h|}{h^2}$. Therefore, the derivative of the Wiener (and consequently $dX(t)$) does not exist in the ordinary sense. Equation (5.3) is solved taking Wiener integrals in the interval $[0, t]$, so that $\int_0^t dX(s) = \alpha \int_0^t dt + \sigma \int_0^t dw(s)$, and we obtain

$$X(t) = X(0) + \alpha t + \sigma w(t) \tag{5.4}$$

This process has the following properties, inherited from the Wiener process:

(i) For any s and t, $s < t$,

$$X(t) - X(s) \sim N(\alpha(t - s), \sigma^2(t - s)) \tag{5.5}$$

(ii) The increments $\{X(t) - X(s) : s < t\}$ are stationary and statistically independent.

(iii) The sample paths of $X(t)$ are everywhere continuous, but nowhere differentiable with probability 1.

By the first property, the value α measures the expected drift rate and σ^2 the variance rate of increments of $X(t)$ per time step. The other two properties imply that the sample paths of the generalized Wiener process are not smooth, jiggling up and down in unpredictable way: due to the independence of increments one can not tell the magnitude of $X(t + h) - X(t)$ from previous observed increment $X(t) - X(t - h)$. These properties make the generalized Wiener process a likely candidate for modeling the behavior of stock prices. But from now on, we will stick to the name Brownian motion in place of Wiener process, following common usage in econometrics and finance.

We can simulate an approximate sample path of an arithmetic Brownian motion over a time interval $[0, T]$, with given α and σ, using the discrete model (5.1) as follows. Partition $[0, T]$ in $n > 0$ uniform parts, each of length T/n, and beginning with $X(0) = 0$ iterate for $i = 1, \ldots, n$ steps the recursive formula

$$X(i) = X(i - 1) + \alpha \frac{T}{n} + \sigma \varepsilon_i \sqrt{\frac{T}{n}} \tag{5.6}$$

with ε_i a randomly generated number, normally distributed with mean 0 and variance 1. A plot of the resulting sequence of values, namely $\{(iT/n, X(i)) : i = 0, 1, \ldots, n\}$, will show the erratic trajectory characteristic of Brownian motion. Running the simulation again with same input values will create a different path, due to the random component $\varepsilon_i \sim N(0, 1)$.

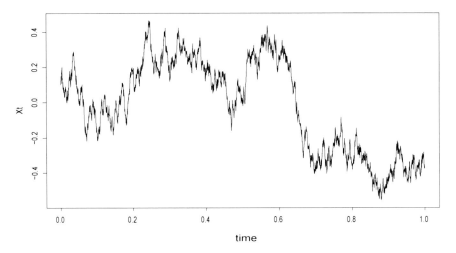

Fig. 5.1 Arithmetic brownian motion for $\alpha = 0, \sigma = 1, X_0 = 0.1, n = 2^{12}$

R Example 5.1 The following R commands creates and plots an approximate sample path of an arithmetic Brownian motion for given α and σ, over the time interval $[0, T]$ and with n points.

```
> ##inputs:
> alpha=0; sigma=1; T=1; n=2^(12); X0=0.1;
> #############Generate 1 trajectory
> dt=T/n
> t=seq(0,T,by=dt)
> x=c(X0,alpha*dt+sigma*sqrt(dt)*rnorm(n,mean=0,sd=1))
> Xt=cumsum(x)
> plot(t,Xt,type='l',xlab="time")
```

Increasing the value of n, so that the time steps T/n become smaller, will improve the quality of the approximate sample path to a true Brownian path. Figure 5.1 shows the plot resulting from running the above simulation of Brownian motion with $n = 2^{12}$.   □

The discrete-time process $X_{t_i} = X(i)$, for $t_i = iT/n$, $i = 0, 1, \ldots, n$, defined by Eq. (5.6), is a random walk with constant drift rate α and constant variance rate σ^2 per unit of time, and normal distribution, which we have seen is not stationary (Example 2.6), although its increments are stationary and independent. These facts implies that any past information is irrelevant to the future values of the process, and so this process complies with the weak form of the Efficient Market Hypothesis.

5.1.2 Itô's Lemma and Geometric Brownian Motion

A limitation for considering the arithmetic Brownian motion as a model for the price of a stock is that in this model the variation of prices over any time period is normally distributed, and this implies that prices can take negative value with positive probability, since the support of the normal distribution is all the reals (observe in Fig. 5.1 that the process takes negative values). This can be fixed, as argued in Chap. 2, by considering prices log normally distributed instead. Therefore, if $\{P(t)\}_{t \geq 0}$ is a continuous process naturally extending discrete observations of the price of a stock, we want to model $P(t)$ as $e^{X(t)}$, with $X(t)$ an arithmetic Brownian motion. This gives $X(t) = \ln P(t)$, and so the price is now being considered log-normally distributed, and furthermore, the log returns

$$r_t = \ln \left(\frac{P(t)}{P(t-1)} \right) = X(t) - X(t-1)$$

are stationary and independent. The continuous model $P(t) = e^{X(t)}$, with $X(t)$ an arithmetic Brownian motion, is called *geometric Brownian motion*, and in order to obtain a SDE that describes the dynamics of this process, obtained by composition of functions, we need an important tool in stochastic calculus known as *Itô's Lemma*. We state this lemma without proof and show how to use it in the case of $P(t) = e^{X(t)}$.

Lemma 5.1 *(Itô's Lemma (1951)) If $X(t)$ is a continuous time stochastic process satisfying*

$$dX(t) = a(X)dt + b(X)dB(t),\qquad(5.7)$$

where $B(t)$ is a Brownian motion, and if $G = G(X, t)$ is a differentiable function of $X(t)$ and t, then

$$dG = \left(a(X) \frac{\partial G}{\partial X} + \frac{\partial G}{\partial t} + b(X)^2 \frac{1}{2} \frac{\partial^2 G}{\partial X^2} \right) dt + b(X) \frac{\partial G}{\partial X} dB(t) \qquad (5.8)$$

The differential equation (5.7) extends the arithmetic Brownian motion by allowing the expected drift rate and variance rate ($a(X)$ and $b(X)$) to depend on X, and in such a form is known as Itô's differential equation. A process $X(t)$ satisfying Eq. (5.7) is called an Itô's process. Thus, an arithmetic Brownian motion $X(t)$ is an Itô's process satisfying $dX = \alpha dt + \sigma dB$, and considering $P(t) = e^{X(t)}$ as the differentiable function $G(X, t)$ in Itô's Lemma, we have that

$$\frac{\partial P}{\partial X} = \frac{\partial^2 P}{\partial X^2} = P \text{ and } \frac{\partial P}{\partial t} = 0$$

and consequently

$$dP = \left(\alpha \frac{\partial P}{\partial X} + \frac{\partial P}{\partial t} + \sigma^2 \frac{1}{2} \frac{\partial^2 P}{\partial X^2} \right) dt + \sigma \frac{\partial P}{\partial X} dB$$

$$= \alpha P dt + \sigma^2 \frac{1}{2} P dt + \sigma P dB = \left(\alpha + \frac{1}{2} \sigma^2 \right) P dt + \sigma P dB$$

Thus the geometric Brownian motion model for the price, $P(t) = e^{X(t)}$, satisfies the SDE

$$dP = \mu P dt + \sigma P dB \qquad (5.9)$$

with $\mu = \alpha + \frac{1}{2}\sigma^2$, while the logarithm of $P(t)$, namely $\ln P(t) = X(t)$, follows the arithmetic Brownian motion dynamics

$$d \ln P = dX = \left(\mu - \frac{1}{2}\sigma^2 \right) dt + \sigma dB \qquad (5.10)$$

Rewrite Eq. (5.9) in the form

$$\frac{dP}{P} = \mu dt + \sigma dB \qquad (5.11)$$

and this makes evident that the instantaneous return dP/P behaves like an arithmetic Brownian motion. Furthermore, if we consider a discrete version of Eq. (5.11),

$$\frac{\Delta P}{P} = \frac{P_{t+\Delta t} - P_t}{P_t} = \mu \Delta t + \sigma \varepsilon_t \sqrt{\Delta t}, \quad \varepsilon_t \sim N(0, 1) \qquad (5.12)$$

then $E(\frac{\Delta P}{P}) = \mu \Delta t$ and $Var(\frac{\Delta P}{P}) = \sigma^2 \Delta t$, so it is clear that the parameter μ represents the *expected rate of return* per unit of time, and σ is the standard deviation of the return per unit of time, that is, the *volatility* of the asset. Both parameters are expressed in decimal form. Now, consider a discrete version of Eq. (5.10) which has the form

$$\Delta \ln P = \ln \left(\frac{P_{t+\Delta t}}{P_t} \right) = \left(\mu - \frac{\sigma^2}{2} \right) \Delta t + \sigma \varepsilon_t \sqrt{\Delta t}, \quad \varepsilon_t \sim N(0, 1) \qquad (5.13)$$

This shows that $\Delta \ln P$ is normally distributed (i.e. the return is log normal) with mean $(\mu - \sigma^2/2)\Delta t$, variance $\sigma^2 \Delta t$ and standard deviation $\sigma \sqrt{\Delta t}$. If $t = 0$ and $t + \Delta t = T$ is some future time, then

$$\ln P_T \sim N \left(\ln P_0 + \left(\mu - \frac{\sigma^2}{2} \right) T, \sigma^2 T \right) \qquad (5.14)$$

which implies that $\ln P_T$ has mean $\ln P_0 + \left(\mu - \frac{\sigma^2}{2} \right) T$ and standard deviation $\sigma \sqrt{T}$. By the same argument as in Example 2.4 based on the Central Limit Theorem, we can find equations for the bounds of the price P_T which hold with a 95 % confidence. These bounds are given by

$$\left(\mu - \frac{\sigma^2}{2}\right) T - 1.96\sigma\sqrt{T} < \ln\left(\frac{P_T}{P_0}\right) < \left(\mu - \frac{\sigma^2}{2}\right) T + 1.96\sigma\sqrt{T}$$

or, equivalently

$$P_0 \exp\left(\left(\mu - \frac{\sigma^2}{2}\right) T - 1.96\sigma\sqrt{T}\right) < P_T < P_0 \exp\left(\left(\mu - \frac{\sigma^2}{2}\right) T + 1.96\sigma\sqrt{T}\right)$$

$$(5.15)$$

Remark 5.1 The reader should be aware of the subtle difference between Eq. (5.15) and Eq. 2.34 of Chap. 2. Here μ and σ^2 are the mean and variance of the simple return, while there μ_r and σ_r are the mean and variance of the continuously compounded return. The relations between each pair of moments are given by Eq. 2.32; for example, $\mu_r = \ln(\mu + 1) - \sigma_r^2/2$. Hull (2009, Sect. 13.3) provides a practical illustration from the mutual funds industry on the difference between μ and μ_r. In practice it is preferable to use Eq. (5.15) for the estimation of price bounds, since we are usually given the information on expected returns and volatility. □

Example 5.1 Consider a stock with an initial price $P_0 = 40$, expected return of 16 % per annum, and volatility of 20 % per annum. Using Eq. (5.15) we can estimate the range of values for the stock price in 1 month time, with 95 % confidence:

the lower bound: $L_T = 40 \cdot \exp\left(\left(0.16 - \frac{0.2^2}{2}\right)(1/12) - 1.96 \cdot 0.2\sqrt{1/12}\right) = 36.14$;

the upper bound: $U_T = 40 \cdot \exp\left(\left(0.16 - \frac{0.2^2}{2}\right)(1/12) + 1.96 \cdot 0.2\sqrt{1/12}\right) = 45.32$. □

Additionally, from the discretization (5.13) we get, for $t = 0$ and $t + \Delta t = T$,

$$P_T = P_0 \exp\left(\left(\mu - \frac{\sigma^2}{2}\right) T + \sigma\varepsilon_t\sqrt{\Delta t}\right), \quad \varepsilon_t \sim N(0, 1) \qquad (5.16)$$

This is the discrete version of the geometric Brownian motion continuous model for the price in the time interval $[0, T]$, which is

$$P_T = P_0 \exp\left(\left(\mu - \frac{\sigma^2}{2}\right) T + \sigma B(T)\right) \qquad (5.17)$$

with $B(t)$ a Brownian motion. Note that Eq. (5.17) is the integral solution in $[0, T]$ of the SDE (5.10).

The discrete geometric Brownian motion model (5.16) allows us to simulate possible paths for the price of a stock, beginning at some known initial price P_0, with known expected return μ and volatility σ, and forward to a given time horizon T, in the same way as the simulations of the arithmetic Brownian motion were done in R Example 5.1. Fortunately, R has a function that encodes a simulation with the discrete model defined by (5.16), so we do not need to program from scratch.

R Example 5.2 (Modeling prices as a Geometric Brownian Motion) For a given stock with expected rate of return μ and volatility σ, and initial price P_0 and a time horizon T, simulate in R n_t many trajectories of the price $\{P_t\}$ from time $t = 0$ up until $t = T$ through n many time periods, each of length $\Delta t = T/n$, assuming the geometric Brownian motion model. This is done with the function GBM in the package sde. The following set of instructions shows how to do it.

```
> library(sde)
> mu=0.16; sigma=0.2; P0=40; T = 1/12 ##1 month
> nt=50; n=2^(8)
>#############Generate nt trajectories
> dt=T/n; t=seq(0,T,by=dt)
> X=matrix(rep(0,length(t)*nt), nrow=nt)
> for (i in 1:nt) {X[i,]= GBM(x=P0,r=mu,sigma=sigma,T=T,N=n)}
>##Plot
> ymax=max(X); ymin=min(X) #bounds for simulated prices
> plot(t,X[1,],t='l',ylim=c(ymin, ymax), col=1,
+     ylab="Price P(t)",xlab="time t")
> for(i in 2:nt){lines(t,X[i,], t='l',ylim=c(ymin, ymax),col=i)}
```

Figure 5.2 shows a simulation of 50 possible paths of a price, assumed to follow a geometric Brownian motion process, with expected rate of return of 16 %, volatility of 20 %, initial price $P_0 = 40$ and one month time horizon. In these simulations the maximum price attained was 47.41, and the minimum 34.71, given us confidence bounds in which prices move for the time horizon (they differ a bit from the bounds computed in Example 5.1 due to the random component of the simulations). Computing the mean of the final results of the 50 simulated paths, and discounting it back to present at the rate of μ with

```
> mean(X[,n+1])*exp(-mu*T)
```

we obtain an approximation to the present value of the stock. This is an example of a straightforward application of the Monte Carlo simulation method for valuation of a stock. We will see more on Monte Carlo simulation in Sect. 5.3.   □

Estimating μ and σ. If we do not know before hand the expected return μ and volatility σ of the price, we can approximate these values from sample data as follows. Let $r = \{r_1, \ldots, r_n\}$ be a sample of n log returns, taken at equally spaced time period τ. Assuming the price P_t follows a geometric Brownian motion then the return r_t is normally distributed with mean $(\mu - \sigma^2/2)\tau$ and variance $\sigma^2\tau$; that is,

$$E(r_t) = (\mu - \sigma^2/2)\tau \quad \text{and} \quad Var(r_t) = \sigma^2\tau \tag{5.18}$$

If $m_r = \frac{1}{n}\sum_{t=1}^{n} r_t$ is the sample mean and $s_r^2 = \frac{1}{n-1}\sum_{t=1}^{n}(r_t - m_r)^2$ is the sample variance of the data, then estimates $\widehat{\mu}$ and $\widehat{\sigma}$ of μ and σ can be obtained from Eq. (5.18) as follows,

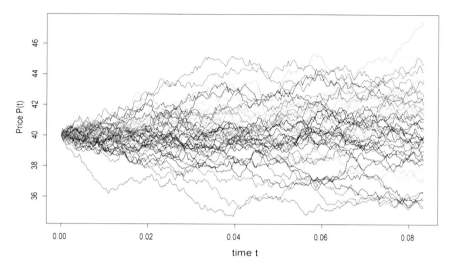

Fig. 5.2 GBM simulated paths for $\mu = 0.16$, $\sigma = 0.2$, $T = 0.83$ and $P_0 = 40$

$$\widehat{\sigma} = \frac{s_r}{\sqrt{\tau}} = \frac{1}{\sqrt{\tau}} \left(\frac{1}{n-1} \sum_{t=1}^{n} (r_t - m_r)^2 \right)^{1/2} \tag{5.19}$$

$$\widehat{\mu} = \frac{1}{\tau} m_r + \frac{\widehat{\sigma}^2}{2} = \frac{1}{\tau} \left(\frac{1}{n} \sum_{t=1}^{n} r_t + \frac{s_r^2}{2} \right)$$

Finally, keep in mind that since the factors μ and σ, under continuous compounding, represent annualized quantities, then if we are observing returns every month, we set $\tau = 1/12$; every week, $\tau = 1/52$; daily, $\tau = 1/252$. We propose as an exercise to do a GBM simulation of the price of a stock from real data, estimating beforehand the expected return and volatility with the equations above (see Exercise 5.4.7).

5.2 Option Pricing Models: Continuous and Discrete Time

We now go on to review two very important and influential option pricing models. The first one is the Black-Scholes model for pricing European options, based on the assumption that the underlying stock prices follow a continuous time geometrical Brownian motion. This model is also refer as the Black-Scholes-Merton model as a recognition to the important contribution made by Robert Merton to the original formula for pricing options proposed by Fisher Black and Myron Scholes in 1973 (Black and Scholes 1973; Merton 1973).

The second option pricing model to be studied is the binomial tree of John Cox, Stephen Ross and Mark Rubinstein, published in 1979 (Cox et al. 1979), and which can be seen as a discrete time counterpart of the Black-Scholes model, as will be shown at the end of this section.

5.2.1 The Black-Scholes Formula for Valuing European Options

Let $G(P_t, t)$ be the price at time t of an option or other derivative written on a stock with current price P_t, with delivery price K and expiration date $T > t$. Let r be the risk-free constant interest rate for the period $[0, T]$. Note that in this notation of $G(P_t, t)$ it is implicitly assumed that the price of the derivative depends on the current price P_t of the stock and not on past prices. Additionally the Black-Scholes model relies on the following hypotheses:

(1) The extended no arbitrage assumptions. Recall that these include the possibility of unlimited borrowing and lending at the constant interest rate of r, that there are no transaction costs, no taxes, no storage fees, no dividends, and most importantly no arbitrage (see Chap. 1, Definition 1.1).
(2) The stock price P_t follows a geometric Brownian motion with constant drift μ and volatility σ; i.e., it is modeled by Eq. (5.9): $dP = \mu P dt + \sigma P dB$.

Originally Black and Scholes (1973) presented a SDE for G that needed further hypotheses, but almost immediately Merton (1973) derived the same equation of Black and Scholes, founded only on the above list of assumptions. The approach by Merton is based on showing that it is possible to create a *self-financing* portfolio where long positions in the stock are completely financed by short positions in the derivative, and that the value of this perfectly hedged portfolio does not depend on the price of the stock. These arguments lead to the differential equation

$$\frac{\partial G}{\partial t} + rP\frac{\partial G}{\partial P} + \frac{1}{2}\sigma^2 P^2 \frac{\partial^2 G}{\partial P^2} = rG \tag{5.20}$$

which has unique solutions under the constraints imposed above, and for each of the boundary conditions related to the derivative that can be defined on a stock. The case of interest is the derivative being a European call or put option. In this case, the boundary conditions are $G(0, t) = 0$ and $G(P_T, T) = \max(P_T - K, 0)$, for a call, or $G(P_T, T) = \max(K - P_T, 0)$, for a put, and the solutions to (5.20) constitutes the Black-Scholes formulas for the price at time t of a European call option (denote this $G_c(P_t, t)$) and a European put option ($G_p(P_t, t)$), for a stock paying no dividends. These are:

$$G_c(P_t, t) = P_t \Phi(d_1) - Ke^{-r(T-t)}\Phi(d_2) \quad \text{for the call,} \tag{5.21}$$

and

$$G_p(P_t, t) = Ke^{-r(T-t)}\Phi(-d_2) - P_t \Phi(-d_1) \quad \text{for the put,} \tag{5.22}$$

where

$$d_1 = \frac{\ln(P_t/K) + (r + \sigma^2/2)(T - t)}{\sigma\sqrt{T - t}}$$

$$d_2 = \frac{\ln(P_t/K) + (r - \sigma^2/2)(T - t)}{\sigma\sqrt{T - t}} = d_1 - \sigma\sqrt{T - t}$$

and $T - t$ is the time to maturity of the option. The function $\Phi(\cdot)$ is the standard normal cumulative distribution function; that is, $\Phi(x) = \int_{-\infty}^{x} \frac{1}{\sqrt{2\pi}} \exp(-t^2/2)dt$.

We will not do here the full derivation of the Black-Scholes and Merton differential equation (5.20) and its solution subject to the appropriate boundary conditions to obtain the European option pricing formulas; for the details read (Campbell et al. 1997, Chap. 9) or (Hull 2009, Chap. 13).

Remark 5.2 (**Risk-neutral valuation revisited**) The Black-Scholes formulas for valuation of European options can also be obtained by application of the risk-neutral valuation principle, as an alternative to solving the SDE (5.20). Recall from Chap. 1, Sect. 1.2.3, that in a risk neutral world the following assertions holds:

(1) The expected return on all securities is the risk-free interest rate r.
(2) The present value of any cash flow can be obtained by discounting its expected value at the risk-free interest rate.

Therefore to compute the present value of, say, a European call option on a stock, in a risk neutral world this amounts to computing, by (1) and (2) above,

$$C = E(e^{-r(T-t)} \cdot \max(P_T - K, 0))$$

where r is the risk-free interest rate, K is the exercise price, $T - t$ is the time to maturity, and S_T the price of the stock at expiration date. If the underlying stock is assumed to follow a geometric Brownian motion (just as for the Black-Scholes formulas), then by risk-neutrality the price of the stock satisfies the equation: $dP = rPdt + \sigma PdB$, where the parameter μ has been replaced by r. This implies that the expectation $E(e^{-r(T-t)} \cdot \max(P_T - K, 0))$ is an integral with respect to the lognormal density of P_T. This integral can be evaluated in terms of the standard normal distribution function Φ, and the solution obtained is identical to Eq. (5.21). The same argument applies to the deduction of the formula for the put (Eq. (5.22)) from risk-neutral valuation.   □

The Black-Scholes option pricing formulas (5.21) and (5.22), and extensions, are implemented in R. These extensions include dealing with stocks that pay dividends and have storage costs. Both of these market imperfections are summarize in one parameter, the cost-of-carry, which is the storage cost plus the interest that is paid to finance the stock discounting the dividend earned. For a non-dividend paying stock the cost-of-carry is r (the risk-free interest rate), because there are no storage costs and no income is earned.

R Example 5.3 (European options without dividends) In R the Black-Scholes
formulas for computing European options is implemented in the function GBSOption
from the package Rmetrics:fOptions. The parameters for this function are:
TypeFlag = c or p (call or put option); S is the security price; X is the exercise
price of option; Time is the time to maturity measured in years; r is the annualized
rate of interest, a numeric value (e.g. 0.25 means 25 %); b is the annualized cost-of-
carry rate, a numeric value (e.g. 0.1 means 10 %); sigma is the annualized volatility
of the underlying security, a numeric value.

   Consider a European call option on a stock with three months to maturity (i.e.
a quarter of a year), a price for the stock of 60, the strike price of 65, the risk-free
interest rate is 8 % per annum, and with annual volatility of 30 %. Thus, in GBSOption
we set $S = 60$, $X = 65$, $T = 1/4$, $r = 0.08$, $\sigma = 0.3$, and execute:

```
> library(fOptions)
> GBSOption(TypeFlag = "c", S = 60, X = 65, Time = 1/4, r = 0.08,
+    b = 0.08, sigma = 0.30)
```

The output is the price of the option at time T:

```
...
Option Price:
  2.133372
```

To obtain the price of the put corresponding to the same inputs set TypeFlag="p".

□

R Example 5.4 (European options with cash dividends) To compute the value of
European options on a stock that pays cash dividends just discount the dividends
from the stock price and proceed as if there were no dividends. As an example,
consider the case of a call option on a stock whose initial price is $S_0 = 100$, the
time to maturity is 9 months, and the stock pays cash dividends at two time instances
t_1 and t_2, once at one quarter of the period and the other at one half of the period,
and for the same amount $D_1 = D_2 = 2$. The rest of the parameters are the strike
price of 90, the risk-free interest rate r is 10 % per annum, and an annual volatility
of 25 %. Then, first compute the price of the stock discounting the dividends with
$S = S_0 - D_1 e^{-rt_1} - D_2 e^{-rt_2}$ and then apply GBSOption to the discounted price S:

```
> S0=100; r=0.10; D1=D2=2; t1=1/4; t2=1/2
> ##apply discount of dividend to final stock price
> S = S0 - D1*exp(-r*t1) - D2*exp(-r*t2)
> GBSOption(TypeFlag ="c", S=S, X=90, Time=3/4, r=r, b=r,
    sigma=0.25)
```

Option Sensitives or Greeks

The partial derivatives of G with respect to its arguments in Eq. (5.20) are commonly
called "Greeks", because of the greek letters used to denote them, and measure

different sensitivities of the option value to changes in the value of its parameters. The option sensitives are defined as follows:

$$\text{Delta}\quad \Delta := \frac{\partial G}{\partial P}, \quad \text{Gamma}\quad \Gamma := \frac{\partial^2 G}{\partial P^2},$$

$$\text{Theta}\quad \Theta := \frac{\partial G}{\partial t}, \quad \text{Rho}\quad R := \frac{\partial G}{\partial r}, \quad \text{Vega}\quad \mathscr{V} := \frac{\partial G}{\partial \sigma}$$

These options sensitives are widely used by professional investors for risk management and portfolio rebalancing. For example, one of the most popular of the option sensitives is the option delta (Δ) which measures the sensitivity of the option value to changes in the value of the underlying security, and it is use to compute the right proportion of calls on the underlying stock to compensate for any imbalances produce by changes in the price of the stock, a strategy known as *delta hedging*, and which we have already implemented in Chap. 1, Example 1.11. For the Black-Scholes formula of a European call option we can evaluate these sensitives explicitly as:

$$\Delta = \Phi(d_1)$$
$$\Gamma = \frac{\phi(d_1)}{P\sigma\sqrt{T-t}}$$
$$\Theta = -\frac{P\sigma}{2\sqrt{T-t}}\phi(d_1) - Kre^{-r(T-t)}\Phi(d_2)$$
$$R = (T-t)Ke^{-r(T-t)}\Phi(d_2)$$
$$\mathscr{V} = P\sqrt{T-t}\phi(d_1)$$

Similar close expressions can be obtained for the Greeks of the Black-Scholes European put options.

R Example 5.5 (Computing Greeks in R) Option sensitives can be computed in R with a couple of functions from the fOptions package. To get the value of a single Greek use GBSGreeks with the parameter Selection set to one of "delta", "gamma", "theta", "vega", or "rho". Let us compute the delta for the European call option considered in Example 5.3:

```
> library(fOptions)
> GBSGreeks(Selection="delta",TypeFlag="c",S=60,X=65,
    Time=1/4,r=0.08,b=0.08,sigma=0.30)
```

The second method is to use GBSCharacteristics which gives the premium plus all the sensitives:

```
> GBSCharacteristics(TypeFlag="c",S=60,X=65,Time=1/4,r=0.08,
    b=0.08,sigma=0.30)
```

5.2.2 The Binomial Tree Option Pricing Model

The binomial tree model for pricing option was proposed by Cox, Ross and Rubin-stein in Cox et al. (1979), and we shall refer to it as the CRR model. The general idea of the CRR model is to construct a binomial tree lattice to simulate the movement of the stock's price, which is assumed that it can only go up or down, at different time steps from current date to expiration of the option on the stock. Each level of the tree is labeled by a time step and contains all possible prices for the stock at that time. Then, working backwards from the leaves of the tree of prices to the root, compute the option prices from exercise date (at the leaves) to valuation date (at the root). For the computation of option prices the CRR model relies on the following assumptions:

- The stock price follows a multiplicative binomial process over discrete periods.
- There are no arbitrage opportunities and, moreover, extended no arbitrage holds.
- The investor's risk preferences are irrelevant for pricing securities; that is, the model assumes a risk-neutral measure.

The inputs to the model are: the current stock price S; the strike price K; the expiration time T measured in years; the number of time periods (or height of the tree) n; the annualized volatility σ and the annualized rate of interest r of a risk-free asset. The time to maturity T is partitioned in periods of equal length $\tau = T/n$. We now expand on the details of the construction of the CRR model.

The first step is to construct the binomial tree lattice from the data. The stock price is assumed to move either up, with probability p, or down, with probability $1 - p$, by a factor of $u > 1$ or $0 < d < 1$ respectively. Given the current stock price S, the price at the next time step is either uS (for an up move) or dS (for a down move); for the following time step uS increases to u^2S or decreases to duS, and so on. This multiplicative process is repeated until the expiration date is reached. Figure 5.3 presents a 3-period Binomial pricing tree.

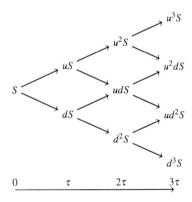

Fig. 5.3 A 3-period binomial price tree

By the no arbitrage assumption, the factors u and d must satisfy the relation

$$d < e^{r\tau} < u \qquad (5.23)$$

since u (respectively, d) represent the rate of return on the stock in the up (down) state. (The reader should think about why is this so, see Problem 5.4.3.)

The CRR model was conceived by their authors as a discrete time model for valuing options which should contain the Black-Scholes model as a limiting case. Therefore, for a smooth transition from discrete time to continuous, one should assume that the interest rate r is continuously compounded (so is independent of the time period), and choose values for u, d and the probability p, such that the mean and variance of the future stock price for the discrete binomial process coincide with the mean and variance of the stock price for the continuous log normal process. With that idea in mind the following equations for computing u, d and p are obtained:

$$u = e^{\sigma\sqrt{\tau}}$$
$$d = 1/u = e^{-\sigma\sqrt{\tau}} \qquad (5.24)$$

and

$$p = \frac{e^{r\tau} - d}{u - d} \quad \text{and} \quad 1 - p = \frac{u - e^{r\tau}}{u - d} \qquad (5.25)$$

An important property of the binomial price tree is that it is *recombining*. This means that at any node the stock price obtained after an up-down sequence is the same as after a down-up sequence. This is what makes the tree look like a lattice of diamond shapes, and hence reduces the number of nodes at the leaves (i.e., at the expiration date) to $n + 1$, where n is the number of time periods. This makes the model computationally feasible since the computation time is linear in n, as opposed to exponential if we had to work with a full branching binary tree. The recombining property allow us to compute the price of a node with a direct formula, instead of recursively from previous valuations, hence dispensing us from actually building the tree. The ith price of the stock at the jth period of time ($1 \le i \le j+1, 0 < j \le n$) is

$$S_{ij} = S \cdot u^{n_u} d^{n_d} \qquad (5.26)$$

where n_u is the number of up-moves and n_d is the number of down-moves, and these quantities verify the equation $n_u + n_d = j$.

Next, we work our way backwards from the leaves to the root to iteratively compute the price of the option. Here is where we use the no arbitrage assumption to equate the value of the option with that of the stock plus some cash earning the interest rate r. We explain how it is done in the case of a call option and one time period to maturity, and then show how to generalized to arbitrarily many time periods until maturity and include the case of put options.

Let C be the current value of the call for the stock with current price S, the strike price is K, and maturity one period of time τ. Let C_u (resp. C_d) be the value of the call

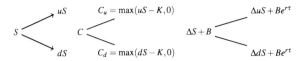

Fig. 5.4 Stock, call option and portfolio evolution of values for a 1-period binomial tree

at the end of the period if the stock price goes to uS (resp. dS). The event C_u occurs with probability p, and the event C_d with probability $1 - p$. Since in either case we are at the end of the contract the exercise policy imply that $C_u = \max(uS - K, 0)$ and $C_d = \max(dS - K, 0)$.

We form a portfolio with Δ many shares of the stock and an amount B of a riskless bond. The cost of this investment at current time is $\Delta S + B$, and at the option's maturity time the value of our portfolio will be either $\Delta uS + rB$ with probability p, or $\Delta dS + rB$ with probability $1 - p$. The evolution of the value of the stock, the value of the call and the value of the portfolio just described is pictured in Fig. 5.4.

The purpose of the portfolio is to cover our investment in the call. This means that at maturity date the value of the portfolio and the option payoff should be equal. Therefore, we must have

$$\Delta uS + Be^{rt} = C_u,$$
$$\Delta dS + Be^{rt} = C_d \tag{5.27}$$

Solving these equations one obtains

$$\Delta = \frac{C_u - C_d}{(u - d)S} \quad \text{and} \quad B = \frac{uC_d - dC_u}{(u - d)e^{rt}}$$

The Δ and B chosen so that the Eq. (5.27) hold makes the portfolio an hedging portfolio. By extended no arbitrage the value of the hedging portfolio and the value of the call should be equal at all past times (Proposition 1.1); that is

$$C = \Delta S + B = \left[\frac{e^{rt} - d}{u - d}C_u + \frac{u - e^{rt}}{u - d}C_d\right] \bigg/ e^{rt}$$

Using Eq. (5.25) we can compute the price of the call option in terms of the two possible exercise values, C_u and C_d, by the equation

$$C = (pC_u + (1 - p)C_d)e^{-rt} \tag{5.28}$$

The generalization to n time periods is a simple iteration of the hedging portfolio construction at each step. Let $C_{i,j}$ be the ith price of the option at jth time period, $0 < i \le j + 1, 0 \le j \le n$. The price of the option at the leaf nodes (i.e., the price at the expiration date) is computed simply as its local payoff. Thus,

$$C_{i,n} = \max(S_{in} - K, 0) \quad \text{for a call option,} \tag{5.29}$$

$$C_{i,n} = \max(K - S_{in}, 0) \quad \text{for a put option.} \tag{5.30}$$

The price of the option at intermediate nodes is computed, as in the one step case (Eq.(5.28)), using the option values from the two children nodes (*option up* and *option down*) weighted by their respective probabilities of p, for going up, or $(1 - p)$ for going down, and discounting their values by the fixed risk-free rate r. Thus, for $j = 0, 1, \ldots, n$, and $0 < i \leq j + 1$,

$$C_{i,j-1} = e^{-r\tau}(pC_{i+1,j} + (1 - p)C_{i,j}) \tag{5.31}$$

Let's named the value of the option (call or put) given for each time period by Eq.(5.31) the *Binomial Value* of the option.

Now depending on the style of the option (European, American or Bermudan) we can evaluate the possibility of early exercise at each node by comparing the exercise value with the node's Binomial Value. For European options the value at all time intervals of the option is the Binomial Value, since there is no possibility of early exercise. For an American option, since the option may either be held or exercised prior to expiry, the value at each node is computed by max (Binomial Value, Exercise Value). For a Bermudan option, the value at nodes where early exercise is allowed is computed by max (Binomial Value, Exercise Value). One can now appreciate the flexibility of this model for computing various type of options: for European we find the value at the root node; for American we find the best time to exercise along any of the intermediate nodes; and for Bermudan we find the best time to exercise along some previously determined intermediate nodes.

The construction of the CRR binomial option pricing tree just described, and summarized in Eqs.(5.29)–(5.31), and the parameters u, d and p given by (5.24) and (5.25), can be assembled together in an algorithm for pricing any vanilla option, written on a security that pays no dividends. This is shown as Algorithm 5.1.[1]

R Example 5.6 The fOptions package has the function CRRBinomialTreeOption that computes the price of vanilla options according to the CRR Binomial model, and whose pseudo code is given in Algorithm 5.1. Additionally, the package provides with the functions BinomialTreeOption that returns the binomial tree for the option, and BinomialTreePlot to make a plot of the binomial tree. We illustrate the usage of these functions by first computing the price of an American put option on a stock with current value of 50, exercise price also of 50, the time to maturity is 5 months, the annualized rate of interest r is of 10 %, the annualized volatility σ of the stock is of 40 %, the annualized cost-of-carry rate b in this case equals the rate of interest r, and the number of time steps n is 5. Then give the instructions to plot the CRR binomial tree for this case.

---

[1] This is in fact a pseudo code of the actual program for the R function CRRBinomialTreeOption in the Rmetrics package due to Diethelm Wuertz.

Algorithm 5.1 CRR Binomial Tree option valuation

---

function CRRBinomialTreeOpt $(\tau, S, K, T, r, \sigma, n)$ {
  # set factor for call or put
  if $(\tau ==$ "ce" **or** $\tau ==$ "ca") { $z = 1$;}
  if $(\tau ==$ "pe" **or** $\tau ==$ "pa") { $z = -1$;}
  else { τ="ce" ; $z = 1$; } # default values
  $dt = T/n$;
  $u = \mathbf{exp}(\sigma *\mathbf{sqrt}(dt)); \quad d = 1/u;$
  $p = (\mathbf{exp}(r * dt) - d)/(u - d);$
  $Df = \mathbf{exp}(r * dt);$ # discount factor
  # initial values at time T
  for $i = 0$ to n {
    OptionValue[i] $= z * (S * u^i * d^{n-i} - K)$;
    OptionValue[i] $= (\mathbf{abs}(\text{OptionValue}[i]) + \text{OptionValue}[i])/2$; }
  # case: European option
  if $(\tau ==$ "ce" **or** $\tau ==$ "pe") {
    for $j = n - 1$ down to 0 {
      for $i = 0$ to j {
        OptionValue[i] $= (p*\text{OptionValue}[i + 1] +$
        $(1 - p)*\text{OptionValue}[i])*Df$; }}}
  # case: American option
  if $(\tau ==$ "ca" **or** $\tau ==$ "pa") {
    for $j = n - 1$ down to 0 {
      for $i = 0$ to j {
        OptionValue[i] $= \mathbf{max}(z * (S * u^i * d^{\mathbf{abs}(i-j)} - K), (p*\text{OptionValue}[i + 1] +$
        $(1 - p)*\text{OptionValue}[i])*Df$); }}}
  return OptionValue[0] ; }

```
> library("fOptions")
> CRRBinomialTreeOption(TypeFlag = "pa", S = 50, X = 50,
    Time = 5/12, r = 0.1, b = 0.1, sigma = 0.4, n = 5)
## output : Option Price: 4.488459

##to plot an option tree (for put american)
> CRRTree = BinomialTreeOption(TypeFlag = "pa", S = 50, X = 50,
    Time = 0.4167, r = 0.1, b = 0.1, sigma = 0.4, n = 5)
> BinomialTreePlot(CRRTree, dy = 1, cex = 0.8, ylim = c(-6, 7),
    xlab = "n", ylab = "Option Value")
> title(main = "Option Tree")
```

Running again the above commands with `TypeFlag = "ca"` gives the result for the corresponding American call. Figure 5.5 shows the plot for both cases. □

Convergence of the CRR model to the Black-Scholes model. We summarize the main ideas of the proof given in Cox et al. (1979, Sect. 5) of the convergence of the CRR binomial tree option pricing model to the continuous time Black-Scholes (BS)

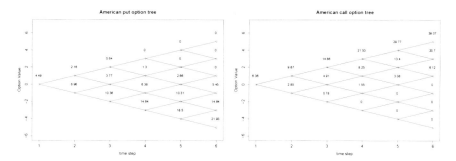

Fig. 5.5 American put and call option trees

model. We centered our discussion around the valuation of a call option. One can show by repeated substitution in Eq. (5.31) of the successive values $C_{i,j}$ of a call, beginning with the values at the leaf nodes of the binomial tree (Eq. (5.29)), that for $n > 1$ periods the value of the call is given by the formula

$$C = \left(\sum_{j=a}^{n} \frac{n!}{(n-j)!j!} p^j (1-p)^{n-j} [u^j d^{n-j} S - K] \right) e^{-rn\tau} \qquad (5.32)$$

where a is the minimum number of upward moves that the stock price must make over the next n periods to finish above the strike price, and so that the call option can be exercised. Formally, a must be the smallest non-negative integer such that $u^a d^{n-a} S > K$. Rewrite (5.32) as

$$C = S \left(\sum_{j=a}^{n} \frac{n!}{(n-j)!j!} p^j (1-p)^{n-j} \frac{u^j d^{n-j}}{e^{rn\tau}} \right) - Ke^{-rn\tau} \left(\sum_{j=a}^{n} \frac{n!}{(n-j)!j!} p^j (1-p)^{n-j} \right)$$

The expression in parentheses multiplying the term $Ke^{-rn\tau}$ is the complementary binomial distribution function $\phi(a; n, p)$; that is $\phi(a; n, p) = 1 - \mathbb{P}(j \leq a - 1)$, the probability that the sum of n random variables, each of which can take on the value 1 with probability p and 0 with probability $1 - p$, will be greater than or equal to a. The expression in parentheses accompanying S can also be represented by $\phi(a; n, p')$ where $p' = ue^{-r\tau}p$ (and hence, $1 - p' = de^{-r\tau}(1 - p)$). Thus,

$$C = S\phi(a; n, p') - Ke^{-rn\tau} \phi(a; n, p)$$

Cox, Ross and Rubinstein then show that

$$\lim_{n \to \infty} \phi(a; n, p') = \Phi(d_1) \quad \text{and} \quad \lim_{n \to \infty} \phi(a; n, p) = \Phi(d_2) \qquad (5.33)$$

Thus, the CRR binomial tree model contains as a special limiting case the BS model. For sufficiently large n it will give almost the same value for a European option as that computed with the exact formula provided by Black-Scholes. However, the CRR model has the advantage over BS that it can efficiently compute the value of options for which earlier exercise is desirable.

A key element for the computational feasibility of the CRR model is its recombining property, as we have already commented. This limit the application of CRR to path-independent style options, like European, American or Bermudan, since for those exotic options that are intrinsically path dependent, like Asian or Barrier options, the computational saving implied by the recombining property is not possible because necessarily all paths must be explore, and the number of distinct paths through a tree is 2^n. Therefore, other option pricing models must be explored.

5.3 Monte Carlo Valuation of Derivatives

A popular numerical approach to valuing securities is the Monte Carlo (MC) simulation method. We have seen already in R Example 5.2 an illustration of the method. In its simpler form it consists of simulating many finite trajectories of a random process that models the price of the asset, and then take the average of the results at a given point as an estimate of the value of the process at that point. This numerical method has been proven to be very useful when applied to pricing options.

The application of the MC method to the valuation of options was first proposed by Boyle (1977), and after this seminal paper followed a long list of publications showing the advantage of the method in finding good approximations to the present value of almost any type of options, which included path dependent options such as Asian, and even further, options with multiple uncertainties as, for example, American arithmetic Asian options on a single underlying asset (Grant et al. 1997). There are different variants of the MC method, depending on the model assumed for the behavior of the underlying asset and the technique employed for speeding up the convergence of the simulations. We will give a very brief account of Monte Carlo discretization methods applied to options whose underlying asset prices follow a continuous-time model. Our exposition will be terse in the technical details, as our purpose is to provide a primer to the important and extensive field of Monte Carlo methods in finance. As a follow up, or better as a companion to these notes, we propose the survey by Boyle et al. (1997) and the textbook by Glasserman (2004).

The general idea of the Monte Carlo method for computing the present value of a security is to numerically approximate the discounted expected payoff of the security, assuming a risk-neutral world. Here is again the risk-neutral valuation principle in action, which we first encounter in Sect. 1.2.3, and then again in Remark 5.2 on the Black-Scholes formulas, and in the construction of the CRR binomial tree model. The MC method implements the risk-neutral valuation of securities through the following steps:

(1) Simulate sample paths of the underlying state variable (e.g., asset prices and interest rates) over the relevant time horizon and according to the risk-neutral measure.
(2) Evaluate the discounted cash flows of a security on each sample path, as determined by the structure of the security in question.
(3) Average the discounted cash flows over a given number of sample paths. The result is taken as an approximation to the value of the security.

When the underlying state variable is assumed to follow a continuous time model, such as the arithmetic Brownian motion, the MC method described above consists formally in computing an approximation to $E(e^{-rT}f(X_T))$, where r is the constant risk-free interest rate used to discount the value $f(X_T)$, $f : \mathbb{R}^d \to \mathbb{R}$ (the payoff of the security) and X_T is the value at time T of the random process $X(t)$ that is solution of the SDE

$$dX(t) = a(X(t))dt + b(X(t))dB(t), \quad X(0) = X_0 \in \mathbb{R}^d \quad (5.34)$$

with $B(t)$ a Brownian motion, and $a, b : \mathbb{R}^d \to \mathbb{R}^d$ functions satisfying certain regularity conditions that ensure the existence of unique solutions to Eq. (5.34). To obtain an approximate solution of (5.34) over a time interval $[0, T]$, and where functions a and b are not necessarily constants, one effective method is to apply a discretization scheme. Begin by dividing the time interval $[0, T]$ in n parts of equal length, such as $0 = t_0 < t_1 < \cdots < t_n = T$, with $t_j = j\Delta t$ and $\Delta t = T/n = t_j - t_{j-1}$, for all $j > 0$. A solution observed at discrete time $t \in \{t_0, t_1, \ldots, t_n\}$ is obtained by integrating (5.34) from t to $t + \Delta t$ which results in

$$X_{t+\Delta t} = X_t + \int_t^{t+\Delta t} a(X_s)ds + \int_t^{t+\Delta t} b(X_s)dB(s) \quad (5.35)$$

The remaining task is to approximate the two integrals in (5.35). We do this with a simple, yet effective, discretization scheme due to Euler.

Euler approximation. The Euler method approximates the integrals in Eq. (5.35) as products of their integrands at time t and the size of the integration domain, that is,

$$\int_t^{t+\Delta t} a(X_s)ds \approx a(X_t) \int_t^{t+\Delta t} ds = a(X_t)\Delta t$$

and

$$\int_t^{t+\Delta t} b(X_s)dB(s) \approx b(X_t) \int_t^{t+\Delta t} dB(s) = b(X_t)\sqrt{\Delta t}\varepsilon_t, \quad \varepsilon_t \sim N(0, 1)$$

Putting all together we get Euler's discrete solution of (5.34), which is

$$X_0 = X(0), \quad X_{t+\Delta t} = X_t + a(X_t)\Delta t + b(X_t)\sqrt{\Delta t}\varepsilon_t \tag{5.36}$$

Now that we have a way to simulate sample paths of the arithmetic Brownian motion, we employ it in a straightforward application of the Monte Carlo method. This example has the twofold purpose of illustrating the method, as well as being a reference upon which we will develop further complicated applications of the method.

Consider the problem of estimating the price of a European option (call or put) on a common stock. Thus, the underlying state variable is the stock's price $\{S_t\}$, which is assumed to follow a geometric Brownian motion, and hence its continuous version is solution of the equation

$$dS(t) = \mu S(t)dt + \sigma S(t)dB \tag{5.37}$$

where μ is the drift rate and σ the volatility of $S(t)$. Assuming a risk-neutral measure, we put $\mu = r$, where r is the risk-free interest rate. The function f is the European option's payoff, which is defined as

$$f(S_T) = \begin{cases} \max(S_T - K, 0) & \text{for a call} \\ \max(K - S_T, 0) & \text{for a put} \end{cases}$$

where K is the strike price and S_T is the stock's price at expiration time T. Then the steps of the MC method adapted to the valuation of this option are as follows:

(1) To simulate the ith sample path of $S(t)$ we use the Euler discretization with $X_t = S_t$, $a(X_t) = rS_t$ and $b(X_t) = \sigma S_t$. Divide the time interval $[0, T]$ into n time periods $\Delta t = T/n$, and recursively compute the approximated path $\{\widehat{S}_{i,j}\}$ of length n as

$$\begin{aligned} &\widehat{S}_{i,0} = S(0); \\ &\widehat{S}_{i,j} = \widehat{S}_{i,j-1} + r\widehat{S}_{i,j-1}\Delta t + \sigma\widehat{S}_{i,j-1}\varepsilon_j\sqrt{\Delta t}, \quad \varepsilon_j \sim N(0,1), j = 1, \ldots, n \end{aligned} \tag{5.38}$$

(2) The discounted payoff of sample path $\omega_i = (\widehat{S}_{i,0}, \widehat{S}_{i,1}, \ldots, \widehat{S}_{i,n})$ is $f(\widehat{S}_{i,n}) \cdot e^{-rT}$.

(3) Considering m sample paths $\omega_1, \ldots, \omega_m$, the average is given by $\dfrac{1}{m}\sum_{i=1}^{m} f(\widehat{S}_{i,n})e^{-rT}$.

    This is the estimated price of the option.

The pseudo-code for this MC simulation based on Euler discretization is presented in Algorithm 5.2. We have adapted the mathematical nomenclature to the language of R, so that the reader can readily copy this code into his R console and put it to work. The default inputs are set to: a call option, stock's initial price and option's strike price equal to 100, expiration time one month, risk-free interest rate of 10 %, and volatility of 25 %.

Algorithm 5.2 Euler MC to calculate the price of European option

---

```
0.   EulerMC = function( Type = "c", S = 100, K = 100, T = 1/12,
            r = 0.1, sigma = 0.25, n = 300, m = 100, dt = T/n ){
1.   sum = 0;
2.   for (i in 1 : m) { # number of simulated paths
3.        for (j in 1 : n) { # length of path
4.             E = rnorm(0, 1);
5.             S = S + r * S * dt + sigma * S * sqrt(dt) * E; }
6.        if (Type == "c" ){ payoff = max(S − K, 0) }
7.        else if (Type == "p" ){ payoff = max(K − S, 0) }
8.        else { payoff = max(S − K, 0) }   # default
9.        sum = sum + payoff }
10.  OptionValue = (sum * exp(−r * T))/m;
11.  return (OptionValue)}
```

Remark 5.3 Observe that dividing both sides of Eq. (5.38) by $\widehat{S}_{i,j-1}$ and subtracting 1 we get a simulation of the return $\widehat{R}_{i,j} = \dfrac{\widehat{S}_{i,j}}{\widehat{S}_{i,j-1}} - 1 = r\Delta t + \sigma \varepsilon_j \sqrt{\Delta t}$, which is the discrete version of the arithmetic Brownian motion model for the instantaneous return. $\qquad\square$

Convergence. How good is the Monte Carlo approximation $\widehat{C}_m^n = \dfrac{1}{m}\sum_{i=1}^{m} f(\widehat{S}_{i,n})e^{-rT}$ to $E(e^{-rT}f(X_T))$? We need to analyze the approximation error

$$\varepsilon(m, n) = \widehat{C}_m^n - E(e^{-rT}f(X_T)),$$

a measure that depends on both m and n (the number of simulated paths and the length of each path). Let's first consider m variable (so that n is fixed). Set $C = e^{-rT}f(X_T)$, and let $\mu_C = E(C)$ and $\sigma_C^2 = Var(C)$. Each $\widehat{C}_i = f(\widehat{S}_{i,n})e^{-rT}$ is iid sample of C, so that $\widehat{C}_m^n = \dfrac{1}{m}\sum_{i=1}^{m} \widehat{C}_i$ is a sample mean $\widehat{\mu}_C$, and the sample variance $\widehat{\sigma}_C^2 = \dfrac{1}{m-1}\sum_{i=1}^{m}(\widehat{C}_i - \widehat{\mu}_C)^2$. By the Central Limit theorem the standardized estimator $\dfrac{\sqrt{m}(\widehat{C}_m^n - C)}{\sigma_C}$ converges in distribution to the standard normal. This also holds for the sample standard deviation $\widehat{\sigma}_C$ in the place of σ_C. This means that

$$\lim_{m \to \infty} \mathbb{P} \left(\frac{\sqrt{m}}{\widehat{\sigma}_C} (\widehat{C}_m^n - C) \le x \right) = \Phi(x)$$

for all x, where $\Phi(x)$ is the cumulative normal distribution. This implies that the approximation error $\varepsilon(m, n)$, with n fixed, has a rate of convergence $O(1/\sqrt{m})$. Thus, increasing the number of simulated paths reduces the approximation error, and at a rate of $1/\sqrt{m}$.

On the other hand, with respect to n, and keeping m fixed, it can be shown that $\lim_{n \to \infty} \varepsilon(m, n) = 0$, and even further that $\varepsilon(m, n)$ is $O(n^{-1})$ (see Kloeden and Platen (1995)). Thus, increasing the number of steps to construct the simulated paths the more accurate the approximations $\widehat{C}_i$, and the rate of convergence is as $1/n$. This means that the Euler approximation is of order 1 convergence. This can be improved. A discretization method that achieves order 2 convergence, i.e., convergence is $O(n^{-2})$, is given by Milstein discretization scheme.

Milstein approximation. An improvement to the Euler approximation to solutions of the arithmetic Brownian motion was obtained by Milstein.[2] The Milstein discretization scheme to approximate solutions to Eq. (5.34) is given by

$$
X_{t+h} = X_t + ah + b\sqrt{h}\varepsilon_t + \frac{1}{2}b'b \left((\sqrt{h}\varepsilon_t)^2 - h \right) \tag{5.39}
$$
$$
+ \frac{h^2}{2} \left(aa' + \frac{1}{2}b^2 a'' \right) + \frac{h^{3/2}}{2} \left(a'b + ab' + \frac{1}{2}b^2 b'' \right) \varepsilon_t
$$

where $h = \Delta t$ (the time step), $a = a(X_t)$, $b = b(X_t)$ and derivatives a', a'', b', b'' are with respect to X, and $\varepsilon_t \sim N(0, 1)$. This approximation has a second-order convergence rate, meaning that the approximation error is $O(h^{-2})$, under some assumptions on the continuity of the coefficient functions a and b. Thus, with the same time step $h = \Delta t$ the Milstein approximation converges faster than Euler approximation. For a deduction of the Milstein discretization scheme and a discussion on its convergence see Glasserman (2004, Chap. 6).

To apply the Milstein approximation to sample paths of the price of a stock S that follows a geometric Brownian motion, we proceed as before with Euler and set $X_t = S_t$, $a = a(X_t) = rS_t$ and $b = b(X_t) = \sigma S_t$ (hence $a' = r$, $a'' = 0$, $b' = \sigma$, $b'' = 0$). Divide the interval $[0, T]$ into n time periods $h = T/n$, and thus, an approximated Milstein path $\{\widehat{S}_j\}$ of length n is given by

$$\widehat{S}_0 = S(0);$$
$$\widehat{S}_j = \widehat{S}_{j-1} + r\widehat{S}_{j-1}h + \sigma \widehat{S}_{j-1}\sqrt{h}\varepsilon_j + \frac{1}{2}\sigma^2 \widehat{S}_{j-1} \left((\sqrt{h}\varepsilon_j)^2 - ht \right)$$
$$+ \widehat{S}_{j-1} \left(\frac{(rh)^2}{2} + r\sigma h^{3/2} \right), \quad \varepsilon_j \sim N(0, 1), \, j = 1, \ldots, n$$

---

[2] Milstein, G. N. (1978). A method of second order accuracy integration of stochastic differential equations. *Theory of Probability & its Applications* 19 (3): 557-562.

You can substitute line 5 of Algorithm 5.2 for this equation and you have a Monte Carlo method for computing the price of a European option with the Milstein discretization scheme. There are several others discretization schemes, some of higher order of convergence; for a review see Kloeden and Platen (1995). We have presented the two most popular discretization schemes.

Path dependent options. Now that we have understood the MC method through a simple application as the valuation of European option, let's see how it applies to the case of valuing path-dependent options. As an example we consider Asian call options on stocks. Recall that the payoff of these type of options is given by the average price of the stock over some predetermined period of time until expiration. Formally, the payoff is defined by $\max(A_S(T_0, T) - K, 0)$, where $A_S(T_0, T) - K, 0) = \frac{1}{n} \sum_{j=0}^{n} S_{t_j}$, for some fixed set of dates $T_0 = t_0 < t_1 < \cdots < t_n = T$, with T the expiration date for the option.

We need to calculate the expected discounted payoff $E(e^{-rT} \max(A_S(T_0, T) - K, 0))$, and to do that it will be sufficient to simulate the path $S_{t_0}, S_{t_1}, \ldots, S_{t_n}$ and compute the average on that path. We can use any of the discretization methods to simulate the path; for simplicity we work with Euler approximation. Algorithm 5.3 shows the steps to do this calculation of an Asian option (which is set by default to a call).

Algorithm 5.3 MC to calculate the price of Asian options

```
0.   MCAsian = function( Type = "c", S = 100, K = 100, T = 1/12,
           r = 0.1, sigma = 0.25, n = 300, m = 100, dt = T/n ){
1.   sum = 0;
2.   for (i in 1 : m) { # number of simulated paths
3.       Asum = 0; # for cumulative sum along the path
4.       for (j in 1 : n) { # length of path
5.           E = rnorm(0, 1);
6.           S = S + r * S * dt + sigma * S * sqrt(dt) * E;
7.           Asum = Asum + S; }
8.       if (Type == "c" ){ payoff = max((1/n) * Asum - K, 0) }
9.       else if (Type == "p" ){ payoff = max(K - (1/n) * Asum, 0) }
10.      else { payoff = max((1/n) * Asum - K, 0) }   # default
11.      sum = sum + payoff }
12.  OptionValue = (sum * exp(-r * T))/m;
13.  return (OptionValue)}
```

Analyzing the form of both Algorithms 5.2 and 5.3 we see that a MC simulation algorithm for computing the present value of an option can be structured on three general subroutines:

(1) A function to generate the options innovation, i.e., the random component of simulated path. This function depends mainly on m (the number of simulated paths) and n (the path length), and possibly other variables for initializing the

generating method and other parameters specific of the method. The output is a $m \times n$ matrix $\varepsilon = (\varepsilon_{ij})$ containing the random components for all simulations, which are produced by any pseudo-random number generator or other procedure for generating random sequences that we choose. Let us denote this function by `OptInnovations(m,n, ...)`.

(2) A function to generate the options underlying security price paths (e.g. Euler, Milstein, or other). This function takes as inputs m, n, the options innovations (the output of `OptInnovations`) and other parameters specific to the numerical procedure (e.g., the interest rate r, the volatility σ, the initial price, time periods, and so on). The output is a $m \times n$ matrix where each row represents a simulated path of the underlying security. We denote his function by `SecurityPath(m,n,innovations, ...)`.

(3) A function to compute the option's payoff according to risk-neutral valuation. This function mainly takes as input the simulated paths to calculate the discounted payoff at exercise time and according to exercise value K, and it is denoted by `OptPayoff(path,K, ...)`.

This is how it is structured the function `MonteCarloOptions` for the `fOptions` package in R. We show how to use it in the following R Example.

R Example 5.7 The following step-by-step instructions for doing MC is based on the examples from the on-line manual of `MonteCarloOptions`. We compute the approximated value of an Asian option on a stock, by simulating the path of the returns of such stock which are assumed to follow a Brownian motion (hence the simulation is of the Euler type by Remark 5.3).

```
library(fOptions)
## MC simulation
## Step 1: function to generate the option's innovations.
## Use normal pseudo random numbers
pseudoInnovations = function(mcSteps, pathLength,init){
  #Create normal pseudo random numbers
  innovations = rnorm.pseudo(mcSteps, pathLength,init)
  #Return value
        innovations }

## Step 2: function to generate the option's price paths.
wienerPath = function(eps) {
  # Generate the Paths:
  path = (b-sigma*sigma/2)*delta.t + sigma*sqrt(delta.t)*eps
  # Return Value:
      path }

## Step 3: function for the payoff for an Asian Call or Put:
arithmeticAsianPayoff = function(path) {
  # Compute the Call/Put Payoff Value:
  SM = mean(S*exp(cumsum(path)))
  if (TypeFlag == "c") payoff = exp(-r*Time)*max(SM-X, 0)
  if (TypeFlag == "p") payoff = exp(-r*Time)*max(0, X-SM)
  # Return Value:
```

```
                    payoff }

## Final Step: Set global parameters for the  option:
TypeFlag <<- "c"; S <<- 100; X <<- 100
Time <<- 1/12; sigma <<- 0.4; r <<- 0.10; b <<- 0.1

# Do the Asian Simulation with pseudo random numbers:
mc = MonteCarloOption(delta.t=1/360, pathLength=30, mcSteps=5000,
    mcLoops = 50, init = TRUE,innovations.gen = pseudoInnovations,
    path.gen = wienerPath, payoff.calc = arithmeticAsianPayoff,
    antithetic = TRUE, standardization = FALSE, trace = TRUE)
# Plot the MC Iteration Path:
par(mfrow = c(1, 1))
mcPrice = cumsum(mc)/(1:length(mc))
plot(mcPrice, type = "l", main = "Arithmetic Asian Option",
      xlab = "Monte Carlo Loops", ylab = "Option Price")
```

We trace the output of each simulation (parameter `trace=TRUE`). This prints the value of the option computed (left number) and the average payoff of all samples computed so far. The options price is taken as the final right most value, the average over all samples. In this example the price obtained is 2.934428.

```
Monte Carlo Iteration Path:
Loop:    No
Loop:    1   : 2.413952 2.413952
Loop:    2   : 3.03552 2.724736
Loop:    3   : 2.932311 2.793928
Loop:    4   : 1.367154 2.437234
Loop:    5   : 3.548186 2.659425
...      ...     ...    ...
Loop:    48    : 1.842802 2.947898
Loop:    49    : 3.339307 2.955885
Loop:    50    : 1.883031 2.934428
```

Figure 5.6 shows a plot of the 50 successive iterations of the Monte Carlo simulation (the Monte Carlo Iteration path), as the path made with the average successive payoffs, and we can see it converging to a value close to 2.9. Increasing the path length or the MC steps we will get sharper estimates. To compare the performance of the MC method against other analytical methods, consider for example the Turnbull and Wakeman's approximation method (Haug 2007, Sect. 4.20.2), which is coded in the fExoticOptions package. Execute:

```
> library("fExoticOptions")
> TW=TurnbullWakemanAsianApproxOption(TypeFlag="c",S=100,SA=100,
    X=100, Time=1/12, time=1/12, tau=0, r=0.1, b=0.1, sigma=0.4)
> print(TW)
```

The price obtained with this method is 2.859122.   □

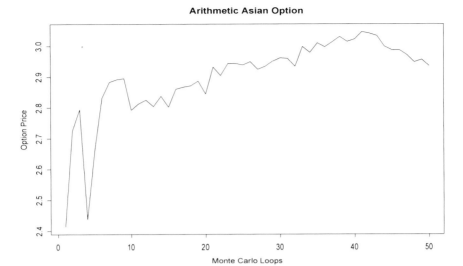

Fig. 5.6 MC simulation of arithmetic Asian option using Brownian paths with pseudo random numbers

5.4 Notes, Computer Lab and Problems

5.4.1 Bibliographic remarks: Three excellent textbooks for more in-depth study of continuous-time models and their applications to option valuation (BS and CRR models) are Campbell et al. (1997), Hull (2009) and Tsay (2010) Our exposition of the CRR model (Sect. 5.2.2) is a summary of the main points of its construction from the original source Cox et al. (1979). We advice the reader to read this paper by Cox, Ross and Rubinstein for it has plenty of motivating examples and clear arguments explaining and sustaining the mathematical details. For a comprehensive reference containing every option pricing tool one may need see Haug (2007). For a primer in Monte Carlo methods see Glasserman (2004) and Boyle et al. (1997).

5.4.2 A brief historical account of financial models: Louis Bachelier (1870–1946) has the credit of being the first person to model stock prices as Brownian motion, in his Ph.D. thesis *Theorié de la Speculation* (1900) presented at the Sorbonne. In 1973, Fischer Black and Myron Scholes, and independently Robert Merton, published their results on the continuous-time option pricing model based on the geometric Brownian motion. The Black–Scholes–Merton model became very popular in the finance engineering industry, and Merton and Scholes received the 1997 Nobel Prize in Economics for this achievement (Black had died by then, and therefore he was not eligible for the prize).

The Black–Scholes–Merton model assumes constant volatility in order to obtain a closed-form formula for pricing only European options. A major breakthrough came in 1979 with the introduction of the first discrete time model for pricing American as well as European options, the binomial tree model proposed by J. C. Cox, S. A. Ross and M. Rubinstein. On the other hand, observations that the volatility of prices may change over time, and even be dependent on the past behavior of prices, were already made by Benoit Mandelbrot (1963), Eugene Fama (1965) and others. Then a further success, in the same line of discrete time modeling, came with the ARCH model and its generalization, GARCH (both seen in Chap. 4), proposed respectively by Robert Engle (1982) and Tim Bollerslev (1986), where the variance of an asset return is considered dependent on past shocks, and on these shocks plus past variance. The ARCH model approach to time-varying variance is an important landmark in the understanding of financial time series behavior, as it is attested by the fact that Robert Engle received the Nobel Prize in 2003 for his work on this model. For more on the pre-history of financial modeling read Bernstein (1992) and Rubinstein (2006). The modern history of models for market behavior is yet to be unraveled and written, and the task is not easy as mutually contradictory views are equally praise. In 2013, Eugene Fama, Lars Peter Hansen and Robert Shiller were awarded the Nobel Prize in Economic Sciences, the first for his orthodox theory of the efficiency of markets, while the other two for their rational theories sustaining the predictability of prices in the long-run.

5.4.3: Give an argument based on no arbitrage to sustain Eq. (5.23).

5.4.4: The *implied volatility* refers to the value of the volatility parameter ("the σ") in an option pricing model, needed to estimate the price of the option. For example, in the Black-Scholes formula for computing the price of an European call (Eq. (5.21)) the parameter σ is the implied volatility of the underlying asset. This is a forward looking measure since it represents the volatility of the asset into the future time until expiration of the option contract. Thus, it is very subjective and differs from historical volatility since the latter is estimated from past returns. Nonetheless, professional traders like to work with implied volatilities estimated from various option calls and puts, for it reflects some sort of general consensus on the future variability of stocks. Note that there exist an index of implied volatilities of options on the S&P 500, named the VIX, and extensively used by traders.[3]

How can we go about computing such implied volatility? Observe carefully Black-Scholes equation (5.21), and you should convinced yourself that if you fix all parameters but σ, you obtain a *monotonically increasing* function in σ for the values of the call. This means that if you increase the values of σ you get higher values of the call. In symbols, $C = f(\sigma)$, where C is the value of the call and f is the function defined by Eq. (5.21) with only σ variable. The bad news is that such f is not invertible, but the monotonicity of f guarantees that for a given value of C there exists a unique value of σ such that $f(\sigma) = C$, and here is where the guessing game starts. Traders offer a likely value for the price of the call, say C, and now you can guess a value σ_i for the implied volatility. If $f(\sigma_i) < C$, by monotonicity take $\sigma_{i+1} > \sigma_i$, and

---

[3] see Hull (2009) for details on this important index.

evaluate $f(\sigma_{i+1})$; otherwise, if $f(\sigma_i) > C$ then take $\sigma_{i+1} < \sigma_i$. Repeat this process until you get a value σ such that $f(\sigma) \approx C$.

5.4.5: Use the procedure explained in the previous note, for estimating the implied volatility in the following European option call contract on a stock that does not pays dividends: the current value of the option is 11.1, the current price of the stock is 100, the strike price is 100, the risk-free interest rate is 3 %, and the time to maturity is 6 months.[4]

5.4.6: Extend Algorithm 5.1 with the case for Bermudan option. The vector of possible exercise dates should be an input to the program, and it is only at these predetermined exercise dates where the value of the option is the max(Binomial Value, Exercise Value), and at all remaining times the value is the Binomial Value.

5.4.7 R Lab: Repeat R Example 5.2 with real data, estimating μ and σ with Eq. (5.19), and tracing 50 possible trajectories of the price one month ahead.

5.4.8: Ornstein-Uhlenbeck process: An important continuous time process that extends the Brownian motion is the Ornstein-Uhlenbeck (OU) process.[5] The OU process $X = \{X(t)\}_{t \geq 0}$ can be defined as the solution of the SDE

$$dX(t) = \gamma(m - X(t))dt + \sigma\, dB(t), \quad t > 0 \tag{5.40}$$

where $\gamma > 0$, m and $\sigma \geq 0$ are real constants, and $\{B(t)\}_{t \geq 0}$ is a standard Brownian motion. The solution of Eq. (5.40) give us an expression for the OU process X in terms of stochastic integrals:

$$X(t) = m + (X(0) - m)e^{-\gamma t} + \sigma e^{-\gamma t} \int_0^t e^{\gamma s} dB(s), \quad t \geq 0 \tag{5.41}$$

$X(0)$ is the initial value of X, taken to be independent of $\{B(t)\}$.

 (i) Verify that X defined by Eq. (5.41) satisfies Eq. (5.40). (Note: we have not done any stochastic differentiation but this is a case where we can get by with ordinary differentiation and obtain same results. Thus, apply the ordinary differentiation rules to Eq. (5.41), with respect to t.)
(ii) Using Eq. (5.41), and assuming $X(0)$ constant, compute the mean and covariances of $X(t)$; i.e. $E(X(t))$ and $Cov(X(s), X(t))$, $s \geq t \geq 0$. Afterwards, note that for t large, $E(X(t)) \approx m$, and observe that the OU process is mean reverting. The mean reversion of X to the level m can be inferred from Eq. (5.40): if $X(t) > m$, then the coefficient of the dt term (the drift of the process) is nega-

---

[4] Answer: 37 %. The method `EuropeanOptionImpliedVolatility` in the R package `RQuantLib` might be of help.

[5] Uhlenbeck, G. E. and Ornstein, L. S. (1930) On the Theory of the Brownian Motion, *Phys. Rev.*, **36**, 823–841.

tive, so X will tend to move downwards and approach m; whereas if $X(t) < m$, the contrary occurs.

(iii) Consider the homogeneous part or zero-mean version of OU process, that is, $X(t) = \sigma e^{-\gamma t} \int_0^t e^{\gamma s} dB(s)$. Show that if this zero-mean OU process is sampled at equally spaced times $\{i\tau : i = 0, 1, 2, \ldots, n\}, \tau > 0$, then the series $X_i = X(i\tau)$ obeys an autoregressive model of order 1, AR(1). Hence, we can consider the OU process as a continuous time interpolation of an AR(1) process. (Hint: Develop the expression $X_{i+1} = \sigma \int_0^{(i+1)\tau} e^{-\gamma((i+1)\tau - s)} dB(s)$.)

For further applications and extensions of the Ornstein-Uhlenbeck process see Maller et al. (2009).

Chapter 6
Trade on Pattern Mining or Value Estimation

The two most popular approaches to investment, although considered as opposite paradigms of financial engineering, are Technical Analysis and Fundamental Analysis. There is a long debate on whether one approach is more mathematically sound than the other, but we shall not be concerned with this dispute. The fact is that both type of analysis are largely used in the finance industry and produce a vast amount of research in numerical methods and computer algorithms, in an attempt to systematize the forecasting techniques proposed by either point of view to investment. What is interesting, from a computational perspective, is that Technical Analysis relies heavily on the recognition of patterns formed by the asset's price history and hence provides grounds for research in sequence labeling, pattern recognition and related data mining techniques. On the other hand, in Fundamental Analysis the main concern is a detailed examination of a company's financial statements to valuate its market shares with respect to its business potentials; hence, here too there is room for data mining applications in information retrieval and data classification. Last but not least, in both fields, there is room for machine learning methods to automatically discover rules of investment, or building artificial intelligent systems for trading.

This chapter deals with both forms of financial analysis, technical and fundamental, their basic principles, methods, and usage in automatic trading systems and forecasting. Also, in between, a mathematical foundation for Technical Analysis is presented.

6.1 Technical Analysis

A definition from a classic in Technical Analysis:

> Technical Analysis is the science of recording, usually in graphic form, the actual history of trading (price changes, volume of transactions, etc.) in a certain stock or in the averages and then deducing from that pictured history the probable future trend (Edwards and Magee 1966).

A. Arratia, *Computational Finance,* Atlantis Studies in Computational Finance and Financial Engineering 1, DOI: 10.2991/978-94-6239-070-6_6, © Atlantis Press and the authors 2014

This definition conveys the basic credo of technical analysts, which is that all information about a stock is contained in the curve drawn by the price with respect to time (the "pictured history"), and forecasting of its future behavior should be based on interpreting recurrent patterns found on that graph, and the general practice of chartism. After reviewing the basic principles and tools (type of charts, graphic marks, trading signals, etc.) we will be concerned with mathematically formalizing the most popular rules of trading of Technical Analysis. This is important for devising trading algorithms. Once we have well-defined rules for trading, our next concern is to determine if these rules are useful for forecasting and to what extend outperform other mathematical models, such as those seen in Chap. 4.

6.1.1 Dow's Theory and Technical Analysis Basic Principles

Charles Henry Dow is recognized as the forefather of Technical Analysis, a title he may not had the intention to bore as he was not apparently trying to forecast stock's prices in his writings as a financial journalist. Dow founded, together with Edward Jones, *The Wall Street Journal*, and throughout the years 1900–1902 he wrote a series of articles in that journal which contain a set of observations on market behavior based primarily on the varying trends of the prices of stocks. It was William Peter Hamilton who after Dow's death in 1902 succeeded to the editorial of *The Wall Street Journal*, and from that tribune, and in a later book (Hamilton 1922), presented Dow's stocks market principles as investment rules and applied them over the period 1902–1929. It is fair then to say that Hamilton was indeed the first chartist in action. The axioms or principles that comprise what is known today as Dow's Theory were laid down by Hamilton, and by a later prominent follower of Dow's investment principles and student of Hamilton, Rhea (1932). Dow's Theory can be summarized in the following principles (Edwards and Magee 1966):

(1) The averages discount everything.
(2) The market has three trends: primary or major, secondary, and minor. Subsequently, primary trends have three phases: (i) accumulation phase; (ii) public participation phase; (iii) distribution phase.
(3) Stock market averages must confirm each other.
(4) The volume confirms the trend.
(5) Trends exist until a definitive reversal signal appears.

By the averages, Dow was referring to the average indices that he composed together with E. Jones: the *Industrial*, comprised of twelve blue-chip stocks, and the *Rail*, comprised of twenty railroad stocks. Since Dow believe that stock's prices reflect the flows of supply and demand, then the market averages discount and reflect everything that is known on stocks by market participants. The primary trends are the broad long-lasting upward movements, referred as Bull market, or the long periods of downward movements known as Bear market. The three phases dividing these primary trends are a reflection of public sentiment towards the market and their course of actions, bearing a lot of subjectivity. For example, for Bull markets Rhea writes (op. cit.): "first [...] a revival of confidence in the future of business [...] second is the response of stock prices to the known improvement in corporation earnings, and the third is the period when speculation is rampant and inflation apparent." From these observations technical analysts infer the following investors actions: in the first phase only informed investors trade; in the second phase the common investors enter the game and speculation begins; and in the third and last phase the informed investors perceived the approach of a peak on prices and shift their investments, leaving the uninformed common investor alone in the market.

Secondary trends are corrective trends, being of opposite direction to the major trends and of shorter duration in time; and the minor trends are much briefer fluctuations within secondary trends, all-in-all conforming to a sort of fractal construction. This minute division of the major trend may seem as an attempt to provide some criteria for removing the chaff from the wheat, the unimportant movements from the major and important one, to prevent investors from making false moves, as jumping out of the market at the wrong time due to a short swing in the trend. However, the recognition of short periods of trend reversal involve many vague proportional comparisons to previous similar trends or past price information, not being any of these informal arguments convincing enough for dispelling any doubt that the investors are possibly confronting a period of pause of the major trend that were recently riding, or are about to completely lose their investment. Thus, the algorithmic recognition of these different trends is an important subject of research.

Among those signals that the specialists perceive as change in the market mood is the divergence of the trends of two different market sectors, as Dow's two average indices, the Industrials and the Rails, and this is what Dow meant by the stock averages confirming each other. Likewise, Dow and later Hamilton (and succeeding technical analysts) have observed a linear relation between the trend of prices and the trend of transaction volume; e.g. an up-trend with low volume is doom to reversal. Later in the chapter we give a table representing the relation between price and volume as established by technical analysts. Finally, the market movements perceived as cyclical will then repeat certain patterns in the price histories of all stocks. The recording of those apparently recurring patterns and the observed correlation of those patterns with the market trend (aka chart analysis) is the bread and butter of Technical Analysis.

The Dow, Hamilton and Rhea practice of investment evolved subsequently into a set of basic guiding principles for all technical investors, which can be summarized as the following facts:

(1) The market price of a stock fully reflects its "real" value (i.e. only supply and demand determines market value).
(2) Prices move in trends.
(3) A trend is sustained through time until the balance of supply and demand changes.
(4) The changes in supply and demand are reflected by meaningful patterns on the charts.
(5) History repeats itself (and so do chart patterns!).

We can appreciate that these investment principles are completely focus on price movements and volume figures of the stock, disregarding any other economic variable pertaining to the company, such as its income or liabilities. Having the focus primarily set on the history of prices, the bulk of the foundations of Dow's Theory and subsequent methods of Technical Analysis revolve around qualitative descriptions of the elements of the market that influence the price and volume of stocks. It is under these foundational beliefs that the principles of Technical Analysis should be understood, and applied.

6.1.2 Charts, Support and Resistance Levels, and Trends

Technical analysts view price histories in three graphic forms, each conveying more or less information about the behavior of the price in a given period of time. The types of charts representation for a price time series are:

Line chart: This is the simplest plot, representing typically the closing price of the security with respect to time. It provides the quickest and easy view of price behavior. Figure 6.1 shows a daily line chart for a stock.

Bars chart: This type of chart give more information than line charts on the evolution of the price. Each bar in a bars chart describes a single session, or period of time, and is depicted as a vertical line where the top indicates the highest value of the price in the period of time, the bottom indicates the lowest value, the closing price is marked on the right side of the bar, and the opening price is marked on the left side of the bar. Figure 6.2 depicts a daily bars chart for a stock.

Candlesticks chart: The candlestick representation of stock prices has its origin in the 1600s in Japan. It is part of a charting technique applied originally to set future prices in the rice market, and then their use extended to all Japanese financial markets as they developed. It was introduced in the Western World around 1990 by a professional technical analysts working at Merrill Lynch, Steve Nison, who wrote a complete manual for understanding and using these patterns (Nison 1991). A candlestick includes information on the high, low, opening and closing price in a period of time (most frequently daily, but also weekly or monthly) plus the relation of opening versus closing. This summaries are depicted in a candle with wicks (or "shadows") at both ends, so that the top (bottom) of the upper (lower) wick marks the high (low), and the real body of the candle shows the opening and closing prices, together with information on the difference between these two. If the closing price is higher than the opening price, then the top (bottom) of the candle body marks the closing (opening) price, and the body is white (or green, for a more colorful representation). If the closing price is lower than the opening price, then the top (bottom) of the candle body marks the opening (closing) price, and the body is black (or red, to best pair with the green). Although their similarities with bars charts, candlesticks have become more popular among technical analysts for their employment in charting trading rules, some of which

Oct 14, 2010 : ▬TEF.MC O: 19.40 H: 19.63 L: 19.35 C: 19.45

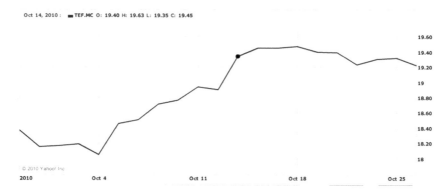

Fig. 6.1 A *line chart* for a stock

Sep 28, 2010 : ▬TEF.MC O: 18.25 H: 18.45 L: 18.09 C: 18.37

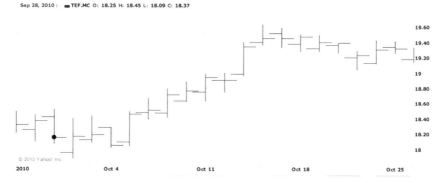

Fig. 6.2 A *bars chart* for a stock

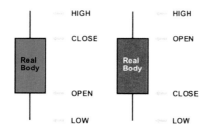

Fig. 6.3 Candlesticks

we discuss in Sect. 6.1.3. Figure 6.3 shows the two forms of candlesticks, and Fig. 6.4 shows a daily candlestick chart for a stock.

Support and resistance levels. On either graphic representation of the series of price, technical analysts draw horizontal lines that mark the balance of forces between sellers and buyers. A *support level* indicates the price where the majority of investors would be buyers (i.e., investors support the price), whereas a *resistance level* indi-

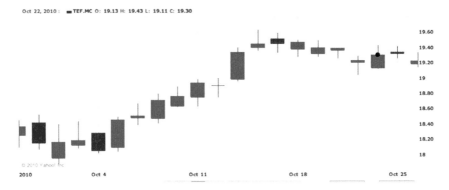

Fig. 6.4 A *candlesticks chart* for a stock

Fig. 6.5 Support and resistance levels

cates the price where the majority of investors would be sellers (i.e., investors are reluctant, or resist, to buy at that price). For a more formal definition, consider that a support (resistance) level of a price series is the lower (upper) end of its recent trading range. We will also see these level lines as special cases of the trend lines to be defined below. Figure 6.5 shows support and resistance levels for a stock's price series. Observe that once the price breaks a resistance level, this becomes a support level, and vice versa.

Trend lines. Let $\{P_t : t = 0, 1, 2, \ldots, \tau\}$ be a stock's price series observed in a time period τ. A trend line can be formally defined by considering two local maxima (or minima) points and tracing the line through those points. Let $P_{t_a} = P_a$ and $P_{t_b} = P_b$

be the highest local maxima (or lowest local minima), found at times t_a and t_b in the period $(0, \tau]$, so that $t_a < t_b < \tau$.

Then the straight line through the points (t_a, P_a) and (t_b, P_b):

$$T(t) = \frac{(P_b - P_a)}{(t_b - t_a)}(t) + \frac{(P_a t_b - P_b t_a)}{(t_b - t_a)} \tag{6.1}$$

defines a trend line. It is an uptrend if, and only if, $P_a < P_b$, or a downtrend if $P_a > P_b$. Note that support or resistance lines can be also considered as trend lines, where in Eq. (6.1) $P_a = P_b$ and thus, $T(t) = P_a$.

6.1.3 Technical Trading Rules

The core of a technical trading rule is a mathematical indicator function that takes as input an asset's price (or price plus other information such as volume), and produces a value use to anticipate a change of price direction.[1] We can classify the various forms of price history-based technical trading rules in three broad groups: moving averages, level crossings and pattern recognition rules. We shall give below succinct descriptions of each one of these groups of rules, and some examples. Then, in Sect. 6.1.4, we study some probability conditions to assess which of these groups of rules are well-defined and useful for predicting successfully exit and entry points. The point that this formalism illustrates is that technical trading rules can be mathematically defined and, consequently, algorithms can be devised for generating the buy and sell signals emitted by these rules. Our list of technical trading rules are far from exhaustive, and should be taken as an initiative to research, that can be expanded with the vast existing literature on this subject, of which we mention a few at the end of the chapter.

Moving averages. The rules in this group consist on considering two or more moving averages of different lengths on the observed values of the price's history of a stock, $\{P_t : t \geq 0\}$, and the intersections of these averages are interpreted as signals for buying or selling. A moving average over n periods of time and up to time t can be defined as

$$MA(n)_t = \frac{1}{n} \sum_{i=0}^{n-1} P_{t-i} \tag{6.2}$$

Depending on the value of n, $MA(n)_t$ gives an indication of the short, mid, or long term tendency of the price series. Here is a classical representative of this group of rules.

Dual Moving Average Crossover: For each time t, consider: (1) a short term Moving Average over n days up to time t, $MA(n)_t = \frac{1}{n} \sum_{i=0}^{n-1} C_{t-i}$, where C_t is the close

---

[1] For the sake of clarity we discuss technical rules in reference only to price history, but most of these rules can apply also to volume.

price at time t; and (2) a long term Moving Average over m days up to time t (with $n < m$), $MA(m)_t = \frac{1}{m} \sum_{i=0}^{m-1} C_{t-i}$. Consider the process

$$X_t = MA(n)_t - MA(m)_t \qquad (6.3)$$

which we begin observing at $t = t_0$, and through a sequence of times $t_0 < t_1 < \cdots < t_i < \cdots$. Then, as soon as for some $i > 0$, X_{t_i} changes sign, we trade according to the following scheme:

(1) If $X_{t_{i-1}} < 0$ and $X_{t_i} > 0$, go long at O_{t_i+1}, where O_{t_i+1} is the open price at time $t_i + 1$ (the following session).
(2) If $X_{t_{i-1}} > 0$ and $X_{t_i} < 0$, go short at O_{t_i+1}.

The general consensus of users of this rule is to take $n = 20$ and $m = 50$. This rule is often extended with a third $MA(r)_t$ of very short term (e.g. $r = 9$ or $r = 12$) and setting up trading rules according to the simultaneous crossings of the three moving averages. We give an example of the use of this rule in R Example 6.1.

Level crossing. The rules in this group consist on considering a boundary (upper or lower) to the price observed through a period of time, and once the price crosses this limit a buy or sell signal is issued. Boundaries can be straight lines as uptrends, downtrends (as defined by Eq. (6.1)), support or resistance levels, or some curve given as a function of price. A common extension of level crossing rules is to jointly consider two boundaries of values enveloping a portion of the price, from the current time back to a specific point in the past: one of these limiting series of values runs above, the other below the price, drawing a band or window around the price values; crossing either of the boundary lines is taken as a signal to buy or sell. The general principles governing the trading signals are: if the crossing is through the upper boundary this means a strengthening in the upward trend, and hence one should go long on the asset; if the crossing is through the lower boundary then a continuation of the down trend will follow and one should go short on the asset. One simple, yet popular, member of this group of rules, the *trend lines crossing*, can be formalized as follows.

Trend lines crossing: Let $\{P_t : t > t_0\}$ be a series of values of an asset's price, which we began observing after a turning point occurred at $t = t_0$. We draw an upper trend, if the desire is to buy the asset (looking for an entry point), or if we have bought the asset then draw a lower trend to alert us of an exit point. Suppose it is the latter case. For each time t consider $T(t)$, the lower trend line drawn through two local minima attained at $t_b > t_a > t_0$ and defined by Eq. (6.1). Along the interval of time $[t_a, t_b]$, $T(t) \le P_t$, but the trend line extends further. As soon as the price P_t crosses $T(t)$ we sell. Formally, consider the process

$$X_t = P_t - T(t) \qquad (6.4)$$

which we monitor through times $t_0 < t_1 < t_2 < \cdots$. At the first $i > 0$ such that

$$X_{t_{i-1}} > 0 \text{ and } X_{t_i} < 0 \tag{6.5}$$

go short in O_{t_i+1}. For the case of buying as our objective, consider $T(t)$ as trend line passing through two local maxima, and the process $X_t = T(t) - P_t$. Then, for $i > 0$ such that Eq. (6.5) holds, go long in O_{t_i+1}.

Technical analysts assert that when either of these crossings occur, then prices will continue beyond the level line for a while; hence, new trend lines should be recomputed after a turning point located near t_{i-1}.

On the subfamily of combined upper and lower crossings, or *trading bands*, we have the following simple rule.

High–Low Envelopes: For a given $n > 0$ and for each time t, consider a window of local extrema around the price P_t, build as follows: in the time interval $[t - n, t]$ take the highest high, $HH(t) = \max(H_{t-n}, \ldots, H_t)$, where H_t is the High at time t; and the lowest low, $LL(t) = \min(L_{t-n}, \ldots, L_t)$, where L_t is the Low at time t. Then:

- Go long when price at time $t + 1$, $P_{t+1} > HH(t)$.
- Go short when $P_{t+1} < LL(t)$.

A variation of this rule is to consider two moving averages in place of the highest high, $HH(t)$, and the lowest low, $LL(t)$. *Bollinger bands*, a well-known rule among professional traders, go a notch further by taking as the window two standard deviations around the 20-day moving average $MA(20)_t$, at each time t.

For all these envelopes or channels, the theory states that movement outside the bands should continue, and this supports the above trading decisions. In Note 6.3.8 at the end of the chapter, we present another interesting and much used technical rule of this family of envelope crossings, the Relative Strength Index (RSI).

R Example 6.1 The quantmod package has many functions for calculating and drawing different technical indicators. The R commands below retrieve the historical quotes of Apple (AAPL) from Yahoo, and charts the price history from June, 2008 to April, 2009, using a candlestick representation for each daily tick, with positive candles in green (up.col='green') and negative candles in blue (dn.col='blue'); two moving averages ($MA(50)_t$ in blue and $MA(10)_t$ in black), Bollinger bands on a 20 day moving average with bands at 2 standard deviations, and RSI indicator are also drawn.

```
> getSymbols("AAPL")
> chartSeries(AAPL,subset='2008-06::2009-04',
+ theme=chartTheme('white',up.col='green',dn.col='red'),
+ TA=c(addBBands(n=20,sd=2,),addSMA(n=50,col="blue"),
+ addSMA(n=10,col="black"),addRSI(n=14)))
```

The resulting chart can be seen in Fig. 6.6. Note that in consistency with the dual moving average crossover rule, as the $MA(10)_t$ crosses down the $MA(50)_t$, for $t > 01/09/2008$, the price drops, while in March 2009 the crossing up of $MA(10)_t$ above $MA(50)_t$ signals a price increase.   □

Fig. 6.6 AAPL *chart* with technical indicators: $MA(10)_t$, $MA(50)_t$, Bollinger, RSI

Pattern recognition. The rules in this group are based on recognizing some pattern drawn by a portion of the price curve ending at the current time, and issuing a buy or sell signal. Most of these patterns can be formally defined using local maxima and minima, a point that we shall illustrate through some meaningful examples. We consider three very popular *western-world* patterns, namely *head and shoulders*, *triangles*, and *gaps*, and two *japanese candlesticks* patterns, the *hammer* and the *morning star*.

Head and Shoulders. A Head-and-Shoulders (HS) price pattern have been regarded among technical analysts as one of the "most important and reliable reversal pattern" (see Edwards and Magee (1966, Chap. VI) and (Achelis 2001, p. 246)). A HS pattern occurs when the price series, being in an uptrend, first describes a stair-step by marking a local maximum E_1, then a local minimum $E_2 < E_1$, followed later by a higher local maximum $E_3 > E_1$; then at this last point the trend reverses falling through a similar stair-step shape with new local minimum E_4, and finalizing at a local maximum $E_5 < E_3$. The resulting figure consists of a left-shoulder, its top marked by E_1, a head, with its top marked by E_3, and a right-shoulder, marked by E_5. Let $t_1^* < t_2^* < t_3^* < t_4^* < t_5^*$ be the time instants (e.g. dates) at which the values E_1, E_2, E_3, E_4 and E_5 occurred. The straight line going through (t_2^*, E_2) and (t_4^*, E_4) is called the *neck line*. The trading rule for this pattern advises to sell (or go short) as soon as the price falls below the neck line. Figure 6.7 presents an example of the HS pattern. Now, in order to completely formalize this pattern (and build an algorithm to recognize it), we need to impose two important restrictions on the extreme values $E_1, \ldots, E_5$.

Fig. 6.7 A Head-and-shoulders pattern

(1) *Bandwidth*: Practitioners of the HS trading rule, advise that the values of E_1 and E_5 (and of E_2 and E_4) should be similar. However, it is highly unlikely that each pair of extrema would be equal, so we consider slightly more relaxed conditions. We ask that

E_1 and E_2 (also E_2 and E_4) be within a b factor of their averages.

Here b should be small; the recommended value being $b = 0.015$ (in percentage terms this corresponds to 1.5%).

(2) *Time-window length*: We should be precise about the width of the shoulders and the head, i.e., the number of dates (or sessions) that each part of the pattern expands throughout. It is sufficient to indicate the total width of all the HS pattern. Thus we must fix before hand the length T of the time window where the pattern should be located, starting at current date t and back T time periods. We are asking then that the times $t_1^*, t_2^*, t_3^*, t_4^*, t_5^*$ at which the extremes E_1, E_2, E_3, E_4, E_5 occur, must be within the T dates preceding the current date t:

$$\{t_1^*, \ldots, t_5^*\} \subset \{t_0, t_1, \ldots, t_T\} \tag{6.6}$$

where $t_0 < t_1 < \cdots < t_T < t$.

With the above provisos we can formally describe Head-and-Shoulders.

HS: Let t be the current time. Let $T > 0$ be the time-window length, and b the bandwidth (for daily observations of price, recommended values are: $T = 35$ and $b = 0.015$). In the time spanned by the sequence, $t_0 < t_1 < \cdots < t_T < t$, find extrema E_1, E_2, E_3, E_4 and E_5, occurring at times $t_1^* < t_2^* < t_3^* < t_4^* < t_5^*$, verifying Eq. (6.6) and such that

Fig. 6.8 A triangle formation by gold prices (*source* StockCharts.com)

- E_1 is a local maximum.
- $E_3 > E_1$ and $E_3 > E_5$.
- For $E = (E_1 + E_5)/2$ and $E' = (E_2 + E_4)/2$, we must have $|E - E_1| < b \cdot E$, $|E - E_5| < b \cdot E$, $|E' - E_2| < b \cdot E'$ and $|E' - E_4| < b \cdot E'$

(Observe that a local maximum, with value P_s occurring at time s, can be determined by checking if $P_{s-1} < P_s$ and $P_s > P_{s+1}$.)

Triangles. Figure 6.8 shows two trend lines converging to a common point describing what technicians know as a triangle. This pattern can be generated by considering a fixed number $k > 1$ of consecutive local maxima appearing in descending order, and k consecutive local minima, in the same time span, appearing in ascending order. Let $M_1 > M_2 > \cdots > M_k$ be local maxima appearing at times $t_1^M < t_2^M < \cdots < t_k^M$, and $m_1 < m_2 < \cdots < m_k$ be local minima appearing at times $t_1^m < t_2^m < \cdots < t_k^m$. The rule for this triangle pattern is that for $t > \max(t_k^m, t_k^M)$, buy (go long) if $P_t > M_k$, or sell (go short) if $P_t < m_k$.

Gaps. An up-gap occurs when the stock's lowest price of the date is above the highest of previous date; i.e., $L_t > H_{t-1}$. A down-gap is when $H_t < L_{t-1}$. Some technical investors buy (sell) after observing one up-gap (down-gap), others prefer to see two or three gaps of the same kind before trading. The fact is that one must fix a $k \in \{1, 2, 3\}$ (k must be small), and the rule for gaps says:

- After observing k up-gaps buy (or go long).
- After observing k down-gaps sell (or go short).

In Fig. 6.6 one can observe a down-gap near Oct. 1, 2008.

Candlesticks' charts. These chart patterns are composed of one or up to five consecutive candlesticks, corresponding to a sequence of days or sessions observed, where by comparing the sizes of their real bodies, the sizes of their upper and lower shadows, the colors, and their High, Low, Close and Open values, and observing certain relations among these parameters and the trend of past days, one obtains either a confirmation of the current trend or a signal of an imminent turning point. To algebraically express candlesticks patterns we need to mathematically formalize their parameters and set of relations. Let us fix some notation first. To simplify the exposition we consider the price observed in a daily basis. Let C_t, O_t, L_t, H_t be the Close, Open, Low, and High of the price at day t.

- The color of a candlestick real body is white (w) or black (b), and if the color is irrelevant we use bw. We use $col(t)$ to indicate the color of the candlestick at day t. The real body is white (resp. black) if, and only if, $C_t > O_t$ (resp. $C_t < O_t$).
- The size of the real body is calculated as $|C_t - O_t|$. However, technical analysts verbally classify the size into four categories: "doji" (d), "small" (s), "medium" (m), and "tall" (t). These qualitative descriptions of size can be given a numerical form as quartiles of the set of size of candlesticks of the k most recent days (e.g. $k = 22$) verifying $d < s < m < t$, and such that a candlestick of size d has $C_t \approx O_t$ (i.e. the Close and Open are almost equal, and so the body looks almost flat); the remaining sizes $> d$ are assigned to the remaining three quantiles: s for the first 25 % of the sample, m for the 50 %, and t for the 75 %. We use $size(t)$ to indicate the size of the real body at day t.
- The size of the upper (resp. lower) shadow is given by $H_t - C_t$ (resp. $O_t - L_t$) if $col(t) = w$; otherwise is $H_t - O_t$ (resp. $C_t - L_t$). We use $us(t)$ (resp. $ls(t)$) for the size of the upper (resp. lower) shadow at day t, and it should be clear from the context which equation we should use to compute its value.
- The trend of the price for the past short term, say 12 days, previous to the candlesticks pattern, can be described by Eq. (6.1) and its slope be computed from that equation.

With these general provisos, we can now define some particular patterns.

Hammer. This pattern consist of a single candlestick, which should be observed at the current day t, although it needs a confirmation by observing the next day candlestick. The previous trend is downward (formalize this using the instructions above). The appearance of the hammer is an indication of a possible reversal of trend. Verbally, the hammer is defined as a candle where the color of its real body can be either black or white, with its lower shadow at least twice the size of its real body and its upper shadow at most one quarter of its real body.[2] Next day confirmation must be a candlestick with closing price greater than the hammer's Close. This description can be formalized by the following equations:

$$col(t) = bw, \quad ls(t) > 2 \cdot size(t), \quad us(t) \leq 0.25 \cdot size(t),$$

$$C_{t+1} \geq C_t.$$

---

[2] see Nison (1991, pp. 28–29) for an even more mathematically vague description.

Morning Star. This pattern consist of three candlesticks observed at times $t - 2$, $t - 1$ and t. The previous trend is downward, and needs no confirmation; so it is a sure signal of a coming turning point, a possible change to an uptrend. Here is the formal definition of the morning star.

$$col(t - 2) = b, \ size(t - 2) = t, \quad col(t - 1) = bw, \ size(t - 2) = s,$$

$$col(t) = w, \ size(t) = t$$

$$\text{and} \quad C_{t-2} > \max(O_{t-1}, \ C_{t-1}), \quad O_t > \max(O_{t-1}, \ C_{t-1}),$$

$$C_t > C_{t-2}.$$

(As an exercise the reader should write this formal description verbally.)

We believe that with this formal framework and the given examples, the reader would be able to program his own candlesticks patterns recognizer. We had done such a programming exercise in Arratia (2010).

6.1.4 A Mathematical Foundation for Technical Analysis

We present in this section some formal criteria to asses if a technical trading rule is mathematically sound and the extent to which these rules, although well-defined, are better forecasters than, say, some of the time series models presented in Chap. 4. These criteria were developed by Salih Neftci in Neftci (1991) and are based on the theory of optimal stopping of stochastic processes (see Shiryaev (2007) for the necessary background). The key concept is the Markov time random variable.

Definition 6.1 (*Markov time*) Let $\{X_t : t \geq 0\}$ be a stochastic process. A non-negative integer-valued random variable τ is a *Markov time* for $\{X_t\}$ if for every $t \geq 0$, the event $\{\tau = t\}$ depends only on $\{X_0, X_1, \ldots, X_t\}$, and does not depend on $\{X_{t+s} : s \geq 1\}$.

Thus, Markov times are random time periods independent of the future, as their values are determined only by current information.[3]

Example 6.1 Let τ_a be the first time that a stochastic process $\{X_t : t \geq 0\}$ hits or passes the upper boundary $y(t) = a$, for some constant $a > X_0$. For example, X_t

---

[3] A word should be said about the rather loose statement "the event $\{\tau = t\}$ depends only on $\{X_0, X_1, \ldots, X_t\}$". By this we mean that to determine $\{\tau = t\}$ it is enough to know the past values X_s ($s \leq t$), and anything that can be said about them, in a definite way; that is, all definite information about X_s, for $s \leq t$. This underlying notion of "current information set relative to the process" can be made mathematically precise using the notion of *sigma algebras* (as it is done in Shiryaev (2007)). But we feel that for our present use of Markov times we can get by with the given informal version, and spare the reader of the mathematical rigmarole.

represents the price of a stock and $y(t) = a$ a resistance level. Then τ_a is a Markov time since, for any time $t \geq 0$,

$$\tau_a = t \iff X_t = a \text{ and } X_{t-1} \neq a, \ldots, X_1 \neq a \qquad (6.7)$$

and it is irrelevant to know the sign of $X_{t+s} - a$, for $s \geq 1$, to decide the event $\{\tau_a = t\}$.

□

Example 6.2 Let τ_b be the beginning of a Bull market; that is, τ_b is that random point in time where a market index (e.g. DJIA or S&P 500) starts on an uptrend. This is not a Markov time, since in order to know $\tau_b = t$ one needs to know future events of any process X_t indicating the possible Bull trend. For example, that a trend line drawn starting at $\tau_b = t$ extends with positive slope for a long period. To compute this slope we need to know a local extrema at some point $s > \tau_b$.   □

It is then reasonable to say that a technical rule is well-defined if the sequence of buy or sell signals that generate is a Markov time, since this implies that the rule is using only current information. If a technical rule fails to be a Markov time, then it means that it is using future information to determine some (or all) of its signals, an unrealistic conclusion.

Let us elaborate further on a theoretical framework for these sequences of buy or sell signals, produced by technical rules, to see how we can sort out their possible Markovian property. When we attempt to automatize a technical rule, we produce a program that on input a price series and other information (all together conforming a random process depending on time) outputs a sequence of signals at certain intervals of times. These are the occurring times for a technical rule.

Definition 6.2 Given a real-valued random process X_t, and an interval of real numbers A, we define the times of X_t occurring at A, or simply the *occurring times* if context is clear, as the sequence of positive integers

$$\tau_0 = 0;$$
$$\tau_{i+1} = \min\{t : X_t \in A, t > \tau_i\}, \text{ for } i \geq 0. \qquad □$$

Thus, to show that a technical rule is well-defined, it is sufficient to show that its occurring times are Markov times. In fact, we prove below that it would be sufficient to show that the event of X_t occurring at A (i.e. $\{X_t \in A\}$) can be determined using only current information.

Proposition 6.1 Let $\{X_t\}$ be a real-valued random process, and A an interval of real numbers. If, for all $t \geq 0$, the event $\{X_t \in A\}$ depends on $\{X_s : 0 \leq s \leq t\} \cap A$ and not on $\{X_{t+s} : s \geq 1\}$, then the occurring times $\tau_0, \tau_1, \ldots$, of X_t at A are Markov times.[4]

---

[4] Recall footnote 3. Hence, to depend on $\{X_s : 0 \leq s \leq t\} \cap A$ means all posible definitive information about X_s, for $s \leq t$, and A.

Proof Proceed by induction. Assume that for $i = 0, \ldots, n$, τ_i is a Markov time. Fix $t > 0$. Then

$$\{\tau_{n+1} = t\} = \{X_t \in A\} \cup \{\tau_n < t\}$$
$$= \{X_t \in A\} \cup \{\tau_n \leq t - 1\}$$
$$= \{X_t \in A\} \cup \bigcup_{s=0}^{t-1} \{\tau_n = s\}.$$

The event $\{X_t \in A\}$ depends only on X_t and A. By induction, each event $\{\tau_n = s\}$ depends only on A and $\{X_0, \ldots, X_s\}$, for $s = 0, \ldots, t - 1$. The union of these sets of information is A and $\{X_0, \ldots, X_{t-1}\}$. Therefore, $\{\tau_{n+1} = t\}$ depends only on A and $\{X_0, \cdots, X_t\}$.   □

As an example note that in Example 6.1 τ_a can be written as

$$\tau_a = \min\{t : t \geq 0,\ X_t \in [a, \infty)\}$$

and the event $X_t \in [a, \infty)$ can be determined by comparing the value of X_t with a, and no extra information ahead of time t is needed; hence τ_a, which is the occurring time τ_1, is Markov time. Observe that without making any further assumptions about the process $\{X_t\}$ in Example 6.1, we can not guarantee that it will eventually hit the line $y(t) = a$, since the best we can say about the probability of the event $\{\tau_a = t\}$ occurring, for some t, is that is less than 1 (and so it could be 0). Then, what good is a technical rule, like "crossing a resistance line", that although being a Markov time (hence well-defined), it may never be realized? We need to impose a further test for the adequacy of technical rules.

Definition 6.3 *(Stopping time)* A Markov time τ is a *stopping time* if it has probability one of being finite, i.e., $\mathbb{P}(\tau < \infty) = 1$. Thus, a stopping time is a Markov time that almost surely has a finite sample value.

Thus, a *well-defined* technical rule is one whose occurring times are Markov times, and it is *adequate* if further the occurring times are stopping times. We shall first sort out which of the technical rules presented in Sect. 6.1.3 are Markov times or not; afterwards we shall try to characterize the adequacy of some of these rules.

The occurring times of technical trading rules and the Markov property.

1. **Moving averages rule**. The process under consideration, given by Eq. (6.3), is

$$X_t = MA(n)_t - MA(m)_t$$

   and a buy or sell signal is generated when the process changes sign from one time instant to the next. The change of sign is captured by the test $X_t \cdot X_{t-1} < 0$. Thus, the occurring times for the moving average rule can be defined as

$$\tau_0 = 0, \ \tau_i = \min\{t : X_t \cdot X_{t-1} < 0, \ t > \tau_{i-1}\}, \ i > 0. \qquad (6.8)$$

The event $X_t \cdot X_{t-1} < 0$ is determined from computing $MA(n)_t$ and $MA(m)_t$, which uses $\{X_t, X_{t-1}, \cdots, X_{t-n}\}$ as information set. By Prop. 6.1, the occurring times of moving averages rule are Markov times.

2. **Level crossings**. Let us analyze first the trend lines crossing rule for taking short positions; the case for long positions being similar. The process considered is $X_t = P_t - T(t)$, and the signal is given by Eq. (6.5). The occurring times for this rule are also given by

$$\tau_i = \min\{t : X_t \cdot X_{t-1} < 0, \ t > \tau_{i-1}\}$$

However, we argue that this sequence of times is not Markov. The problem is that whenever a signal occurs at time τ_{i-1}, the trend $T(t)$ have to be recomputed with local minima attained at times t_a, t_b in $(\tau_{i-1}, t]$. We have not specified a recipe for determining if the values P_{t_a} and P_{t_b} are local minima, since the "correct" application of this rule by experts considers, more often than not, the lowest of the low possible values found in the range of $(\tau_{i-1}, t]$, and to find these low values one probably needs to make comparisons with values attained beyond t. A deterministic solution like "consider P_s to be a local minimum if $P_{s-1} > P_s$ and $P_s < P_{s+1}$, and then take the two consecutive local minima thus formed", is likely to produce a non desirable trend line for this rule.

Another possible definite algorithm for computing $T(t)$, using only past information, could be: "take 8 time instants previous to τ_{i-1} and fit a straight line by least squares method". The problem with this fix is that technical analysts believe that as soon as a signal is produced, at the time τ_{i-1}, the preceding trend is to suffer a reversal, then the proposed remedy will have worse effects than the originally ill-defined rule. Therefore, it seems inescapable to use some knowledge of the future to trace a technically correct trend line $T(t)$, and hence, in general the occurring times for this rule are not Markov times.

On the contrary, the high-low envelope rule, and in general the family of envelope crossings, have occurring times that are Markov times, since one can see that the envelopes are computed using information previous to current time t (e.g. the highest high and lowest low of n previous days, or two standard deviations around the past 20-day moving average).

3. **Patterns**. For the head-and-shoulders pattern, the process considered is

$$X_t = P_t - N(t)$$

where P_t is the price and $N(t)$ is the neckline passing through the points (t_2^*, E_2) and (t_4^*, E_4). The rule is to sell once $X_t < 0$, for $t > t_T$. The one time occurrence for this rule can be defined as

$$\tau_1 = \min\{t : X_t < 0, \ t > t_T\}$$

This time τ_1 is a Markov time, since we apply a very simple test for determining local extrema, which consists of comparisons with immediate price values P_{s-1} and P_{s+1}, for each $s < t_T$; in consequence, E_2 and E_4 (and $N(t)$) are computed with information known up to time t_T (see Eq. (6.6)). However, we remark that the neckline obtained by the algorithm HS does not in general coincides with the neckline that technical analysts will draw by observing the price curve. We have not imposed any relation among the lines drawn by $\{(t_2^*, E_2), (t_4^*, E_4)\}$ and $\{(t_1^*, E_1), (t_5^*, E_5)\}$, whereas users of the head-and-shoulders rule feel more confident with it if these lines are parallel and considerably apart. Forcing this parallelism or spread among these lines implies to look at more values ahead to find better fit extrema. This turns the occurring time into a non Markov time.

Triangles and gaps generate occurring times that are Markov time (we leave to the reader to define their occurring times τ_i). For triangles we are looking at k past and consecutive local extrema, and for gaps we are looking at k jumps, defined by comparing the current High or Low with previous day Low or High. Nevertheless, observe that what makes the occurring times of these rules Markov is the a priori specification of the parameter k. For triangles this is a questionable issue, since it is not clear how many local minima and maxima have to be observed, in ascending and descending order respectively, to decide that a real triangle has formed. But if one starts adding requirements to realize some angle of incidence of the meeting lines, or similar qualities, then almost surely the occurring time will not be Markov time. For gaps there are no doubts, since almost every user is comfortable with applying $k = 1$.

Finally, candlesticks patterns do generate occurring times that are Markov times, provided the length of past observations used to compute the four quantiles describing size of real body, and the number of past days for computing the trend preceding the pattern, are previously fixed.

Conditions for some technical rules to generate occurring times that are stopping times. We analyze the possible characteristics that the random process should have for some of the technical rules to be stopping times. The following lemma is adapted from a general result in Breiman (1992, Prop. 6.38), and it is a key tool for showing that certain Markov times produced by strictly stationary processes are stopping times (in Problem 6.35 we give a hint for proving this important fact).

Lemma 6.1 *Let $\{Z_t : t \geq 0\}$ be a strictly stationary process, and A an interval of reals. If $\mathbb{P}(Z_t \in A, \text{ at least once}) = 1$, then the occurring times τ_i, $i \geq 0$, of Z_t at A, are stopping times.*   □

We also need Proposition 2.2 from Chap. 2, so the reader is advise to review that result. With all previous tools we can now show that stationarity and m-dependance are sufficient conditions for moving averages trading signals to be stopping times.

Theorem 6.1 *If $\{Y_t : t \geq 0\}$ is a strictly stationary and m-dependent process, then the Dual Moving Averages rule for $\{Y_t\}$ generates trading signals that are stopping times.*

Proof Let $MA(n)_t$, $MA(m)_t$ moving averages for Y_t given by Eq. (6.2). By Prop. 2.2, both compositions of processes $X_t = MA(n)_t - MA(m)_t$ and $Z_t = X_t \cdot X_{t-1}$ are stationary. By stationarity $E[Z_t] = 0$, hence $0 < \mathbb{P}(Z_t \geq 0) < 1$, for all t. We want to show that $\mathbb{P}(Z_t \in (-\infty, 0]$, at least once$) = 1$. Consider,

$$\mathbb{P}(Z_t \leq 0, \text{ at least once for } t \leq n) = 1 - \mathbb{P}(Z(0) > 0, Z(1) > 0, \ldots, Z(n) > 0)$$

Since Z_t is m-dependent, taking Z_t's sufficiently apart will be independent. Hence, there exist an integer $u > 0$, such that Z_t and Z_{t+u} are independent. Then, for large n,

$$\mathbb{P}(Z_0 > 0, Z_1 > 0, \ldots, Z_n > 0)$$
$$\leq \mathbb{P}(Z_0 > 0) \cdot \mathbb{P}(Z_u > 0) \ldots \mathbb{P}(Z_{ku} > 0) \leq \mathbb{P}(Z_0 > 0)^k$$

by stationarity and m-dependence. As $k \to \infty$, $\mathbb{P}(Z_0 > 0)^k \to 0$, since $\mathbb{P}(Z_0 > 0) < 1$. Thus $\mathbb{P}(Z_t \leq 0$, at least once$) = 1$, as desired. By Lemma 6.1 the signals constitute a sequence of stopping times. □

The assumptions of stationary and m-dependance are necessary for the conclusion of the previous theorem to hold. Indeed, consider as a counter-example the following AR(1) process

$$Y_t = \phi Y_{t-1} + \varepsilon_t, \quad t = 1, 2, \cdots$$

with $\phi > 2$ and ε_t an iid random variable uniformly distributed over $[-1, 1]$. An AR(1) process is not m-dependent for any finite m. Consider the moving averages $MA(1)_t$ and $MA(2)_t$ for Y_t, and the dual moving average rule:

$$X_t = MA(1)_t - MA(2)_t = Y_t - (1/2)(Y_t + Y_{t-1}) = (1/2)[(\phi - 1)Y_{t-1} + \varepsilon_t]$$

Now, the event $\{Y_t > 1$, and tends to $\infty\}$ has positive probability, and in consequence the occurring times for the rule X_t (see Eq. (6.8)) is such that $\tau = \infty$, so it is not a stopping time.

On the predictive power of technical analysis. Our next concern is the usefulness for forecasting of those technical rules which are well-defined and adequate. The question is under what conditions these rules can be more effective than other econometric models. This is an ample research question which admits many answers depending on the characteristics of each of the many classes of time series models. To begin, recall from Sect. 2.4 that for a given stationary process $\{X_t\}$ the best predictor of X_{t+s}, based on past history $\{X_t, X_{t-1}, \ldots, X_{t-k}\}$, is precisely the conditional expectation $E(X_{t+s}|X_t, X_{t-1}, \ldots, X_{t-k})$. Therefore, given a sequence of Markov times $\{\tau_i : i \geq 0\}$ obtained from a finite history of $\{X_t\}$, we have by definition that the values of $\{\tau_i\}$ are determined by $\{X_t, X_{t-1}, \ldots, X_{t-k}\}$, for some $k > 0$, and in consequence

$$E(X_{t+s}|\{X_t, X_{t-1}, \ldots, X_{t-k}\}, \{\tau_i : \tau_i < t\})$$
$$= E(X_{t+s}|\{X_t, X_{t-1}, \ldots, X_{t-k}\})$$

Consider now the case of $\{X_t\}$ being a Gaussian process. We know from Proposition 2.3 that the conditional expectation

$$E(X_{t+s}|\{X_t, X_{t-1}, \ldots, X_{t-k}\}) = \alpha_0 X_t + \alpha_1 X_{t-1} + \cdots + \alpha_k X_{t-k}, \qquad (6.9)$$

and so the best predictor can be obtained by linear regression on $\{X_t, X_{t-1}, \ldots, X_{t-k}\}$, and in this case an autoregressive model does a better job in predicting than the technical rules producing the stopping times. There are other classes of processes that verify Eq. (6.9) and, in fact, the sub-Gaussian processes are characterize by this equation, as shown by Hardin (1982). So, in general, for any process whose best predictor (or conditional expectation on past history) has a linear form (e.g. Martingales, which are characterize by $E(X_{t+1}|\{X_t, X_{t-1}, \ldots, X_{t-k}\}) = X_t$), no sequence of finite Markov times produced by some technical analysis rules can be better in prediction than an autoregressive model.

Thus, if Technical Analysis is to have any success in forecasting this might be plausible on securities whose price time series is such that its conditional expectation, based on its past history, is a non linear function.

6.2 Fundamental Analysis

Consider the following *raison d'être* for the security analyst

> The security analysts develops and applies standards of safety by which we can conclude whether a given bond or preferred stock may be termed sound enough to justify purchase for investment. These standards relate primarily to [*the company's*] past average earnings, but they are also concerned with capital structure, working capital, asset values, and other matters (Graham 2003, p. 281)

Benjamin Graham can be rightly considered as a founding father of Fundamental Analysis, and elaborating further from his investment policy, summarized in the previous quotation, a fundamental analyst is someone who should strive to determine a security's value by focusing on the economic fundamentals that affect a company's actual business and its future prospects, such as its revenue growth, its profits and debts, and other business indicators, as opposed to just analyzing its price movements in the stock market. In the next sections we study the core elements and methods of investment of Fundamental Analysis.

6.2.1 Fundamental Analysis Basic Principles

Those applying Fundamental Analysis in their investments (aka fundamental investors), act according to three basic principles inherited from Benjamin Graham and David Dodd's approach to investment (Graham and Dodd 1934; Graham 2003):

(1) The market price of a stock does not fully reflect its "real" value. The real value of a stock is intrinsic to the value of the issuing company, termed the intrinsic value.

(2) The market price and the intrinsic value of a stock often diverge, but in the long run will align to each other.

(3) Shares of a stock should be bought with a "margin of safety". This means to buy a stock when its market price is significantly below its intrinsic value (ideally, about a factor of one-half).

By following these principles, a fundamental analysts general investment procedure amounts to compute the intrinsic value of a stock, by valuing the financial standing of the stock's company, and compare that value to the current market price. If the market price is lower than the intrinsic value, then there is a potential profit, since eventually the market price should equal the intrinsic value. Furthermore, buying with a margin of safety is an almost surely guarantee for not loosing money, as this acts as a safeguard from market capricious behavior.

Now, the generally agreed econometric model to estimate the intrinsic value of any of the company's assets is the *discounted cash flow* (DCF) model, that considers the sum of the different cash flows that the asset will produce for investors into the future, discounted back to the present (cf. Sect. 1.2.2). Regardless of its theoretical importance, the DCF model gives such gross approximation to present value that it can hardly be a reliable tool for investment decisions, because it depends on two intangible parameters, the future dividend growth and future rate of return, and both must be forecasted by some model of price. A slight deviation of the projected values for any of these parameters from their future realization can have disastrous consequences for investments (see Remark 1.1).

The fundamental analysis promoted by Graham and Dodd avoids the pitfalls of discounted cash flows, and similar models of valuation dependent on estimating unknown future values, by basing their valuation model on presently known economic information and real figures describing the financial situation of the company. As Graham describes it, the valuation of a common stock should be a two-part appraisal process, where one first analyze the associated business "past-performance value, which is based solely on the past record [...] This process could be carried out mechanically by applying a formula that gives individual weights to past figures for profitability, stability and growth, and also for current financial condition. The second part of the analysis should consider to what extend the value based solely on past performance should be modified because of new conditions expected in the future." (Graham 2003, p. 299). Let us then review the most used past-performance business value indicators for quantifying the stock's current value.

6.2.2 Business Indicators

There are three basic financial statements that a company publishes on a quarterly or annual basis, that serve as different sources for estimating the intrinsic value of its stock. These are:

- The balance sheet.
- The income statement.
- The cash flow statement.

The *balance sheet* gives a snapshot of a company's assets, liabilities and share-holders' equity at a particular point in time. The assets refer to resources own by the company, and are grouped into *current assets* (cash, marketable securities and inventories, and in general anything that can be quickly converted to cash), and *fixed assets* (properties, equipment, and any tangible valuable asset, not easily converted to cash). The liabilities are basically the unpaid bills, also grouped into two categories: *current*, for short termed bills, and *non-current*, for bills with no payment deadline or time negotiable. The shareholders (or ownership) equity is cash contributed to the company by the owner's or shareholders. The figures in this financial statement have to balance according to the equation:

$$\text{All Assets} = \text{All Liabilities} + \text{Shareholder equity}$$

This shows the amount of debt incurred and owner's money used to finance the assets. A fundamental measure that can be extracted from this information is the company's *book value* (BV): the cash that would be obtained by selling all assets free of liabilities; that is, the result of subtracting liabilities from assets. By comparing this quantity with shareholders' equity, investors can estimate their benefits or losses if the company were to be liquidated.

The *income statement* provides information on the company's revenues and expenses, during an specific period (usually quarterly or yearly). The revenue is basically the amount of money received from sales in the given period. The expenses are divided into: production costs (cost of revenue), operational costs (for marketing, administration, or any non-recurring event), loan interests and taxes, and deprecia-tion (the lost of value of fixed assets used in production). From these figures one gets the company's net income over the period, computed as:

$$\text{Net Income} = \text{Revenues} - \text{All Expenses}$$

However, for a more accurate measure of how profitable is the company, financial analysts look at the income without deducting the payments for loan interests and taxes. This is the EBIT, or *earnings before interests and taxes*, given by

$$\text{EBIT} = \text{Revenues} - (\text{production} + \text{operational} + \text{depretiation})$$

Put another way, Net Income = EBIT − (loan interests + taxes).

The *cash flow statement* represents a record of a company's cash inflows and outflows over a period of time. These flows of cash are reported from the investing, operating and financing activities of the company. The main purpose of this financial statement is to report the company's liquidity and solvency, and provides a measure of the ability of the company to change cash flows to meet future expenses.

6.2.3 Value Indicators

What do we do with all the business' financial data? We use it to compute financial ratios to gauge the value of a stock and, by and large, the value of the company behind. Let us begin reviewing some common fundamental ratios to value a stock with respect to the business.

Earnings per Share *(EPS)*: is the portion of the company's profit allocated to each outstanding share of its stock, for a given period of time τ (which by default is the last twelve months). Calculated as:

$$EPS_\tau = \frac{NI_\tau - DivP_\tau}{WNSO_\tau}$$

where NI_τ is the net income perceived in the period of time τ, $DivP_\tau$ is the total amount of dividends on *preferred* stock, and $WNSO_\tau$ is the weighted average number of shares outstanding for the period considered.[5] Thus, *EPS* represent that portion of the profits that a company could use to pay dividends to *common* stockholders or to reinvest in itself. A positive *EPS* is an indication of a profitable company.

Price to Earnings *(P/E or PER)*: is the ratio of a company's current share price with respect to its earnings per share for the trailing twelve months (the period by default). Formally, if P_t is the price per share of a company's stock at time t, τ is the twelve months time period ending at t, then the price to earnings at time t is

$$P/E_t = \frac{P_t}{EPS_\tau}$$

This is the most used factor by financial specialists.[6] It gives a measure of the multiple amount per euro of earnings that an investor pays for a share of the company. In this regard, it is useful to compare the values of two stocks in the same industry. For example, if the P/E of the shares of bank A is lower than the shares of bank B, and the remaining fundamentals for both banks are quite similar, then the shares of A are the better deal. Benjamin Graham recommends as a reasonable P/E for a value stock an amount that is 20 % below the multiplicative

---

[5] This is the "official" formula for the (basic) EPS as established by the Financial Accounting Standards Board, and issued in its *Statements of Accounting Standard No. 128* of March 3, 1997. Nonetheless, many financial traders and firms apply a simpler formula where *WNSO* is just the number of shares outstanding at the end of the period. Observe the difference with a simple example. Assume that over a year a company has 1 million shares outstanding the first 3 months, and for the rest of the year increases to 2 millions shares outstanding; so the final number of shares is 2 millions, while $WNSO_{year} = 0.25 \times 1 + 0.75 \times 2 = 1.75$.

[6] There is another popular form known as the *forward P/E* which is the current price of the share divided by the *projected* EPS. This is not a value indicator under the fundamentalist standards for it uses an estimation of future information. The P/E presented here is also known as *trailing P/E ratio*.

inverse of the current high grade bond rate (Graham 2003, p. 350). If quantities are given in percentage terms this can be expressed mathematically by

$$\text{good } P/E \approx 80/AAbond \tag{6.10}$$

where *AAbond* equals a AA corporate bond yield. For example, if the one year yield of a AA-rated corporate bond (or high-grade corporate bond index) is of 6.43 %, then a good P/E for a stock (that is, for the stock to be considered of some value) is $80/6.43 = 12.44$. The P/E makes sense for a company with positive earnings. If $EPS \leq 0$ then financial analysts give a value of 0 or NA (Not Applicable) to the P/E.

Price to Book (P/B): is the ratio of a company's current share price with respect to the company's book value per share. In mathematical terms, the price to book at time t is

$$\text{P/B}_t = \frac{P_t}{BVPS}$$

where *BVPS* is the most recent quarter's book value of the company for each outstanding share, calculated as $BVPS = BV/(\text{number of outstanding shares}) = (\text{Assets} - \text{Liabilities})/(\text{number of outstanding shares})$, and P_t is the price of a share at time t. Note that this is the same as

$$(\text{market capitalization})/(\text{book value})$$

and that is why is also known as the Market to Book ratio (M/B). We have seen that for a financially well-balanced company, its book value represents the shareholders equity; hence, a P/B close to 1 means a fair price for the share since it is close to its "price in books". A P/B much higher than 1 means that market investors are overvaluing the company by paying for its stocks more than the price in reference to the company's value. Conversely, a low P/B (lower than 1) means that the market is undervaluing the company, and either the stock is a bargain with respect to its book value or has been oversold for reasons extrinsic to the company and perceived by the market.

Regarding the value assessment of business, the most commonly used measurements factors are organized into four groups pertaining to the following attributes: financial performance, the efficient use of resources, financial leverage, and liquidity.[7]

Performance measures: We seek to know, How well is the company doing? How profitable it is? Besides the raw Net Income and EBIT, there are also the following metrics.

---

[7] In each class may enter many more different measures than the ones listed here, not all applicable to all type of business; so we only indicate the most common few. For a more extensive account see Brealey et al. (2011).

Return on equity (ROE): (net income)/(shareholder's equity)

It measures a company's profitability by revealing how much profit it generates with the money shareholders have invested. Note that for a company that pays dividend only for common stock, this measure becomes $ROE = EPS/$Equity, which is the most frequent used form. Many analysts use this ROE to estimate the long term dividend growth rate: the parameter g in the perpetual growth DCF model for computing present value (Chap. 1, Eq. (1.120)). The estimation is done with the equation $g = \left(1 - \dfrac{DIV}{EPS}\right) \cdot ROE$, which expresses the dividend growth rate g as the ratio of reinvesment after payment of dividends per share's earnings (DIV/EPS) multiply by the factor of return on equity.

Return on capital (ROC): (net income + after-tax interest)/(total capital)

The total capital is the sum of shareholder's equity and long-term debt. Thus, the ROC measures the proportion of return obtained from investing the money borrowed or owned. It is also useful for calculating the *economic value added* (EVA) for a company, which is the profit obtained after deducting all costs, including the expected return on all the company's securities (known as *cost of capital*). This is the case, since

$$EVA = (ROC - (\text{cost of capital})) \times (\text{total capital})$$

Return on assets (ROA): (net income + after-tax interest)/(total assets)

The proportion of the return obtained from operations financed with the company's total liabilities plus shareholder's equity (i.e. its total assets).

Efficiency measures: These measures are answers to the question: Is the company allocating resources efficiently?

Asset turnover: (sales)/(total assets at start of year)

A profit ratio of the amount of sales for each euro invested in resources.

Inventory turnover: (costs of assets sold)/(inventory at start of year)

A factor of how rapidly a company turns inventories into cash. A low value (e.g. <1) means poor sales and excess of inventory.

Profit margin: (net income)/(sales)

The proportion of sales that actually turn into profits.

Leverage measures: with this class of indicators we want to determine the level of debt.

Debt ratio: (total liabilities)/(total assets)

A high ratio means a high leveraged company.

Times-interest-earned ratio: EBIT/(interest payments)

A factor of coverage of interest obligations by earnings. Values below 2 may indicate lack of money to pay interests.

Cash-coverage ratio: (EBIT + depretiation)/(interest payments)

A factor of coverage of interest obligations by all the operating cash.

Liquidity measures: What's the company's capability of changing the flow of cash? (e.g., converting assets into cash, paying short term bills)

Working capital: (current assets) − (current liabilities)

This is the most basic measure of a company's liquidity; it gives the amount of cash that the company can have immediately.

Working-capital to assets: (working capital)/(total assets)

The proportion of all resources that represent working capital.

Current ratio: (current assets)/(current liabilities)

A value <1 shows liquidity problems, a negative working capital.

R Example 6.2 The R package `quantmod` has some functions for retrieving the three basic financial statements (balance sheet, income statement and cash flow) from Google Finance, and for viewing. Let us retrieve and view the financials for Apple Inc. (AAPL):

```
> getFinancials('AAPL')  #returns AAPL.f  to "env"
> viewFin(AAPL.f,"CF","A")  #Annual Cash Flows
> viewFin(AAPL.f,"BS","A","2009/2011") #Annual Balance Sheets
```

The `getFinancials` (or `getFin`) returns a list of class 'financials' containing six individual matrices, and for the last four years: IS a list containing (Q)uarterly and (A)nnual Income Statements, BS a list containing (Q)uarterly and (A)nnual Balance Sheets, and CF a list containing (Q)uarterly and (A)nnual Cash Flow Statements. Each type (financial statement) can be handle as a four column matrix:

```
> applBS = viewFin(AAPL.f, "BS","A")
> colnames(applBS) #give years
> rownames(applBS) #give 42 rows labelled with financial indicators
> CA=applBS["Total Current Assets","2009-09-26"]
> ##returns the value of Total Current Assets in 2009
> ##same as
> applBS[10,4] #since row 10=Total Current Assets, col 4=2009-09-26
> CL=applBS["Total Current Liabilities","2009-09-26"]
```

We can compute some liquidity measure for Apple, like its current ratio as of Sept. 2009:

```
> CurrentRat = CA/CL
> CurrentRat
```

The result obtained is 2.742482 which is a pretty good current ratio.   □

6.2.4 Value Investing

How to use the value indicators to construct a set of criteria for good stock selection? Different value investing professionals have his or her own particular combination of financial ratios, selected from the ones listed in the previous section, or others drawn from their own research and experience, to make a selection of potentially profitable stocks. But all, more or less, will coincide to some extend with Graham's own set of guiding principles for acquiring value stock. So let us review Graham's fundamental investment policies (updated to today's market conditions) from (Graham 2003, Chap. 14):

(1) *Adequate size of the enterprise.* The recommendation is to exclude companies with low revenues. Graham advises to consider companies with more than $100 million of annual sales. But that was back in 1973, so to adjust that lower bound to our times, compound that basic amount with a reasonable yearly interest rate, say 7 %, for the number of years from that date to today, to get $100(1+0.07)^{40} = 1497.45$, that is $1.5 billions in revenues.

(2) *Strong financial condition.* The company's current assets should be at least twice current liabilities (a 2-to-1 current ratio), and must have a positive working capital.

(3) *Earnings stability.* The company should have positive earnings in each of the past ten years.

(4) *Dividend record.* Uninterrupted payments of dividends for at least the past 20 years.

(5) *Earnings growth.* An increase of 33 % on the last three-years average of earnings per share (EPS), with respect to the three-years average EPS of a decade ago. This really amounts to a 3 % average annual increase, which is not a very hard test to pass.

(6) *Moderate P/E ratio.* Graham recommends a P/E in between 10 and 15 (or for a precise quantity apply Eq. (6.10)).

(7) *Moderate price-to-book ratio.* The price-to-book ratio should be no more than 1.5.

(8) The last two tests can be combined, as we shall explain, if it is the case that the P/E ratio is in the lower bound (and possibly below) or the price-to-book ratio is slightly above 1.5. The combined P/E and P/B test suggested by Graham, consists on calculating the product of the P/E times the P/B ratio, and this should not exceed 22.5 (this figure corresponds to 15 times earnings and 1.5 times book value). This alternative admits stocks selling at, for example, 9 times earnings and 2.5 times book value. Note that equating the combined ratio with the upper bound of 22.5 we get $\dfrac{P_t^2}{EPS \cdot BVPS} = P/E \cdot P/B = 22.5$, from where one obtains the *Graham number*

$$P_t = \sqrt{22.5 \cdot EPS \cdot BVPS} \qquad (6.11)$$

an estimate of the price of a stock under the value test by Graham. Any stock that passes the tests (1)–(5) and with a market price below the value calculated by Eq. (6.11) is considered a buying opportunity by Graham's criteria.

By the end of 1970 the DJIA passed Graham's test (1)–(7), but only five individual stocks from this market passed the test: American Can, ATT, Anaconda, Swift, and Woolworth, and this group performed better than the index.[8] This reinforces the advantages of diversification: a large and well diversified portfolio has greater chance of being conservative and profitable; on the other hand, stock picking with a conservative criteria which primes intrinsic value, may optimize the portfolio in size and profits. Doing a likewise screening and valuation of 4700 NYSE and NASDAQ stocks in 2013 gives also a very small amount of issues meeting Graham's test.[9] There are several internet sites that perform stock market valuation based on Graham's criteria, or other combination of indicators established by some of his notable disciples.[10] Alternative the reader can build his or her own stock screener and fundamental valuation in R, using the tools exhibited in R Example 6.2.

6.3 Notes, Computer Lab and Problems

6.3.1 Bibliographic remarks: An initiation to the craft of Technical Analysis can well start with the classics: Edwards and Magee (1966), for TA general principles and methods; Nison (1991) for the specifics of candlesticks patterns analysis. For an entertained discussion on the effectiveness of Dow's theory as applied by William Peter Hamilton read (Brown et al. 1998). There are several encyclopedic collections of technical analysis indicators and trading rules; we have consulted (Achelis 2001; Katz and Mccormick 2000), in these you can find almost all known technical trading rules. A review of the literature on the profitability of technical analysis is given in Park and Irwin (2007). The mathematical foundation of Technical Analysis (Sect. 6.1.4) is based on Neftci (1991), and we do not know of any other attempt to formalizing TA as serious as this work.

Our fundamental reference (and we dare say *the* reference) for Fundamental Analysis is Graham (2003). A complementary reference is Greenwald (2001), which exposes Value Investing, the modern form of Fundamental Analysis, and its implementation by some of its most notorious practitioners, from Warren Buffet, Mario Gabelli, Glenn Greenberg and others. Examples and detailed analysis of income statements, balance sheets, cash flows plus financial ratios calculations for real life companies can be found in Brealey et al. (2011, Chaps. 28, 29).

6.3.2 R Lab (Technical and Fundamental Analysis in R): Besides `quantmod` there are other packages in R with functions for technical and fundamental analysis. The `fTrading` package contained in `Rmetrics` has many tools for computing technical indicators, such as: Bollinger bands (function `bollingerTA`); Relative Strength Index (`rsiTA`); Simple Moving Averages (SMA), and many more. An example of usage: `SMA(x, n = 10)` returns the moving average of on a period of

---

[8] Graham, op. cit. p. 354.

[9] see http://www.gurufocus.com/news/214204/.

[10] see www.serenity.org, and to find more do an internet search with keywords as "stock valuation", "stock selection criterion", or "automatic stock screening".

10 days for the time series object x; $\texttt{rsiTA(close, lag)}$ returns the Relative Strength Index Stochastics indicator for a vector of closing prices, and time lag. The package $\texttt{TTR}$ (Technical Trading Rules) contains much more technical indicators, although has no nice interface for plotting them. This graphical interface is provided by $\texttt{quantmod}$, which uses $\texttt{TTR}$ for calculations and adds the results of the indicators into a plot of the stock's price.

6.3.3 R Lab: Write an R function $\texttt{head-and-shoulder}$ to recognize the Head-and-Shoulders price pattern. This function takes as parameters the price data series x, the bandwidth b (with default value 0.015) and the time-window length T (default value 35). It outputs true if there is a HS in the time span $[t_0, t_T]$, or false otherwise. Use the algebraic description given in this chapter.

6.3.4: Write the mathematical equations for the occurring times τ_i produced by triangles and gaps trading rules.

6.3.5: Prove Proposition 6.1. (*Hint*: From the hypothesis $\mathbb{P}(Z_t \in A,$ at least once$)$ $= 1$ follows immediately that $\mathbb{P}(\tau_1 < \infty) = 1$. Use stationarity of Z_t to show that $\mathbb{P}(\tau_1 < \infty) = 1$ implies $\mathbb{P}(\tau_k < \infty) = 1$.)

6.3.6 Volume indicators: According to one of Dow's principles "the volume confirms the trend". This relation of price trend and volume is summarized in the following table.

Price	Volume	Trend
UP	UP	continues
UP	DOWN	weakens
DOWN	UP	continues
DOWN	DOWN	weakens

One can also apply Moving Averages to volume time series to analyze the evolution of volume just as with prices. Do an R experiment to test serial correlation of trading volume and price for a given stock or market index. (To learn more on the relation of volume to price read: Blume, L., Easley, D. and O'Hara, M., Market Statistics and Technical Analysis: The Role of Volume, *The Journal of Finance*, 49 (1), 153–181, 1994; and Campbell, J., Grossman, S., and Wang, J., Trading volume and serial correlation in stock returns, *The Quarterly J. of Economics*, 108 (4), 905–939,1993.)

6.3.7 ROE, ROC and ROA revisited: To further understand the subtle differences between the three measures of management efficiency, namely, ROE, ROC and ROA, it is perhaps enlightening to view each of these factors as answer to the probably most basic and fundamental investor's query: How much the company would have earned if financed solely by X?, where X is a variable representing resources, and takes the following values. For ROE, X = the owners; for ROC, X = owners and loans; for ROA: X = all equity. For further descriptions and applications of these and other measures see Brealey et al. (2011).

6.3.8 Relative Strength Index (RSI): For each time t, consider

- Average Up Closes over n days up to time t:
  $AUC(t) = \frac{1}{n} \sum_{i=t-n+1}^{t} \{(C_i - C_{i-1}) : C_i > C_{i-1}\}$
- Average Down Closes over n days up to time t: $ADC(t) = \frac{1}{n} \sum_{i=t-n+1}^{t} \{(C_{i-1} - C_i) : C_i < C_{i-1}\}$
- Relative Strength at time t: $RS(t) = AUC(t)/ADC(t)$
- Relative Strength Index at time t: $RSI(t) = 100 - (100/(1 + RS(t)))$
- Entry Thresholds: an interval $(ET, 100 - ET)$, for a constant value $0 < ET < 100$, within which RSI values fluctuates, generating buy or sell signals when crossing.

We observe the process $RSI(t)$, beginning at some $t = t_0$ where $ET < RSI(t_0) < 100 - ET$. Then as soon as $RSI(t_i)$, $i > 0$, crosses either bound we trade as follows:

- Go long when $RSI(t_i)$ falls below ET.
- Go short when $RSI(t_i)$ rises above $100 - ET$.

In practice, the parameter n is taken around 14, and $ET = 30$. The rationale is that when $RSI(t)$ is close to ET (respectively, $100 - ET$) then the stock is oversold (resp. overbought) and so a turning point in the trend of prices is imminent. Note that this rule is in the family of channel crossings, where the channel is defined by horizontal lines of constant values and the relative strength process $(RS(t))$ is projected into this horizontal space. Show that the RSI rule defines a sequence of occurring Markov times.

6.3.9: Using the notation introduced in Sect. 6.1.3 for formally expressing the candlesticks patterns, write the equations for the pattern known as the *Evening Star*. This pattern is the opposite of the morning star. It consist of three candlesticks c_1, c_2, and c_3, observed at days $t - 2, t - 1$ and t, respectively, and preceded by an uptrend. Candlestick c_1 is white and tall, c_2 is small and of either color, c_3 is black and tall. The closing of c_1 and the opening of c_3 are both below the body of c_2, the closing of c_3 is below the closing of c_1.

6.3.10: Elaborate a report of investment opportunity on a stock of your choice, using technical and fundamental methods.

Chapter 7
Optimization Heuristics in Finance

In this book optimization heuristics refers to algorithmic methods for finding approximate solutions to the basic optimization problem of minimizing (or maximizing) a function subject to certain constraints. The problems of interest are typically of big size, or admitting several local optimum solutions, for which exact deterministic approaches are inadequate. Then, an appropriate heuristic for solving these hard problems must, in principle, offer an improvement over the classical local search procedure. We review in this chapter three of the most popular heuristics that extend local search with some Nature-inspired technique. These are: *simulated annealing*, based on the process of crystallization during the cooling or annealing of melted iron; *genetic programming*, based on the evolution of species governed by the laws of survival of the fittest; *ant colony systems*, based on the collective exchanged of information by populations of organisms in order to find best routes to resources. Applications of these heuristics to financial problems are given in this and the next chapter. We end discussing some ways to mix these schemes to obtain *hybrid systems*, and outlining some general guidelines for the use of optimization heuristics.

7.1 Combinatorial Optimization Problems

Definition 7.1 An optimization problem is a tuple $\mathscr{P} = (\mathscr{S}, \Omega, F)$, where $\mathscr{S}$ is a set of *feasible solutions*, Ω is a set of *constraints* over a fixed number of parameters, and $F : \mathscr{S} \rightarrow \mathbb{R}^+$ is an *objective function* to be minimized. The set of constraints Ω consists of all the variables in a given set of equations. The set of feasible solutions $\mathscr{S}$ consists of all arrays of values for the variables that satisfies the constraints in Ω. A solution for the optimization problem $\mathscr{P}$ is an element s^* of $\mathscr{S}$ such that $F(s^*) \leq F(s)$ for all $s \in \mathscr{S}$; such a s^* is called a *globally optimal solution*. $\qquad \square$

In a strict sense we have defined above a minimization problem. For a maximization problem just ask for the global solution s^* to satisfy $F(s^*) \geq F(s)$ for all $s \in \mathscr{S}$.

A. Arratia, *Computational Finance*, Atlantis Studies in Computational Finance and Financial Engineering 1, DOI: 10.2991/978-94-6239-070-6_7, © Atlantis Press and the authors 2014

However, note that maximizing F is the same as minimizing $-F$, and hence, we can formally treat any optimization problem as a minimization problem.

Example 7.1 Consider a GARCH(1,1) model for a log return series $\{r_t\}_{t=1}^{T}$,

$$r_t = \mu + a_t, \quad a_t \sim N(0, \sigma_t^2) \tag{7.1}$$
$$\sigma_t^2 = \omega + \alpha_1 \cdot a_{t-1}^2 + \beta_1 \cdot \sigma_{t-1}^2$$

with $-1 \le \mu \le 1$, $\omega > 0$, $\alpha_1 \ge 0$, $\beta_1 \ge 0$, $\alpha_1 + \beta_1 < 1$. Here we want to estimate the parameters μ, ω, α_1 and β_1, so that we can recursively obtain σ_t and a_t. We have seen in Chap. 4 that this estimation can be done by the method of maximum likelihood. Thus, the underlying combinatorial optimization problem $(\mathscr{S}, \Omega, F)$ can be explicitly define as follows. The set of constraints Ω consist of Eq. (7.1) plus the inequalities involving tuples of four variables $\psi = (\omega, \alpha_1, \beta_1, \mu)$; the set of feasible solutions $\mathscr{S}$ are the values for $\psi = (\omega, \alpha_1, \beta_1, \mu)$ that satisfy the Eq. (7.1) and inequalities, and we seek for those values that maximizes the log likelihood function (without additive constants)

$$F(\psi) = -\frac{1}{2} \sum_{t=1}^{T} \left(\ln \sigma_t^2 + \frac{a_t^2}{\sigma_t^2} \right) \tag{7.2}$$

This is the objective function to consider for this problem, where one seeks to maximize its value with respect to all $\psi \in \mathscr{S}$. This function can have various local maxima, and therefore a traditional iterative numerical method (e.g. the local search described below as Algorithm 7.1) will often fail to find a global optimum. Hence, the need of an optimization heuristic is justified.   □

Example 7.2 Another maximization problem we consider in this chapter is to find trading rules from Technical Analysis that best perform in a period of time $[1, T]$ and for an asset's return time series $\{r_t\}_{t=1}^{T}$. A description of this problem as a tuple $(\mathscr{S}, \Omega, F)$ is given by:

Ω as a set of production rules which describe the process of constructing technical trading rules (or investment recipes as those presented in Sect. 6.1), from some predefined set of constants, variables and functions.

$\mathscr{S}$ consists of the trading rules conforming to the specifications given by Ω.

F could be the compounded benefit of a trading rule acting on $\{r_t\}_{t=1}^{T}$ (for an absolute measure of performance), or the excess return over the whole period $[1, T]$ with respect to a buy-and-hold strategy, which is a more realistic measure of performance as it is a test against the most common passive investment attitude.

The objective is to find the trading rule in $\mathscr{S}$ with maximum F-value when acting on $\{r_t\}_{t=1}^{T}$. Note that the search space of all possible trading rules is exponential in the number of basic elements used for defining trading rules; hence an exhaustive deterministic search is out of the question, and we can justly turn to heuristic optimization techniques.   □

The classical local search algorithm for solving an optimization problem $(\mathscr{S}, \Omega, F)$ starts with an initial solution (usually generated at random), and by making small perturbations or changes to the current solution creates another; that is, new solutions are obtained "in the neighborhood of current solutions". If the new solution is better (i.e. has a smaller F-value) then the current solution is substituted by the new one. This process is repeated until some stopping criteria is met, which can be (and often is) that there is no further significant improvement in the values of F, or that a predetermined number of repetitions have been accomplished. Algorithm 7.1 presents the classical local search procedure for solving $(\mathscr{S}, \Omega, F)$. Observe that with this procedure we can only guarantee that the search ends in a *local* optimum. In other words, if our optimization problem has more than one local minimum, once the search routine reaches one of these local solutions, it is likely to get trap in its vicinity because small perturbations will not produce better solutions, and so it misses the global minimum. A way to escape from this local trap is to allow to enter into the search path some solutions that are worse than those previously obtained. However, these "escape-moves" (also known as *uphill-moves*) must be performed in a controlled manner to avoid jumping out of the vicinity of a potential global solution. The heuristics presented in the following sections are in essence different implementations of these escape-moves.

Algorithm 7.1 Classical local search algorithm for $(\mathscr{S}, \Omega, F)$

1. **initialization**: generate a solution $s \in \mathscr{S}$
2. **while** (stopping criteria not met) **do**
3.   compute a solution s' in $N(s)$ {*the neighbor of s*}
4.   **if** $(F(s') < F(s))$ **then** $s = s'$
5. **end while**
6. **output**: s {*best solution*}

7.2 Simulated Annealing

Simulated annealing (SA) falls into the category of stochastic *threshold* methods. This means that the algorithm starts with a randomly generated instance of a solution, and in successive iterations a neighbor of a current solution is randomly generated and their costs difference is compared to a given threshold. For simulated annealing the threshold at each iteration is either 0 (so that solutions that optimize the cost function F are accepted), or a positive random number that follows a probability distribution skewed towards solutions corresponding to smaller increases in cost, and depending on the difference of costs ΔF, and a global parameter T. In practice, these random thresholds control the frequency of making escape-moves in the following way. The parameter T is a non-increasing function which is slowly decremented at each time step, and this produces a reduction in the probability of accepting a worse solution, as this probability is governed by the negative exponential distribution of ΔF, with coefficient $1/T$. The process is repeated until some stopping criteria holds, that is usually a combination of reaching a predefined number of iterations, or making no further moves (for better or worse solutions) for some number of steps. The general scheme of simulated annealing just described, for solving an optimization problem $(\mathscr{S}, \Omega, F)$, is shown in Algorithm 7.2.

Algorithm 7.2 Simulated annealing basic algorithm for $(\mathscr{S}, \Omega, F)$

---

1. **initialization**: generate a random solution s_0 ($s_0 \in \mathscr{S}$)
2. $s^* = s_0$; $F^* = F(s_0)$; $s^c = s_0$ {*current solution*}
3. $T = T_0$ {*initial temperature*}
4. **while** (stopping criteria not met) **do**
5.   compute a solution s at random in $N(s^c)$ {*neighbor of s^c*}
6. $\Delta F = F(s) - F(s^c)$
7.   **if** ($\Delta F \le 0$) **then** $s^c = s$
8.   **if** ($F(s) \le F^*$) **then** $(s^*, F^*) = (s, F(s))$
9.   **else** generate number p at random in $[0, 1]$ and
10.   **if** ($p \le \exp(-\Delta F/T)$) **then** $s^c = s$
11.   reduce T
12. **end while**
13. **output**: s^* {*best solution*}

7.2.1 The Basics of Simulated Annealing

Simulated annealing was first proposed as an optimization heuristic in Computer Sciences by Kirkpatrick et al. (1983) for the design of VLSI circuits, and independently by Černý (1985) for a solution to the Traveling Salesman Problem. The idea for this heuristic comes from an analogy with the work of Metropolis et al. (1953) in statistical thermodynamics, who gave a method for simulating the energy changes of a solid in a heat bath from melting point down to a state of thermal equilibrium (i.e. cool state), by means of a cooling, or annealing, process. According to the laws of thermodynamics, the probability of having an energy increment ΔE at a certain temperature t can be estimated by the equation

$$\mathbb{P}(\Delta E) = \exp(-\Delta E/kt) \tag{7.3}$$

where k is the Boltzmann constant. The Metropolis algorithm is a Monte Carlo simulation that successively makes random perturbations to the system, and computes the change in energy ΔE when passing from one state to the next. If there is a drop of energy (i.e. $\Delta E < 0$) then the new state is accepted, otherwise the new state is accepted with a probability given by Eq. (7.3). The process is repeated several times through a series of temperature decrements, which will lower the energy of the system until thermal equilibrium is reached.

The analogy of simulated annealing in combinatorial optimization to the thermodynamical annealing process of Metropolis should be now clear: think of instances of solutions as states of the physical system; cost function as energy; control parameter T as the temperature; and an optimal solution as a cool (or low energy) state. The eventual convergence of the Metropolis algorithm to a state of thermal equilibrium is characterized in terms of the Boltzmann distribution, which relates the probability of the system of being at a certain state with a certain energy to the current temperature. From this fact can be derived an analysis of the convergence of simulated annealing, which we shall not do here and rather refer the reader to Aarts and Lenstra (2003, Chap. 4) for all the details of this analysis (see also Laarhoven and Aarts (1987)).

A more important issue for practical purposes is the setting up of the parameters that control the simulated annealing heuristic. There are basically three parameters to consider: the stopping criterion, the cooling schedule and the neighborhood range. All three parameters are interdependent and the determination of their values depend on the nature of the optimization problem. However, observe that in general, from Eq. (7.3) setting $k = 1$, $p = \mathbb{P}(\Delta E)$, and solving for the temperature t, we get

$$t = -\Delta E / \ln p.$$

This leads to the following simple idea for determining an initial value for the temperature and a cooling schedule: For an arbitrary large number of steps, compute solutions at random in the neighborhood of initial solution s_0 and determine the average increase $\overline{\Delta F}$ of the objective function over this set of solutions. Fix a high probability $p_0 = 0.8$ or 0.9 and set initial temperature to $T_0 = -\overline{\Delta F} / \ln p_0$. Then apply simulated annealing algorithm with initial temperature as T_0 and successive decrements given by the scheme $T_k = \gamma T_{k-1}$, for $k \geq 1$, and γ a constant positive number less than but closer to 1 (the *cooling parameter*). A standard value for γ is 0.95. In applications of simulated annealing this *static geometric cooling schedule* is usually adopted for setting up the values for the parameter T_k. The stopping criterion can now be taken to be the moment the temperature reaches a very low value $T_k < \epsilon$. However, more often, and simpler, is to repeat the search for a large (fixed) number $I > 0$ of iterations. A typical default value for I is 10,000. Finally, the neighborhood range can be adjusted beginning with a large value, so as to explore a large set of solutions, but successively reduce the radio for allowing escape-moves. We put this heuristic to work for estimation of the parameters of a $GARCH(1, 1)$ process, in the next section.

7.2.2 Estimating a GARCH(1, 1) with Simulated Annealing

We solve the optimization problem proposed in Example 7.1, that is, the estimation of parameters for a GARCH(1,1) process, with the SA method. Recall that this is a maximization problem and therefore we must adapt the general SA heuristic (Algorithm 7.2) stated for a minimization problem. This is attained simply by turning

$-\Delta F$ to ΔF, and tests of negativity $\Delta F \leq 0$ to $\Delta F \geq 0$. The objective function is the likelihood estimator $F(\psi)$ of a tuple $\psi = (\omega, \alpha_1, \beta_1, \mu)$, and given by Eq. (7.2). Algorithm 7.3 presents the SA solution to this GARCH estimation. One can set the initial temperature T_0, and the cooling parameter γ, to values calculated with the heuristics explained in the previous section. The stopping criterion is determined by reaching a maximum number I of iterations. At each iteration, for a current solution $\psi = (\psi_1, \psi_2, \psi_3, \psi_4)$, a new candidate solution ψ' is randomly generated by selecting one element $\psi_j \in \psi$ and defining a perturbation $\psi'_j = \psi_j + u \cdot z$, where $z \in [-1, 1]$ is a i.i.d random number and $u \in (0, 1)$ is the neighborhood range; for $k \neq j$, $\psi'_k = \psi_k$.

We can program Algorithm 7.3 in R, or an equivalent algorithm without much programming effort, using already existing R-packages for general purposes optimization.

Algorithm 7.3 Simulated annealing for estimating $GARCH(1, 1)$

---

1. **initialization**: $\psi := (\psi_1, \psi_2, \psi_3, \psi_4)$ with random values
2. $\psi^* := \psi$; T_0 {*initial temperature*}
3. γ {*cooling parameter*}; u_0 {*neighborhood range*}
4. **for** $i = 1$ **to** I **do**
5. $\psi' := \psi$; $z_i = randomValue \in [-1, 1]$; $u_i = z_i \cdot u_{i-1}$;
6. $j = randomInteger \in [1, \ldots, narg(\psi)]$
7. $\psi'_j := \psi_j + u_i \cdot z_i$; $\Delta F = F(\psi') - F(\psi)$
8.   **if** $(\Delta F > 0)$ **then** $\{\psi := \psi'$
9.   **if** $(F(\psi) > F(\psi^*))$ **then** $\psi^* = \psi$ $\}$
10.   **else** $\{p = randomNumber \in [0, 1]$
11.   **if** $p \leq exp(\Delta F / T_{i-1})$ **then** $\psi = \psi'$ $\}$
12. $T_i := \gamma \cdot T_{i-1}$ {*lower the temperature*}
13. **end for**
14. **output:** ψ^* {*best solution*}

R Example 7.1 In R the function `optim` includes an option for performing optimization based on simulated annealing. The implemented SA heuristic is one given in (Bélisle 1992), which uses the Metropolis function for the acceptance probability, temperatures are decreased according to the logarithmic cooling schedule, and the stopping criterion is a maximum number of iterations. To use it, set `method="SANN"`. Then define the function to be minimized (or maximized), and pass this function to `optim` together with a vector of initial values for parameters to be optimized over, and appropriate `control` settings (i.e. the type of optimization -if a maximization or a minimization-, the maximum number of iterations, etc.). Let `fn <- function(psi,r){…}` be the objective function given by Eq. (7.2) with `psi` a 4-component vector, where `psi [1]` = omega, `psi [2]` = a1, `psi [3]` = b1, `psi [4]` = mu, and `r` is the sample log return series. Set up an indicator function for the $GARCH(1, 1)$ parameters constraints:

```
I <- function(psi){
  if(psi[1] > 0 && psi[2]>=0 && psi[3]>=0 && psi[2]+psi[3]<1
          && psi[4]>-1 && psi[4]<1) {return(1)}
  else {return(0)}  }
```

and put together the log likelihood with the constraints for optimization

```
LI <- function(psi,r) return(fn(psi,r)*I(psi))
```

We run an experiment with r = SAN, the log return of stock SAN (Bank Santander, trading at IBEX), data ranging from 09-02-2009 to 29-01-2010. The command to do the *GARCH*(1,1) fitting with SA is

```
> parinit = c(0.1,0.1,0.1,0.1)
> opt = optim(parinit,LI,r=r,method="SANN",
+    control=list(fnscale=-1,maxit=5000))
> opt$par
```

The coefficients of the *GARCH*(1, 1) are obtained from opt$par[1] = ω, opt$par[2] = α_1, and opt$par[3] = β_1, and the estimated mean μ =opt$par[4]. To compare the quality of this result with the one obtained by the garchFit method, we computed the long-run volatility[1]

```
> Vl = sqrt(opt$par[1]/(1-opt$par[2]-opt$par[3]))
```

and got a value of 0.02283154, close to the standard deviation obtained with coefficients computed by the method garchFit. Note that we can estimate μ separately (with mu=mean(r)) and feed the estimated value to the objective function fn, hence reducing the likelihood estimation to three parameters (psi[1],psi[2] and psi[3]), and probably gain in speed and precision. The reader should try all these variants. □

Be aware that a different run of optim in the previous experiment (or any SA algorithm) will yield different values for the target parameters, due to the random component of the algorithm. Thus, to have some confidence in the goodness of the solution, a convergence analysis is needed. The theory behind this type of performance analysis is beyond this book, and hence we recommend, as a starter on this subject, Chap. 4 in the collection (Aarts and Lenstra, 2003) and the work by Lundy and Mees (1986). In Chap. 8, Sect. 8.3.3, we present an application of simulated annealing to the problem of portfolio selection.

7.3 Genetic Programming

The genetic programming (GP) heuristic belongs to a class of evolutionary strategies characterized by being computer-intensive, stepwise constructors of subsets of feasible solutions, using operators that model sexual reproduction and governed by the

---

[1] cf. Eq. (4.38), Chap. 4.

Darwinian principle of survival of the fittest. The general structure of this heuristic is as follows:

> Beginning with a randomly generated set of *individuals* (i.e. feasible solutions for the problem), which constitutes an *initial population*, allow them to evolve over many successive generations under a given *fitness* (or performance) criterion, and by application of operators that mimic procreation in nature, such as, *reproduction*, *crossover* and *mutation*, until no significantly better individuals are produced.

The optimal solutions are then found among the best fitted individuals of the last generation.

Note that, in the context of genetic programming, the formal notation $(\mathscr{S}, \Omega, F)$ for an optimization problem is interpreted as: $\mathscr{S}$ the set of individuals, Ω the constraints to guarantee well-formed individuals, and F the fitness function. The individuals are *programs* which are encoded as *trees*; that is, hierarchical structures where nodes at one level may represent variables, constants or other terms, which are arguments for functions represented by nodes at a preceding level. Applying an operator means to remove some nodes or adding new ones to form new trees (or programs) of varying size and shape.

Remark 7.1 From now on we use indistinctly the words "program" and "tree" to refer to individuals generated by a genetic programming heuristic, due to the natural identification between the object and its graphical representation.   □

Remark 7.2 An important feature about the genetic operations is that all perform local perturbations to create new programs. Therefore, in the general scheme of genetic programming outlined above, if for each generation we consecutively select the best fitted programs for procreating new offsprings (incurring in what's known as *over-selection*), then we can at most guarantee that at the end of the process the best programs obtained correspond to local optimum solutions.[2] Therefore we need to make some escape-moves at certain generations to aim for a global optimum. A escape-move in this context can be realized by selecting some of the worse fitted programs, and combine them among themselves or with best fitted programs, for procreating new offsprings.   □

Genetic programming was conceived by John R. Koza (1992) as an extension of the genetic algorithm of John John Holland (1975). The main distinction between both evolutionary schemes is that in Holland's paradigm individuals are coded as *linear bit strings*, usually implemented as an array of fixed length, whereas in Koza's genetic programming the individuals are coded as *non-linear structures* (e.g. trees) that do not have the size limitation in the design, although of course are limited by the memory capabilities of the computer. The representation of individuals in a genetic heuristic as non-linear structures or programs is more suitable for optimization problems that deals with defining and analyzing financial decision rules, since a decision procedure is easily representable as a tree.

---

[2] The reader should think carefully about why is this the best we can get.

7.3.1 The Basics of Genetic Programming

Given an optimization problem $\mathscr{P} = (\mathscr{S}, \Omega, F)$ that we want to solve with the genetic programming paradigm, the first step is to associate to $\mathscr{P}$ two sets of nodes:

$\mathscr{T}$: set of *terminal* nodes which represent constants (e.g. real, integer or boolean) or variables that take values from input data relevant to $\mathscr{P}$.

$\mathscr{F}$: set of *function* nodes which represent functions relevant to $\mathscr{P}$.

In general, the functions in the function set may include:

- Arithmetic operators: $+, -, \times, /$;
- Relational operators: $<, >, =$;
- Boolean operators: AND, OR, NOT;
- Conditional operators: IF-THEN-ELSE;
- Real functions and any domain-specific function appropriate for the problem.

Terminal nodes have no descendant; function nodes have as many descendants as their number of parameters. Each node transmit to its unique predecessor its value (for terminal nodes), or the result of evaluating the function on the values transmitted by its descendants. It is customary to link nodes under this "descendant of" relation in bottom-up style, so a program $\pi \in \mathscr{S}$ is represented as a tree with terminal nodes at the bottom, function nodes at intermediate levels and a single function node (the root) at the top. The result of evaluating this tree, which gives the output of the program π, is the value returned by the root after recursively evaluating every node of the tree from bottom to top.

Example 7.3 Consider the following numerical equation,

$$C_0 \times C_1 < C_0 + C_1 \tag{7.4}$$

with C_0 and C_1 two numerical constants, $\times$ and $+$ the usual arithmetic functions, and $<$ the order relation, which can be turned into a boolean function by a common trick best illustrated by an example: $3 < 5 \iff < (3, 5) = \mathbf{true}$. (We are sure the reader can generalized this idea to turn any relation into a function.) This equation is a program that outputs **true** if the inequality holds, or **false** if the inequality does not holds. It can be encoded with $\mathscr{T} = \{C_0, C_1\}$ and $\mathscr{F} = \{+, \times, <\}$ as the tree in Fig. 7.1. The boolean value of expression (7.4) is obtained by parsing the tree bottom-up. □

Fig. 7.1 A tree encoding the equation $C_0 \times C_1 < C_0 + C_1$

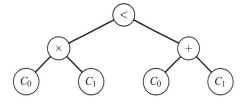

The basic scheme for a genetic programming algorithm is presented in Algorithm 7.4. Details are explained in the following paragraphs.

Algorithm 7.4 Genetic programming basic scheme for $(\mathscr{S}, \Omega, F)$

---

1. **initialization**: generate initial random population $P \subseteq \mathscr{S}$
2. *generation* $= 0$
3. **while** (stopping criterion not met) **do**
4.   evaluate fitness $F(\pi)$ of each program $\pi \in P$
       {*Create a new generation:*}
5.   **while** (Size_of_Population not reached) **do**
6.   select genetic operator Op randomly
7. $n =$ number of arguments for Op
8.   select n programs randomly based on fitness
9.   apply Op to the n programs selected
10.   insert new programs obtained into *NewPopulation*
11.   **end while**
12. $P = NewPopulation$
13.   *generation* $=$ *generation* $+ 1$
14. **end while**
15. **output**: best program π^* based on fitness

Initial individuals. The initial trees conforming the first population are produce by randomly selecting elements from the terminal or function sets, $\mathscr{T} \cup \mathscr{F}$, observing the following recursive rules:

(1) Begin by selecting at random an element of $\mathscr{F}$ to be the root of the tree.
(3) For an already defined node of the tree, if its label is a function $f \in \mathscr{F}$, then define as many descendants nodes as arguments that f takes and of the *appropriate type* from $\mathscr{T} \cup \mathscr{F}$; if the label is a terminal element (from $\mathscr{T}$), then this node has no descendants.
(3) Repeat rule 2 until a predefined depth is reached or until all nodes at a same level are terminal with the proviso that all nodes at the maximum specified depth must be terminal. (The depth of a tree is the longest path from the root to a terminal node.)

There are two variants to this construction which Koza called Full and Grow. In the Full method terminal elements can only be selected for nodes at the last level, which is located at the given depth. Thus all trees are fully branching trees. In the Grow method terminals can be selected at any level, and hence trees has branches of varying lengths (bounded by the predefined length). A sensible generative method is a mix of Full and Grow applied at varying depths ranging from 2 to the maximum established depth. For example, build an equal number of trees with the Full and the Grow method for each depth 2, 3, ..., up to maximum depth. This last strategy is what Koza calls the "ramped half-and-half" method.

Closure property. The provisos in rule (2) for constructing trees refers to a closure restriction stated by Koza in his genetic programming paradigm, where it is required that all functions (and by extension all programs) be well defined and closed for any combination of the appropriate number of arguments that are inputted to it. This guarantees that we only build valid programs, and hence reduces the search space by avoiding to work with malformed programs, which will eventually be discarded by the natural selection process anyway.

Fitness. The fitness of a program is a measure that, from the genetic point of view, should reflect its probabilities of reproduction; that is, its expected surviving time and number of successive descendants. As regards the underlying optimization problem, the fitness of a program should reflect some rate of approximation of its F-value to the F-value of the currently best solution, or a predefined target solution. This is known as *raw fitness*.

Selection. Programs are selected according to their fitness value. But, by Remark 7.2, we must allow some not-so-well fitted programs to be selected also, to ensure some degree of variety and have some hope of reaching a global optimum. We must avoid over-selecting programs. There are various selection strategies proposed in the literature that overcomes the over-selection and guarantees some variety in the population up to certain degree.[3] One effective method is to sort programs from best to worst, with respect to some normalization of their raw fitness that would equally distribute them, and apply a random selection with a probability distribution function skewed towards best fitted programs; so that most of the time the best are chosen but sometimes a not-so-good program is also chosen. This is known as *ranking selection*, a technique attributed to Baker (1985). Note that this probabilistic selection is in fact a sufficient condition to have almost surely convergence to a global optimum (i.e. global convergence with probability 1).[4]

Operations. The three most frequently used genetic operators are: reproduction, crossover and mutation. The first step in applying any of the operations is to select programs, which should be done following some strategy ensuring variety, and as outlined above.

Reproduction: it consist on selecting a program and transfer a copy of it, without any modifications, to the next generation.

Crossover: it consists on randomly selecting two programs (the parents) to generate two new individuals (the offsprings) by a process mimicking sexual recombination.[5] From a copy of each parent randomly select a node (the crossover point), and then exchange the subtrees hanging from the chosen nodes, to obtain two new trees with pieces of material from both parents. After the operation we apply some *survival criteria* (based on fitness) to remove one parent and one child. This

---

[3] See Goldberg and Deb (1991) for an analysis of selection schemes.

[4] See Eiben et al. (1991).

[5] There are several variations of this process in the literature: you may produce one (instead of two) offspring from two parents, and then remove both parents, or keep the best fitted from all parents and offsprings. Whatever the case, it is important to control the population size, which should be bounded to a certain quantity throughout the different generations.

is important for keeping the population size bounded. Figure 7.3 shows a full picture of a crossover operation.

Mutation: it consists on choosing at random some node in the tree, remove this node together with the subtree below it, and then randomly generate a new subtree in its place. Mutation is a unary operator, for it applies to a single program. The purpose of mutation is to restore diversity in a population that may be converging too rapidly, and hence, very likely to a local optimum.[6] However, in practice, when crossover is applied hardly two trees that exchange parts will produce a child equal to one of them. Thus, in general, the algorithm should perform few mutations and rely more on crossover and reproduction.

There are other less popular operators, which include *permutation*, *editing*, *encapsulating*, and *destroying*, but we shall not deal with these and refer the reader to (Koza 1992) for a full description and extensive analysis.

Stopping criterion. A genetic programming algorithm can go on forever, and hence we must force it to stop. The stopping criterion is usually set as a predetermined number of generations or conditioned to the verification of some problem-specific success test, as for example, that the programs do not improve their fitness through some generations.

Setting up the control parameters. Depending on the problem we wish to apply the GP scheme, one should adjust the parameters that controls the algorithm accordingly. The most important parameters to calibrate are: the size of the populations, the number of generations, the number of iterations, the frequency of applying each genetic operator, and the size of the programs. Koza (1992, Chap. 5) provides a set of default values that have been widely validated through the extensive practice of genetic programming. However, other authors refrain from giving a set of exact values and rather provide some general guidelines (Reeves and Rowe, 2003, Chap. 2). Our advise is to take Koza's generic parameter values as a starter, and proceed to adjust them through several trials considering the particular features of the problem and computational resources (e.g., more often than not, one is forced to reduce the size of the population from the recommended 500 individuals, due to machine memory limitations). For the financial application of GP presented in the next section, we present our own table of parameters values obtained following the previously stated advise.

7.3.2 Finding Profitable Trading Rules with Genetic Programming

A widely studied application of genetic programming in finance is the automatic generation of technical trading rules for a given financial market and analysis of their profitability. The classic work on this subject is by Allen and Karjalainen (1999), who applied a GP heuristic to generate trading rules from Technical Analysis adapted to the daily prices of the S&P 500 index from 1928 to 1995. Considering transaction

---

[6] Convergence can be deduced from observing that trees remain unchanged through several generations.

costs, these researchers found that the rules produced by their algorithm failed to make consistent excess returns over a buy-and-hold strategy in out-of-sample test periods. Following the work of Allen and Karjalainen (from now on AK), other researchers have tried similar GP heuristics to produce technical trading rules for different financial instruments. Neely et al. (1997) produced trading rules for foreign exchange markets, and reported positive excess returns when trading with six exchange rates over the period 1981–1995; Wang (2000) applied GP to generate rules to trade in S&P 500 futures markets alone and to trade in spot and futures markets simultaneously, finding that his GP trading rules did not beat buy-and-hold in both cases; Roger (2002) investigated GP generated rules to trade with particular stocks (as opposed to an index as did AK) from the Paris Stock Exchange, and found that in 9 out of 10 experiments the trading rules outperformed buy-and-hold in out-of-sample test periods, and after deducting a 0.25 % per one-way transaction.

There are many others experiences with GP generation of technical trading rules with varying conclusions.[7] Those resulting in negative excess returns over buy-and-hold (as AK or Wang), can be interpreted as being in line with market efficiency. On the other hand, those experiments resulting in GP generated rules that consistently outperform buy-and-hold, should be taken first as an indication of some deficiency in the model, before being considered as a contest to the efficiency of markets hypothesis. For example, profitable GP generated rules could have been obtained because the penalty for trading (represented in the transaction costs) has been set too low; or due to over-fitting the data, which occurs when programs that are trained in a given time period of the data, produce other programs by means of some of the genetic operators, and these new programs are trained in the same time period (and with same data) as their parents, whereby knowledge about the data is transmitted through generations. Another possible cause of flaw in the model is the data snooping bias from the selection of specific time periods for training or testing; e.g., periods where data seems more (or less) volatile could influence the results, and by selecting these periods the researcher could be unwillingly forcing the conclusions. These observations motivate the following provisos that should be attached to the definition of a GP algorithm for finding profitable technical trading rules:

(1) Transaction costs should be part of the fitness function, and trials be made with different values, within realistically reasonable quantities.
(2) Consider separate time periods for generating the trading rules and testing their profitability. As a matter of fact, in the process of generating programs in the GP heuristic, there should be a time period for *training* and a different time period for *validation* (or selecting). In general, the time series data should be split into training, validation and test periods, in order to construct trading rules that generalize beyond the training data.
(3) The data snooping bias can be avoided by applying the *rolling forward* experimental set up. This consist on defining two or more partially overlapping sequences of training–validation–test time periods, as in the example shown in Table 7.1, and make a trial of the GP algorithm on each of these sequences.

---

[7] For a survey see Park and Irwin (2007).

Table 7.1 Rolling forward method with 3 sequences of training, validation, and testing, from 1995 to 2006

	1995	1996	1997	1998	1999	2000	2001	2002	2003	2004	2005	2006
1.	training				validation		test					
2.		training				validation			test			
3.			training					validation			test	

The GP algorithm by AK (and Wang's too) work with these provisos. AK tested their rules with one-way transaction costs of 0.1, 0.25 and 0.5 %, to find that only with the rather low fee of 0.1 % a few trading rules gave positive excess returns. They also applied different periods of time for training and validation, and used the rolling forward method.

We have conducted our own experiment using GP to find rules for trading in the Spanish Stock Market index (IBEX) and the Dow Jones Industrial (DJI) index (Llorente-Lopez and Arratia 2012). We followed the AK methodology, and based on our own experience we shall provide a general description of the experimental set up, that hopefully can guide the reader into conducting her own GP experiment for this financial application.

We first observe that the problem of finding profitable trading rules from Technical Analysis has the following global parameters:

- the essential features of the financial time series that is used as information for trading in the given financial asset; for example, daily or monthly price history, volume history, returns, or the time series of any other quantifiable fact of the asset;
- the trading strategy that is carry out.

These parameters condition and determine the building blocks to use for constructing trading rules and, by extension, the genetic programming scheme to learn these trading rules. Indeed, we argue that as trading rules are in general functions that take as input financial data (i.e., financial time series) and output a sequence of signals that trigger some trading action, it is the trading strategy that determines the nature of these signals. For example, if the strategy is to always buy and sell the same fixed quantity of the asset, then the signals emitted by the trading rules should be either "buy" or "sell", that is a binary valued output. But for a more sophisticated strategy that buys and sells varying proportions of the asset, then the trading rules should be functions that output signals from a possibly infinite set of real values (say, in [0, 1]) to quantify different degrees of buy or sell. Therefore, for the sake of precision, when applying a genetic programming scheme to learn technical trading rules one must indicate the underlying financial asset and the trading strategy, and be conscious that results are relative to these parameters. In this section a genetic programming solution is given for:

Finding technical trading rules for *common stocks*, using as information the *daily price history* and applying the simple, yet common, trading strategy of *buying and selling the same fixed proportion of the stock the day following the trading signal*.

Thus, all trading rules give binary valued signals (**false** for sell or **true** for buy), and the trading strategy consist on specifying the position to take the following day in the market (long or out), for a given trading rule signal and current position. More specifically: if we are currently long in the market, and the trading rule signals "sell", then we sell and move out of the market; if the current position is out, and the trading rule signals "buy", then we buy and move in to the market; in the other two cases, the current position is preserved. Each time we trade an equal amount of stock (e.g., one unit). This trading strategy is fully describe by Table 7.2.

Table 7.2 Trading strategy based on 2- valued trading rule signals

		Position	
		in	out
Signal	sell	out	out
	buy	in	in

The full GP algorithm for finding a profitable trading rule from Technical Analysis is given as Algorithm 7.5. The technical details of this GP solution follows below.

Algorithm 7.5 GP algorithm for finding a technical trading rule

---

1. **initialization**: generate a random set P of N rules; best rule $\rho^* := $ NULL;
     $generation := 0$; $S = P$ {*unselected group*}
     $M := \mathbb{P}(crossover) \times N$; $p_m = \mathbb{P}(mutation)$
2. **while** (stopping criterion not met) **do**
3.   evaluate fitness of each rule in P during *training* period $\{fit_T\}$
4.   set ρ as the rule with best fitness in *training* period
5.   evaluate fitness of ρ in *validation* period $\{i.e.$ *compute* $fit_S(\rho)\}$
6.   **if** $(fit_S(\rho) > fit_S(\rho^*))$ **then** $\rho^* := \rho$
7. $i := 0$ {*number of new individuals*}
8.   **while** $(i \leq M)$ **do**
9.   select two rules ρ_1 and ρ_2 at random, using a probability
             distribution skewed towards the best rule (w.r.to fit_T)
10. $S := S - \{\rho_1, \rho_2\}$
11.   apply *crossover* to ρ_1, ρ_2 to generate offsprings ξ_1, ξ_2
12.   apply *survival criterion* to $\rho_1, \rho_2, \xi_1, \xi_2$ to remove one parent and one child.
             Let ρ and ξ be the surviving parent and descendent
13.   with probability p_m apply mutation to the child ξ. Let ξ' be the result
14.   insert ρ and ξ' into *NewPopulation*
15. $i := i + 2$
16.   **end while**
17. $P := NewPopulation \cup S$
18. $generation := generation + 1$
19. **end while**
20. **output**: ρ^*, best individual based on fitness fit_S

The general experimental set up. Following provisos (2) and (3) for the experimental set up of a GP algorithm for finding rules for trading with certain financial instrument, we must train the rules in a period different from the validation (where the best rules are selected), and then test their profitability in another period. To avoid the data snooping bias we applied the rolling forward method, which consists on repeating the experiment with different overlapping sequences of training–valuation–test periods. The number of these sequences and the length of each period conforming each sequence is determined by the amount of data available. In the case of DJI and IBEX we have plenty of data, and study these time series from 1995 to 2011. This allow us to define five time sequences, each divided in a training period of 4 years long, a validation period of 2 years, and a testing period of 2 years. Beginning in 1995 we can set the sequences overlapping for 6 years (see back Table 7.1, where the first 3 sequences are presented). For each data sequence, we made 50 runs of the GP algorithm (the parameter "number of iterations" is set to 50). This give us 50 "best rules" to be tested for profitability. In each run, the three data periods are used in the following manner: In each generation of rules, the best fitted rule against the *training* period is selected and its fitness is evaluated against the *validation* period. If this rule has better validation fitness than the previous candidate to "best rule", then this is the new "best rule". The process is repeated until the stopping criterion is met. The resulting "best rule" is tested for profitability against the *test* period. We denote the result of applying the fitness function *fit* to a rule ρ in the training period by $fit_T(\rho)$, and in the valuation period by $fit_S(\rho)$.

Individuals are trading rules. Since each trading rule is a boolean function that takes as input a given price history, and returns **true** to signal a "buy", or **false** to signal a "sell", it can be easily represented by the parsing tree of a boolean expression. Take, for example, a triple moving average crossover rule for periods of 9, 20 and 50 market days (cf. Sect. 6.1.3). This rule can be expressed by the following boolean function

$$\left(\frac{1}{9}\sum_{i=0}^{8}P_{t-i} > \frac{1}{20}\sum_{i=0}^{19}P_{t-i}\right) \text{ AND } \left(\frac{1}{20}\sum_{i=1}^{19}P_{t-i} > \frac{1}{50}\sum_{i=0}^{49}P_{t-i}\right)$$

We use $MA(n)_t$ to denote the average of the price history considered over the recent past $n-1$ days, and up to current time t; i.e., $MA(n)_t = \frac{1}{n}\sum_{i=0}^{n-1}P_{t-i}$. Figure 7.2 presents a tree representation of the triple moving average crossover rule (9, 20, 50)-periods.

Formally, the trading rules are built from the following set of terminals and functions of different types and varying number of parameters:

Terminals: constants of type real, integer and boolean. Reals are fixed to a finite set of values around 1 to represent a set of normalize prices; integers range from 0 to 252, to represent working days up to a year; boolean constants have value **true** or **false**.

Variables: *price* that takes as its value the stock price of current day (a real number).

Fig. 7.2 Triple moving average crossover rule coded as tree

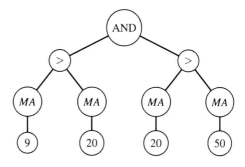

Functions: AND, OR, NOT are of type boolean, the first two received two parameters, while NOT has one parameter, all of type boolean.

$<, >$ and $=$, are of type boolean and of two parameters, both of type real.

IF-THEN–ELSE, another boolean function, but of three boolean parameters; if the boolean valued of the first parameter is true then return the value of second parameter, otherwise return value of third parameter.
$+, -, /, \times$, are real type functions of two real parameters.
Finally, the domain-specific real functions with one parameter of type integer: $MA(n)_t = \sum_{i=0}^{n-1} P_{t-i}/n$, $\max(n)$, $\min(n)$ and $lag(n)$, interpreted respectively as moving average, maximum and minimum of the series of prices over the past n days, and the price lagged by n days from the current day.

Now, in order to comply with the closure property of any genetic program and at the same time with our intended binary valued trading signals, we impose the following restrictions in the construction of trading rules:

- only allow boolean type functions as the root of the tree (this ensures that the output of the rule is **true** or **false**);
- every function node must have as many descendants as the number of its parameters, and each descendant is a terminal or function node of a type matching the type of the corresponding parameter in the function.

Initial population. We apply the ramped half-and-half method described in the general initialization procedure in Sect. 7.3.1. Each trading rule is limited to at most 80 nodes (terminals and functions), which sets a bound to the depth of the associated tree around 7. The size of each population is limited to 200.

Fitness. The fitness function measures excess return over a buy-and-hold strategy, taking into account transaction costs. To define it, we shall assume that the cost of a one-way transaction is always a fixed positive fraction (or percentage) α of the current price. Recall from Problem 2.7.2 that the return for a single trade, say to buy at date $t = b$ and sell at date $t = s$, considering transaction costs at the rate of α, is given by

$$R_{bs} = \frac{P_s}{P_b} \cdot \left(\frac{1-\alpha}{1+\alpha}\right) - 1$$

The log return (with transaction costs) is

$$r_{bs} = \ln\left(\frac{P_s}{P_b}\right) + \ln\left(\frac{1-\alpha}{1+\alpha}\right) \qquad (7.5)$$

Let T be the number of trading days, and $n \leq T$ be the number of trades made by the rule within the trading period $[1, T]$. This is a sequence of n pairs of "buy" followed by "sell" signals (or an empty sequence in case the rule never signals to go in the market).

Now in order to compute the total continuously compounded return for the trading rule ρ generating this sequence of n pairs of signals, we can mark and save explicitly the entry dates and the exit dates, and at the end of the period only sum over all corresponding terms given by Eq. (7.5). Alternatively, for a less intensive use of computer memory, we can compute, on each day t, the daily return r_t and check if the rule ρ is signaling to be in or out, in which case multiply r_t by 1 or 0 respectively. We take this latter approach, which is formalize as follows. Consider the indicator function:

$$I_\rho(t) = \begin{cases} 1 & \text{if } \rho \text{ signals "buy"} \\ 0 & \text{otherwise} \end{cases}$$

Then the continuously compounded return for the trading rule ρ throughout the trading period $[1, T]$ is

$$r = \sum_{t=1}^{T} I_\rho(t) \cdot r_t + n \ln\left(\frac{1-\alpha}{1+\alpha}\right) \qquad (7.6)$$

and the return for the buy-and-hold strategy (buy the first day, sell the last day) is

$$r_{1T} = \ln\left(\frac{P_T}{P_1}\right) + \ln\left(\frac{1-\alpha}{1+\alpha}\right)$$

The excess return (or fitness) for the trading rule is given by

$$fit(\rho) = r - r_{1T} \qquad (7.7)$$

Selection. In selecting programs (potential trading rules) we avoid over-selection by using a rank-based selection. This is done as follows. The current population of N programs is sorted, based on their raw fitness value ($fit(\rho)$), from best fitted (which has rank 1) to worst fitted (which has rank N). Let f_b (resp. f_w) be the fitness value of the best (resp. worst) fitted program. Then the rank-based fitness value f_i for the program of rank i is

$$f_i = f_b - (f_b - f_w) \left(\frac{i-1}{N-1} \right)$$

Note that $f_1 = f_b$ and $f_N = f_w$. Using f_i as an adjusted fitness, only for the purposes of selection, programs which have close raw fitness are now separated and have similar chance of being randomly selected, while programs with very high fitness are bound for the f_b value, reducing the chance of being the most selected.

Genetic operations. We apply crossover with high probability. For that the parameter $\mathbb{P}(crossover) = p$ is set to $p = 0.9$, and define the variable $M = p \cdot N$ that sets the number of programs that should be obtained by crossover from the current population. Crossover is done as explained in the general scheme: two trading rules are recombine by joining their tree representations at some compatible node, and two offsprings are thus generated. An example is shown in Fig. 7.3. To keep the size of the population bounded by N, we eliminate one parent and one offspring applying the following

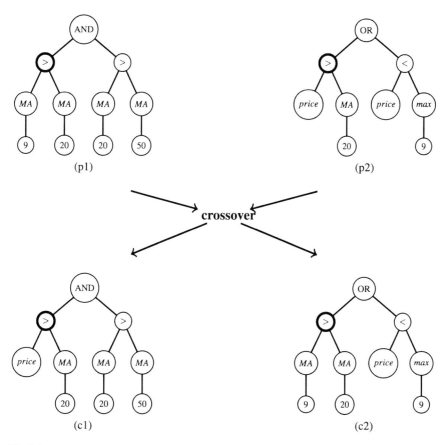

Fig. 7.3 Crossover operation: Parent trees (**p1**) and (**p2**) exchange their subtrees rooted at leftmost (>) node (highlighted); the resulting children are (**c1**) and (**c2**)

survival criterion: the best fitted offspring replaces one of the parents, randomly selected using a probability distribution skewed towards the worst fitted parent. In this way we keep the best fitted offspring, plus one of the parents that is not always the best fitted.

Mutation is applied to the surviving offspring from the crossover operation (following a general advise given by Koza (1992)). This is to guarantee certain degree of diversity, in case parents are too similar; although this is a rare event, and hence this should be done with low probability. We set the probability of mutation $p_m = 0.1$.

We remark that the algorithm by AK does crossover always; i.e. the authors have set $p = 1$. Their crossover produces only one offspring, which replaces one of the parents in the manner we have described. Also AK does not perform any mutation. **Stopping criterion**. This is set to a limited number of generations (in our experiments we used 100), or if the best rule does not improves for a number of generations (we used 50).

Table 7.3 summarizes the values used for the parameters of our GP algorithm for finding trading rules.

Table 7.3 Parameters values for GP

Crossover probability	90 %
Mutation probability	5 %
Number of iterations	50
Size of population	200
Maximum number of generations	100
Maximum number of nodes per tree	80
Number of generations w/o improvement	50

A note on implementation. The GP heuristic makes an intensive use of memory space, so it is important to use a compact representation of individuals with fast access to the information. The article by Keith and Martin (1994) studies various ways of implementing the nodes and storing the trees (programs) generated by a GP, efficiently with respect to memory space and execution time. Although the analysis they do is geared for a C++ platform, it can be adapted to other programming language.

7.4 Ant Colony Optimization

The ant colony optimization (ACO) heuristic belongs to the class of swarm intelligence based methods where a collection of individuals, having a common goal, exchange information about their local actions to build collectively a global procedure that would lead everyone of the participants to the goal. Originally proposed by M. Dorigo in his Doctoral Dissertation Dissertation (Dorigo 1992), and Colorni et al. (1992a, b), as an optimization heuristic in Computer Sciences, the ant colony optimization paradigm has been since then successfully applied to many optimization problems, ranging from purely combinatorial, to problems in telecommunication networks, scheduling, machine learning, bioinformatics and finance (Dorigo and Stützle 2004; Stützle et al. 2011). We give a brief account of this heuristic and show an application to option valuation in the following sections.

7.4.1 The Basics of Ant Colony Optimization

The inspiration for the ACO heuristic comes from the way a colony of ants search for food and traces their way back to the nest. In this search various ants go out of the colony searching for food, and once an ant finds food, it takes a portion back to the nest leaving traces of pheromone on the ground, a chemical trail to guide other ants to the food source. The first ants to arrive with food to the nest have found shortest paths to the food sources, and the quantity of pheromone deposited in each trail is an indication of the quantity and quality of the food. All this information is gathered by other ants from the nest, which quickly go out for food through the currently best routes (those with higher pheromone traces and made by the first arriving comrades). This immediate reaction updates the importance of some trails, since previously laid pheromone slowly evaporates. As the food at certain sources reduces in quantity or quality the trail slowly disappear, as the intensity of pheromone diminishes, and ants start searching for other food sources.

It is clear that this is an appropriate heuristic for solving optimization problems which can be reduced to finding routes in graphs. Hence, the crux in the design of an ACO solution for a given optimization problem $(\mathscr{S}, \Omega, F)$ is to devise a constructive heuristic $\mathscr{H}$ for building solutions (elements of $\mathscr{S}$), as sequences of elements from a finite set $\mathscr{C}$ of *solution components*, which one must determined. We may view $\mathscr{H}$ as a map from sequences of elements of $\mathscr{C}$ onto $\mathscr{S}$. Any solution is then seen as a path $S = \langle u_1, u_2, \ldots, u_n \rangle$ in a totally connected graph formed by the elements of $\mathscr{C}$, and a solution component u_i is added to the partial solution $S^{i-1} = \langle u_1, u_2, \ldots, u_{i-1} \rangle$ by the constructive heuristic $\mathscr{H}$, which makes sure the constrains Ω are met.

Example 7.4 To illustrate the crucial step just described for the design of an ACO solution, consider the asymmetric traveling salesman problem (ATSP), where an instance is a completely connected directed graph $G = (V, A)$ with a positive weight w_{ij} associated to each arc $a_{ij} \in A$. The vertices of the graph (elements of V) represent cities and the arc weights represent distances between the cities. The goal in this problem is to find the minimum weighted directed Hamiltonian cycle, that is, a Hamiltonian cycle in G for which the sum of the weights of its arcs is minimal. The ATSP is an NP-hard problem (Garey and Johnson 1979) Then to find approximate solutions to this problem with the ACO heuristic, consider as solution components the set of arcs, so $\mathscr{C} = A$, and the constructive heuristic $\mathscr{H}$ consist on assembling arcs together in such a way that a path is built verifying the Hamiltonian restriction. □

Algorithm 7.6 ACO basic algorithm for $(\mathcal{S}, \Omega, F)$

---

1. **initialization:** $\mathcal{C}$ {*finite set of solution components*}
     $\forall u \in \mathcal{C}, \tau_u = c > 0$ {*pheromone values*}
     $\alpha, \beta, \rho, J(0) = \emptyset$
     $S^* = $ NULL {*best path*}
     $t = 0$ {*iteration counter*}
2. **while** (stopping criterion not met) **do**
3. $t = t + 1$
     {*Phase 1: ants build their paths*}
4.   **for** $k = 1, \ldots, N_a$ **do**
5. $S_k \leftarrow BuildPath(k)$
6.   **if** $(F(S_k) < F(S^*)$ or $S^* = $ NULL$)$ **then** $S^* = S_k$
7. $J(t) = J(t-1) \cup \{S_k\}$
8.   **end for**
     {*Phase 2: pheromones are updated on the paths*}
9. $UpdatePheromone(J(t), \langle \tau_u \rangle_{u \in \mathcal{C}})$
10. **end while**
11. **output:** S^* {*best path*}

The basic ACO scheme is presented in Algorithm 7.6. The input is an instance of an optimization problem $(\mathcal{S}, \Omega, F)$, and it is assumed that we have a constructive heuristic $\mathcal{H}$ with the characteristics described above, which builds candidate solutions $S \in \mathcal{S}$ as paths formed with solution components from a finite set $\mathcal{C}$. Additionally, for a true ACO solution, in this constructive process for building solutions as paths, ants choose next solution component $u \in \mathcal{C}$ randomly with probabilities subjected to the pheromone model. These probabilities of moving from one solution component to another (or transition probabilities), are defined for the kth ant, acting at time t, by the equation

$$P^k(u, t) = \begin{cases} \dfrac{(\tau(u,t))^\alpha \cdot (\eta(u,t))^\beta}{\sum_{w \in \mathcal{R}(k)} (\tau(w,t))^\alpha \cdot (\eta(w,t))^\beta} & \text{if } u \in \mathcal{R}(k) \\ 0 & \text{otherwise} \end{cases} \qquad (7.8)$$

where $\tau(u, t)$ is the pheromone value present in component u at time t, $\eta(u, t)$ is the *heuristic information* value[8] of the solution component u at time t, $\mathcal{R}(k)$ is the set of solution components that ant k has not yet visited, and the parameters α and β, $\alpha > 0$ and $\beta > 0$, measure the relative importance of pheromone value and heuristic information.

This process of building a path by an ant, using the constructive heuristic $\mathcal{H}$ to build solutions from the set $\mathcal{C}$ of solution components, subjected to the constraints Ω,

---

[8] $\eta(u, t)$ is a measure of how much the component u is desired to be part of the solution. For the ATSP problem, where the solution components are arcs, is usual to take $\eta(a_{ij}, t) = 1/w_{ij}$, the multiplicative inverse of the weight, for any time t.

and selecting a next component in the path using the transition probabilities given by Eq. (7.8), is encoded in the algorithm (line 5) as function $BuildPath(k)$. This function returns a path S_k, built by the kth ant. The quality of each path S_k is evaluated by the objective function F and the best path (or best solution) is updated accordingly (line 6). Then the path is saved in the set $J(t)$, which contains already constructed paths verifying the constraints up to time (or iteration) t (line 7).

The pheromone values are initially set to some positive constant c, and uniformly for all solution components. Afterwards, at each iteration $t > 0$, when all ants have built their paths, pheromone values are updated for all components u that belong to a path in the set $J(t)$. This is what the subroutine $UpdatePheromone(J(t), \langle \tau_u \rangle_{u \in \mathscr{C}})$ does (line 9). The pheromone value of component u is updated at time t, according to the following rule

$$\tau(u, t) \leftarrow (1 - \rho)\tau(u, t) + \rho \sum_{k=1}^{N_a} \Delta \tau^k(u, t) \qquad (7.9)$$

where $0 < \rho \leq 1$ measures the pheromone decay (or evaporation rate), and

$$\Delta \tau^k(u, t) = \begin{cases} \frac{\gamma}{1+F(S_{|u})} & \text{if } k\text{th ant takes component } u \\ 0 & \text{otherwise} \end{cases}$$

with $S_{|u}$ a path (candidate solution) containing component u, γ some positive constant, and N_a is the total number of ants. Thus, Eq. (7.9) gives an update of the value of the pheromone present at time t in solution component u, as a convex combination of current value (which if it hasn't been visited is the initial constant c) and the cumulative sum of the pheromone deposited in u by all ants. In this manner, with the parameter ρ, one can control the velocity of convergence of the algorithm towards local optima, by setting, for example, a high evaporation rate. A stopping criterion for ACO algorithm is usually set as an upper bound in the number of iterations. When this bound is reached the best path found so far is given as output. The next section describes an application of the ACO heuristic to the valuation of American put options.

7.4.2 Valuing Options with Ant Colony Optimization

Keber and Schuster (2002) give an ant colony based heuristic to learn analytical expressions for the valuation of American put options on non dividend paying stocks. Their main idea is to define a context-free grammar[9] $\mathscr{G} = (T, N, P, S)$ to generate analytical expressions that are to be interpreted as functional approximations for

---

[9] For details on the theory of formal grammars see Lewis and Papadimitriou (1998, Chap. 3).

the valuation of stock options. Then an ACO heuristic is used as search engine for producing the derivations of the grammar.

Let us begin fixing the notation. In a context-free grammar $\mathcal{G} = (T, N, P, S)$, the set T is the vocabulary of the grammar and its elements are the *terminal* symbols, N is the set of *non-terminal* symbols, S is the *start* symbol, which is a member of N, and P is the set of *production rules*. All the sets T, N and P are finite. The language $\mathcal{L}(\mathcal{G})$ produced by the grammar $\mathcal{G}$, consists of all finite concatenations (strings) of elements of T, which can be derived from the start symbol S by successive application of the production rules in P acting on the non-terminal symbols to produce elements in N or T. Formally,

$$\mathcal{L}(\mathcal{G}) = \{w : S \overset{*}{\Longrightarrow} w, \ w \in T^*\}$$

where T^* is the set of all finite strings over the vocabulary T, and $S \overset{*}{\Longrightarrow} w$ means that w is derived from S by finitely many successive applications of production rules from P. The following example illustrates how a context-free grammar could be defined to obtain analytical expressions like the Black-Scholes option pricing formulas (Eqs. 5.21 and 5.22).

Example 7.5 We require a grammar (T, N, P, S) where the set T of terminals should contain the basic items found in a real valued analytical equation (e.g. mathematical operators, real numbers, and some predefined variables). The following T should fit our purpose:

$$T = \{S_0, K, r, T, \sigma, 0, 1, 2, 3, 4, 5, 6, 7, 8, 9,$$
$$+, -, *, /, x^y, \sqrt{x}, \ln(x), \exp(x), \Phi(x)\}$$

The first five symbols represent, respectively, initial stock price (i.e. at time $t = 0$), option's exercise price, annual continuous risk-free interest rate, annualized time to expiration, and annual volatility. We have the ten first natural digits that are used to construct real numbers in a manner that will be explained shortly. We have symbols for the basic arithmetic operations, and some analytical functions, including the standard normal distribution $\Phi(x)$.

The set N of non-terminals contain symbols that can be expanded to terminals or other non-terminals, and the start symbol S is a member of N. Let

$$N = \{expr, var, biop, uop, digit, int, real\}$$
$$S = expr$$

The production rules in P describe how the non-terminals are mapped to other non-terminals or terminals to create analytical expressions. Consider the following rules:

$$\langle expr \rangle \rightarrow (\langle expr \rangle \langle biop \rangle \langle expr \rangle)$$
$$| \langle uop \rangle (\langle expr \rangle)$$
$$| \langle real \rangle * (\langle expr \rangle)$$
$$| \langle var \rangle$$
$$\langle biop \rangle \rightarrow + | * | / | \hat{} $$
$$\langle uop \rangle \rightarrow - | \exp | \ln | \sqrt{} | \Phi$$
$$\langle var \rangle \rightarrow S_0 | K | r | T | \sigma$$
$$\langle real \rangle \rightarrow \langle int \rangle . \langle int \rangle$$
$$\langle int \rangle \rightarrow \langle int \rangle \langle digit \rangle | \langle digit \rangle$$
$$\langle digit \rangle \rightarrow 0 | 1 | 2 | 3 | 4 | 5 | 6 | 7 | 8 | 9$$

The "|" (or) separates the different alternatives for a rule acting on a non-terminal. A derivation consist in a sequence of replacements of a non-terminal symbol by another symbol, until a string containing only terminal symbols is obtained. Each step in the derivation conforms to a mapping from a non-terminal to another symbol defined by one of the production rules. Special attention deserves the set of rules for $\langle real \rangle$, $\langle int \rangle$ and $\langle digit \rangle$. These conform an efficient method to create numerical constants, integer or real, of any size, and which has been termed *digit concatenation* (O'Neill et al. 2003). For example, the real number 3.14 can be derived using these set of rules, as follows:

$$\langle real \rangle \rightarrow \langle int \rangle . \langle int \rangle \rightarrow \langle int \rangle . \langle int \rangle \langle digit \rangle \rightarrow \langle int \rangle . \langle digit \rangle \langle digit \rangle$$
$$\rightarrow \langle digit \rangle . \langle digit \rangle \langle digit \rangle \rightarrow 3 . \langle digit \rangle \langle digit \rangle \rightarrow 3.1 \langle digit \rangle \rightarrow 3.14$$

A derivation in the language $\mathscr{L}(\mathscr{G})$ of the grammar starts with the symbol S, and results in an analytical expression on the variables S_0, K, r, T, and σ. Let us see how the expression $-3.14 * S_0 * \Phi(T)$ could be derived in this grammar.

$$S \rightarrow \langle expr \rangle \langle biop \rangle \langle expr \rangle \rightarrow \langle uop \rangle (\langle expr \rangle) \langle biop \rangle \langle expr \rangle \rightarrow -(\langle expr \rangle) \langle biop \rangle \langle expr \rangle$$
$$\rightarrow -(\langle real \rangle * \langle expr \rangle) \langle biop \rangle \langle expr \rangle \rightarrow \cdots \rightarrow -3.14 * \langle expr \rangle \langle biop \rangle \langle expr \rangle$$
$$\rightarrow -3.14 * \langle expr \rangle * \langle expr \rangle \rightarrow \cdots \rightarrow -3.14 * \langle var \rangle * \langle uop \rangle (\langle expr \rangle)$$
$$\rightarrow \cdots \rightarrow -3.14 * S_0 * \langle uop \rangle (\langle var \rangle) \rightarrow \cdots \rightarrow -3.14 * S_0 * \Phi(T)$$

We have abbreviated some steps of this derivation: the first $\rightarrow \cdots \rightarrow$ replaces the derivation of the number 3.14 done before; the other ellipsis are shortcuts for simultaneous applications of two or more obvious production rules.   □

Having defined the appropriate grammar, the next step is to use the ACO heuristic as search engine in the space of words $\mathscr{L}(\mathscr{G})$. The adaptation of the heuristic is somewhat natural: the strings $w \in \mathscr{L}(\mathscr{G})$ are the feasible solutions (the potential analytical expressions for valuing options), the solution components are the non-terminal and terminal symbols, paths are derivations in the grammar, ants begin building paths from the start symbol S, and for the k-th ant ($k = 1, \ldots, N_a$) the path construction process *BuildPath(k)* is comprised of the production rules of the grammar plus the transition probabilities given by Eq. (7.8) to decide the alternatives of the rules to use in the next derivation.

The fitness function F is defined as follows. Each path built by an ant, or derivation $w \in \mathcal{L}(\mathcal{G})$ (an analytical expression), is a function on the variables contained in the terminals T, so that on input data $\mathcal{E}$, subject to certain domain restrictions, delivers an output data $\mathcal{A}$, $w : \mathcal{E} \to \mathcal{A}$. Therefore, to test the fitness of w, several input data records $\mathcal{E}_i$ are produced respecting the restrictions, for $i = 1, \ldots I$, and each corresponding output record $w(\mathcal{E}_i) = \mathcal{A}_i$ is compared to a target output $\mathcal{A}_i^*$. The fitness function is then defined by $F(w) = \dfrac{1}{1 + Q(w)}$, where $Q(w) = \displaystyle\sum_{i=1}^{I} \delta(\mathcal{A}_i, \mathcal{A}_i^*)$ is the aggregated deviations given by some error function $\delta(\mathcal{A}_i, \mathcal{A}_i^*)$ (e.g. Q could be a sum of squared errors).

Additionally, to the ACO heuristic (Algorithm 7.6) applied to the language $\mathcal{L}(\mathcal{G})$ and fitness function F as described above, Keber and Schuster apply a reduction in size of the set $J(t)$ in the *UpdatePheromone* routine (line 9), by deleting those paths (instances of analytic expressions) where pheromone has evaporated. The motivation for this adjustment is that the number of expressions in $\mathcal{L}(\mathcal{G})$ can grow exponentially, and hence some form of pruning the space search is needed.

For their experimental set up, Keber and Schuster used as training data a randomly generated sample of 1,000 American put options on non-dividend paying stocks, described as tuples $(P_0, S_0, K, r, T, \sigma)$, where P_0 is the "exact" value of an American put option on non-dividend paying stock at $t = 0$, computed by the finite difference method, and $S_0 > S^*$, where S^* represents the killing price (computed by some known numerical procedure). These sample tuples were transformed into an input data record $\mathcal{E}_i (= (S_0, K, T, r, \sigma))$ and a corresponding target output record $\mathcal{A}_i^* (= P_0)$, for $i = 1, \ldots, 1000$, and for the aggregated deviations function they used the sum of squared errors $\sum_{i=1}^{1000}(\mathcal{A}_i - \mathcal{A}_i^*)^2$. The domains of values for the variables were restricted as follows: $2 \leq r \leq 10$, $1/360 \leq T \leq 1/4$, and $10 \leq \sigma \leq 50$; $S_0 = 1$ and $S_0 > S^*$. The number of ants were taken as $N_a = 50$, and for the relative importance of pheromone and heuristic information they used $\alpha = 1$ and $\beta = 1$; ρ was set to 0.5, $\gamma = 1$, and the stopping criterion is to halt after 100,000 iterations.

Keber and Schuster found in their experiments as one of their best analytical approximations derived by their ACO-engined grammar, the following expression

$$P_0 \approx K \cdot e^{-rT} \Phi(-d_2) - S_0 \cdot \Phi(-d_1)$$

where

$$d_1 = \frac{\ln(S_0/K)}{\sigma\sqrt{T}}, \qquad d_2 = d_1 - \sigma\sqrt{T}$$

A formula that has an amazing similarity in its structure to the Black-Scholes formula for valuing European put options. Validation of expressions like these, and other best solutions obtained by their program, against 50,000 randomly generated American put options, resulted in a 96 % of accuracy for some cases, outperforming some of the most frequently used analytical approximations for American put options.[10]

7.5 Hybrid Heuristics

There are different ways in which we can combine the individual heuristics presented in previous sections and obtain a *hybrid heuristic*. Talbi in (Talbi 2002) has compiled an almost exhaustive taxonomy of hybrid heuristics, which the reader may consult as a guide for mixing heuristics according to the problem at hand. In his compilation, Talbi classifies hybrid heuristics in *hierarchical* or *flat* schemes, the latter refers to algorithms whose descriptors can be chosen in arbitrary order. But the most frequent form of hybrid heuristic is hierarchical, and is of this form that we shall limit the examples in this section.

The class of hierarchical hybrid heuristics is divided in *low level* and *high level*, and within each level we may further distinguish between *relay* and *cooperative*.[11] The low level hybridization replaces a given function of a heuristic by another heuristic. In high level hybridization the different heuristics are self contained. The relay form combines heuristics in pipeline fashion: applying one after another, each using the output of the previous input. In cooperative hybridization various heuristics cooperate in parallel.

The following example presents cases of high level relay hybrid heuristic.

Example 7.6 Consider the basic genetic programming algorithm (Algorithm 7.4). A possible hybridization is to use a greedy heuristic to generate the initial population. This greedy strategy could consist on applying the fitness function to the initial individuals that are randomly generated, and keep those ranking high in an ordering by fitness (or discard those individuals with fitness below a certain threshold). The drawback with this extension of the GP algorithm is that it might induce over-fitted succeeding generations, and this, as it has been commented before, could trap the algorithm into a local optimum.

Another hybridization is to implement a simulated annealing-like heuristic after the operation of selection in the GP algorithm, to enhance the fitness of the selected individuals but still considering as solutions some not well fitted individuals, due to the capability of making escape-moves of the SA heuristic. This is in fact an instantiation of the general probabilistic selection strategy outlined in Sect. 7.3.1, namely, to make a random selection on sorted individuals (assumed uniformly distributed) with

---

[10] Those described in: Johnson (1983), An analytic approximation for the American put price, *J. of Financial and Quantitative Analysis*; and Geske , Johnson (1984), The American put option valued analytically, *J. of Finance*.

[11] Talbi originally calls it *teamwork*.

a probability function skewed towards best-fitted individuals. The simulated anneal-ing implementation comes down to do this random selection under the Metropolis probability function. In this case this hybridization does provides an enhancement of the raw GP algorithm (i.e. where selection of individuals consists on simply choos-ing the best-fitted), for as we have commented before this random selection give theoretical guarantees of global convergence.[12]

Both forms presented of hybridization of GP are of the type hierarchical high level relay, since the additional heuristics (greedy in the first, SA in the sec-ond construction) are applied in addition to the operations of GP and working in tandem.   □

Our second example presents a case of low level cooperative hybrid heuristic. These type of hybridizations are motivated by the need of improving the local search for those heuristics, such as GP or ACO, that are good in exploring the search space globally but do not care to refine, or exploit, the local findings.

Example 7.7 In the ACO heuristic (Algorithm 7.6), one can refine the local search of solutions by substituting the *BuildPath*(k) routine by a SA heuristic, so that instead of choosing a next component solution, based on transition probabilities and pheromone values, choose a next component solution as the best solution found in a local search with SA in the neighbor of the current component solution, using as objective func-tion the pheromone valuation, or some similar function. We leave as exercise the formalization of this hybridization.[13]   □

7.6 Practical Considerations on the Use of Optimization Heuristics

The heuristics defined in this chapter, and others alike, are powerful alternatives when classical optimization methods, such as least squares estimation and other algebraic methods, have limited capacity for finding solutions of certain problems of interest. However, before embarking on the design of an optimization heuristic one must assess if it is really needed. The use of these heuristics should be founded, in principle, on some conviction of the computational infeasibility of the problem, for otherwise a more exact deterministic method should be preferred. Once one is convinced of the computational hardness of the problem (maybe by some proof of reduction to an NP-complete problem,[14]) or that the objective function may have several local optimum (by observing some partial plot), one should decide on a

---

[12] See also Hartl 1990, A global convergence proof for a class of genetic algorithms, *Tech. Report, University of Technology, Vienna*, available from the author's webpage.

[13] You might find some hints for this problem in the work by Taillard and Gambardella 1997, An ant approach for structured quadratic assignment problems. In: *Proc. 2nd. Metaheuristics International Conf., Sophia-Antipolis, France*.

[14] cf. Garey and Johnson (1979).

heuristic that best suits the characteristics of the problem. For that, it is essential to have a good definition of the search space. This would indicate if the need is for doing intensive local search, refining singular instances of solutions, hence calling for simulate annealing or similar trajectory method; or if the need is to maintain a pool of solutions to do selective picking, in which case genetic programming or ant systems are appropriate. Whatever the choice, do begin experimenting with a simple general heuristic before turning to some complicated hybrid system.

Do not forget the stochastic nature of every solution obtained with these optimization heuristics, which is due to the random component of the algorithms. This implies that repeated applications of the heuristic with same input do not yield the same result (almost surely). Therefore, every solution obtained must not be considered definitive, but rather a random drawing from some unknown distribution. This distribution can be empirically computed by executing the optimization heuristic several times on the same problem instance, with fixed parameters, and plot the values $F(\theta_r)$ of the objective function over the results of the r-th run, θ_r; from this plot one can observe lower (or upper) quantiles which serve as estimates of the minimum (or maximum) values expected for the objective function. For all the optimization heuristics presented in this chapter, global convergence results have been proven that guarantee that for any replication the solution obtained can be close to the global optimum with high probability, provided the heuristic runs through a sufficient number of iterations I (stopping criterion).[15] However, the value of I required by these global convergences to hold is usually computationally unattainable with our current computer resources, so we have to be content with rough approximations of the distribution of values of the objective function, for different values of I.

We now list a few useful recommendations, specific to the parameters controlling an optimization heuristic. For methods involving neighborhood search (e.g. SA):

- The solutions within a neighborhood should be easy to compute (e.g. as in Algorithm 7.3 where a neighbor solution is obtained by perturbing one coordinate of a given solution). This is important for reducing computation time.
- The topology induced by the objective function on the solution space should not be too flat. If we view the values of the objective function as the height of a solution, then if almost all solutions within a neighborhood have similar height (describing a flat surface) it would be difficult for the method to escape from such locality.

For methods involving co-operation among individuals in the form of exchanging information about the solution space (e.g. GP or ACO), it is important that:

- The information transmitted is relevant, so that there is improvement in the generations.
- The combination of two equivalent parent solutions should not produce an offspring that is different from the parents (consistency in information transmission).

---

[15] To learn asymptotic convergence results for simulated annealing see Laarhoven and Aarts (1987); Aarts and Lenstra (2003); for genetic algorithms (Reeves and Rowe 2003); for ant colony (Dorigo and Stützle 2004).

- Diversity in the population should be preserved, to avoid premature convergence to local optimum.

For all the heuristics, due to their stochastic nature, it is recommendable to plot values of the objective function, for different runs, to determined convergence, and with different values for the number of iterations, or other parameters, to tune these accordingly. Further discussion and examples of this empirical tuning of an optimization heuristic can be found in Gilli and Winker (2008). Also, from these authors, it is worth adding their recommendation on properly reporting results of experiments with any optimization heuristic. This includes indicating the amount of computational resources spent by the method; empirical demonstrations (via plots or tables) of the rate of convergence; comparison with other methods, and application to real as well to synthetic data to attest for the robustness of the heuristic.

7.7 Notes, Computer Lab and Problems

7.7.1 Bibliographic remarks: Gilli and Winker (2008) and Maringer (2005, Chap. 2) are two excellent summaries of optimization heuristics and their use in econometrics and portfolio management, respectively. For the general theory of heuristics see Aarts and Lenstra (2003). The development of genetic algorithms is generally attributed to John Holland (1975), although other evolution based strategies were proposed by contemporaries (Rechenberg 1973) and some precursors (Fogel et al. 1966). The book by Koza (1992) contains all you want to know about genetic programming. A step further from genetic programming is Grammatical Evolution (Ryan et al. 1998), where the mechanism for creating programs is described through a formal grammar, and with a linear representation of programs, as opposed to trees. The ACO application of Keber and Schuster (2002) that we have presented in Sect. 7.4.2 can be considered within this class of grammatical evolutionary heuristics; in fact, it is an example of Grammatical Swarm. For more on these biological inspired heuristics and their applications to finance see the textbook by Brabazon and O'Neill (2006).

7.7.2 R references: There is an R project for genetic programming, called R Genetic Programming (RGP), at http://rsymbolic.org/projects/show/rgp. This is a simple modular genetic programming system build in pure R. In addition to general GP tasks, the system supports symbolic regression by GP through the familiar R model formula interface. GP individuals are represented as R expressions, an (optional) type system enables domain–specific function sets containing functions of diverse domain and range types. A basic set of genetic operators for variation (mutation and crossover) and selection is provided.

7.7.3 R Lab: Write the code in R for the function fn <- function(psi,r){…} corresponding to the objective function given by Eq. (7.2), and perform the computations exhibited in R Example 7.1 for estimating the parameters of a *GARCH*(1, 1) model of a return series of your choice.

The following two notes refer to the GP application of Sect. 7.3.2.

7.7.4: Suppose that in the out-of-market periods the trading system automatically puts our money in a risk-free asset earning a return of $r_f(t)$ for each day t. What would be in this case the continuously compounded return for a trading rule which operates through a period $[1, T]$ under transaction costs of some percentage α of the price, such that while in the market earns the market daily return of r_t, and when out of the market earns the risk-free return at the rate $r_f(t)$? Redefine the fitness function adapted to this situation.

7.7.5: In a GP algorithm it is desirable to avoid duplicated individuals (Koza calls these duplicate deadwood). Duplicates affect variety (in proportion of repetitions), so that if there are no duplicates then the variety is 100%. For this duplicate checking and processing, one needs an algorithm for comparing trees. Write such algorithm. Another further enhancement to the GP heuristic for finding technical trading rules is to work with rules involving price and volume (and not just price), or any other statistic about the price history of the asset, such as: k period return, variance in daily returns over past n days, relative strength index, and other indicators. And for a GP trading system based on fundamental analysis see Doherty (2003)

7.7.6: Write the algorithm for the hybrid system that combines the ACO and SA heuristics, and as suggested in Example 7.7.

Chapter 8
Portfolio Optimization

Harry Markowitz presented in 1952 the basic tenet of portfolio selection[1]: to find a combination of assets that in a given period of time produces the highest possible return at the least possible risk. This laid the foundation of portfolio management, where the process of compounding a portfolio for a particular investor reduces to two main steps: first determined the different combinations of assets, or portfolios, that are optimal with respect to their expected return and risk equilibrium equation; second, choose that portfolio that best suits the investor's utility function, which has to do solely with the investor's personal constraints, such as, his tolerance to risk, the type and number of assets desired for his portfolio, and the total capital reserved for the investment.

This chapter presents the Markowitz mean-variance model for portfolio optimization, and some of the tools for portfolio management, such as Sharpe ratio, beta of a security, and the Capital Asset Pricing Model. We look at some of the limitations of the mean-variance model, which are basically due to the constraints imposed to the portfolio selection problem to make it computationally feasible, and then see how to deal with more computationally difficult, although more realistic extensions of the portfolio optimization problem, using the already learned optimization heuristics.

8.1 The Mean-Variance Model

8.1.1 The Mean-Variance Rule and Diversification

The departing hypothesis of Markowitz for his portfolio selection model is that investors should consider expected return a desirable thing and abhor the variance of return. This mean of returns versus variance of returns rule of investors behavior has, as a first important consequence, a clear mathematical explanation to the widely

---

[1] Markowitz (1952). This is the beginning of an important bulk of work in portfolio theory by the author, rewarded in 1990 with the Nobel Prize in economics.

A. Arratia, *Computational Finance,* Atlantis Studies in Computational Finance and Financial Engineering 1, DOI: 10.2991/978-94-6239-070-6_8, © Atlantis Press and the authors 2014

accepted belief among investors on the importance of diversification in portfolio selection. The formalization of diversification can be done as follows.

Let $\mathscr{P}$ be a portfolio of N *risky* assets. We learned in Chap. 2, Sect. 2.1, that the τ-period simple return of $\mathscr{P}$ at time t is

$$R_t^{\mathscr{P}}(\tau) = \sum_{i=1}^{N} w_i R_{i,t}(\tau) \tag{8.1}$$

where $R_{i,t}(\tau)$ is the τ-period return of asset i at time t, and w_i is the weight of asset i in $\mathscr{P}$. It should be clear that a portfolio is totally determined by the vector of weights $w = (w_1, \ldots, w_N)$: a weight $w_i = 0$ means that there are no positions taken on the i-th asset; positive weights signifies long positions, and negative weights stand for short positions in the portfolio. Thus, from now on we shall refer to a portfolio by its vector of weights, and use $R_t^w(\tau)$ instead of $R_t^{\mathscr{P}}(\tau)$ to denote its τ-period return at time t. To simplify notation we will omit τ and t wherever possible, provided both parameters are clear from context.

From Eq. (8.1) the mean value (or expected return) at time t of the portfolio w is

$$\mu_w = E(R_t^w) = \sum_{i=1}^{N} w_i E(R_{i,t}) \tag{8.2}$$

and the variance of portfolio w is (cf. Exercise 3.5.2 of Chap. 3)

$$\sigma_w^2 = Var(R_t^w) = \sum_{i=1}^{N}\sum_{j=1}^{N} w_i w_j \sigma_i \sigma_j \rho_{ij} = \sum_{i=1}^{N}\sum_{j=1}^{N} w_i w_j \sigma_{ij} \tag{8.3}$$

where σ_i is the standard deviation of $R_{i,t}$; ρ_{ij} the correlation coefficient of the returns of assets i and j; $\sigma_{ij} = Cov(R_{i,t}, R_{j,t})$ is the covariance of returns of assets i and j; and note that $\sigma_{ii} = Cov(R_{i,t}, R_{i,t}) = \sigma_i^2$. Now, consider the following two extreme cases:

Case 1: The returns of the N assets in the portfolio are pairwise uncorrelated, i.e., $\forall i \neq j$, $\rho_{ij} = 0$. Then $Var(R_t^w) = \sum_{i=1}^{N} (w_i)^2 \sigma_i^2$. If we further assume that all assets are equally weighted, that is, for all i, $w_i = 1/N$, we have

$$Var(R_t^w) = \frac{1}{N^2} \sum_{i=1}^{N} \sigma_i^2 \leq \frac{\sigma_M^2}{N},$$

where $\sigma_M = \max[\sigma_i : i = 1, \ldots, N]$. We can see that as N grows to infinity, $Var(R_t^w)$ tends to 0. This means that *the greater the number of pairwise uncorrelated assets, the smaller the risk of the portfolio.*

Case 2: The returns of the assets in the portfolio are similarly correlated. This can be realized by considering that all assets' returns have same constant variance σ^2, and the correlation of any pair of distinct returns is some constant $c \leq 1$. Also consider again all assets equally weighted. Then, for all $i \neq j$, $\sigma_{ij} = Cov(R_{i,t}, R_{j,t}) = c\sigma^2$ and

$$Var(R_t^w) = \sum_{i=1}^{N}\sum_{j=1}^{N} w_i w_j \sigma_{ij} = \frac{1}{N^2}\sum_{i=1}^{N}\sum_{j=1}^{N}\sigma_{ij}$$

$$= \frac{1}{N^2}\left(\sum_{1\leq i\leq N}\sigma_{ii} + \sum_{1\leq i<j\leq N}\sigma_{ij}\right)$$

$$= \frac{\sigma^2}{N} + \left(1 - \frac{1}{N}\right)c\sigma^2 = \frac{(1-c)}{N}\sigma^2 + c\sigma^2$$

This shows that no matter how large N is made it is impossible to reduce the variance of the portfolio below $c\sigma^2$. The conclusion is that *the greater the presence of similarly correlated assets, the closer is the risk of the portfolio to a risk common to all assets* (e.g., some weighted average of individual assets' risks).

Therefore, diversification in a mean-variance world is accomplished by considering highly uncorrelated assets in some reasonable number. By plotting the variance of the portfolio as a function of N, given by either of the above cases, one sees that from 15 to 20 are reasonable numbers for the size of a portfolio.

8.1.2 Minimum Risk Mean-Variance Portfolio

Given a portfolio of N risky assets, determined by its vector of weights $\mathbf{w} = (w_1, \ldots, w_N)$, define its covariance matrix $\mathbf{C} = [\sigma_{ij}]_{1\leq i,j\leq N}$, where $\sigma_{ij} = Cov(R_{i,t}, R_{j,t}) = \sigma_i\sigma_j\rho_{ij}$, and define its expected return vector $\boldsymbol{\mu} = (\mu_1, \ldots, \mu_N)$, where $\mu_i = E(R_{i,t})$ is the expected return of the i-th asset in the portfolio. With this vectorial notation we can rewrite the expressions for the expected return and the variance of the portfolio $\mathbf{w}$ as

$$E(R_t^w) = \mathbf{w}'\boldsymbol{\mu} \quad \text{and} \quad Var(R_t^w) = \mathbf{w}'\mathbf{C}\mathbf{w}.$$

Here $\mathbf{w}'$ denotes the transpose of $\mathbf{w}$. According to the Markowitz's mean of returns versus variance of returns rule, investors' main concern is to obtain a certain level of benefits under the smallest possible amount of risk. Therefore the Markowitz portfolio selection problem amounts to:

Find weights $\mathbf{w} = (w_1, \ldots, w_N)$ such that, for a given expected rate of return r^*, the expected return of the portfolio determined by $\mathbf{w}$ is r^*, i.e., $E(R^w) = r^*$, while its variance $Var(R^w)$ is minimal.

This minimum variance for given return portfolio optimization problem can be mathematically formulated as follows:

$$\min_{w} \; w'Cw \qquad (8.4)$$

$$\text{subject to:} \quad w'\mu = r^*, \qquad (8.5)$$

$$\text{and} \; \sum_{i=1}^{N} w_i = 1 \qquad (8.6)$$

This is a quadratic programming problem, since the objective is quadratic and the constraints are linear, which we will see can be reduced to a linear system of equations and solve. Constraint (8.6) means that the investor uses all his budget for the N assets. Observe that the model imposes no restrictions on the values of the weights (i.e. there could be either long or short positions). Under these relaxed conditions on weights (and assuming extended no arbitrage[2]) this portfolio optimization problem can be solved analytically using *Lagrange multipliers*. This method consists in defining the Lagrangian

$$L = w'Cw - \lambda_1(w'\mu - r^*) - \lambda_2(w'\mathbf{1} - 1) \qquad (8.7)$$

$$= \sum_{1 \le i,j \le N} w_i w_j \sigma_{ij} - \lambda_1 \left(\sum_{i=1}^{N} w_i \mu_i - r^* \right) - \lambda_2 \left(\sum_{i=1}^{N} w_i - 1 \right)$$

where λ_1 and λ_2 are the Lagrange multipliers. Then, differentiating L with respect to each w_i and setting these partial derivatives to zero, one obtains the following N linear equations in addition to the two equations for the constrains (Eqs. (8.5) and (8.6)), for a total of $N + 2$ equations for the $N + 2$ unknowns $w_1, \dots, w_N, \lambda_1, \lambda_2$:

$$2 \sum_{j=1}^{N} w_j \sigma_{ij} - \lambda_1 \mu_i - \lambda_2 = 0, \quad \text{for } i = 1, \dots, N,$$

$$\sum_{i=1}^{N} w_i \mu_i = r^* \; \text{and} \; \sum_{i=1}^{N} w_i = 1$$

Solving this linear system one obtains the weights for a portfolio with mean r^* and the smallest possible variance. This solution is termed *efficient* in the sense of being the portfolio with that expected return r^* and minimum variance, and such that any other portfolio on the same set of securities with same expected return r^* must have a higher variance.

---

[2] We are ignoring transaction costs and other nuisances, and assuming securities are divisible in any proportions.

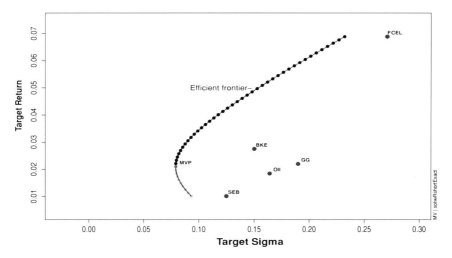

Fig. 8.1 Efficient frontier, the global minimum variance portfolio (MVP), and single stock points from data in R Example 8.1

8.1.3 The Efficient Frontier and the Minimum Variance Portfolio

For a set of N assets for constituting a portfolio, consider different values of r^* for the expected return of the portfolio, and for each of these values solve the system (8.4)–(8.6) to get the efficient solution of weights w^*, from which obtain the corresponding minimum variance, and consequently the standard deviation of the portfolio determined by w^*

$$\sigma^* = std(R^{w^*}) = \sqrt{(w^*)'Cw^*}.$$

The locus of points (σ^*, r^*) in the σ-μ plane, or the *risk-mean* plane, is the right branch of a hyperbola. The vertex of this hyperbola has as coordinates the standard deviation and mean values of the *minimum variance portfolio* (mvp). Let us denote these values by σ_{mvp} and r_{mvp}, respectively. The part of the curve above the point (σ_{mvp}, r_{mvp}), that is, the locus of points (σ^*, r^*) for $r^* > r_{mvp}$ obtained from solutions w^* of (8.4)–(8.6), is called the *efficient frontier* (and the symmetric reflexion curve below is called the *minimum variance locus*). Figure 8.1 shows the efficient frontier, the mvp and the positions in the risk-mean graph of five stocks used to build efficient portfolios in the forthcoming R Example 8.1.

The minimum variance portfolio corresponds to the solution for the particular case where $\lambda_1 = 0$ in Eq. (8.7). This is the Lagrangian for the system of equations (8.4) and (8.6) only, which is asking to find the minimum variance for the portfolio *regardless* of expected returns. In fact, one could have started by solving this case first to get the point (σ_{mvp}, r_{mvp}), and then continue solving the system (8.4)–(8.6) for $r^* > r_{mvp}$ to get the efficient frontier, then draw the lower border by symmetry.

The region enclosed by the right branch of hyperbola (i.e., below and including the efficient frontier and above the minimum variance locus) is the *feasible set* of mean-variance portfolios. Any point in this region is a pair (σ_w, r_w) corresponding to the standard deviation and the expected return of a portfolio w with covariance matrix C and mean vector μ. In other words, only the points (σ_w, r_w) in the feasible set verify the equations

$$\sigma_w = w'Cw, \quad r_w = w'\mu \quad \text{and} \quad \sum_{i=1}^{N} w_i = 1.$$

8.1.4 General Mean-Variance Model and the Maximum Return Portfolio

In the mean-variance model given by Eq. (8.4) we have to choose the desired mean r^*. But if we don't know the mean r_{mvp} for the minimum variance portfolio we might choose r^* below r_{mvp} and get an inefficient portfolio with low mean and unnecessarily high variance. This can be fixed, without recurring to the computation of r_{mvp}, by extending the objective function of minimizing the variance with the objective of maximizing the mean. Since maximizing a function f is the same as minimizing $-f$, and by the convexity of the efficient frontier, we can express the mean-variance portfolio model as the following more general problem. For a given $\gamma \in [0, 1]$, solve

$$\min_{w} (1 - \gamma)w'Cw - \gamma w'\mu \tag{8.8}$$

$$\text{subject to} \quad \sum_{i=1}^{N} w_i = 1 \tag{8.9}$$

Note that when $\gamma = 0$ we are back in the case of minimizing risk regardless of expected returns; that is, we are seeking for the minimum variance portfolio with mean r_{mvp} and standard deviation σ_{mvp}. When $\gamma = 1$ we are now eliminating risk from the equation (hence, completely ignoring it) and seeking for the portfolio with *maximum return*. This can be made more evident by rewriting Eq. (8.8) in the equivalent form:

$$\max_{w} \gamma w'\mu - (1 - \gamma)w'Cw \tag{8.10}$$

Therefore, the global maximum return portfolio corresponds to the solution of the problem

$$\max_{w} w'\mu \quad \text{subject to} \quad \sum_{i=1}^{N} w_i = 1 \tag{8.11}$$

which is a linear programming problem (linear objective and linear constraints), solvable by some efficient optimization method (e.g. the simplex method). In R we can use the function lp contained in the package lpSolve. We show how to do this in R Example 8.1.

The parameter γ can be interpreted as the degree of an investor's tolerance to risk: If $\gamma \mapsto 1$ then the investor seeks to maximize expected return regardless of risk; if $\gamma \mapsto 0$ then the investor seeks to subtract as much variance as possible, hence avoiding as much risk as possible at the expense of obtaining less benefits. The efficient frontier can now be drawn by taking different values of $\gamma \in [0, 1]$ and solving the optimization problem (8.8) for each one of these values.

R Example 8.1 For this example we need to load the packages lpSolve and RMetrics. We analyze five of the stocks contained in the data set SMALLCAP provided by fPortfolio, a sub-package of RMetrics. This data set contains monthly records of small capitalization weighted equities from U.S. markets recorded between 1997 and 2001. The extension SMALLCAP.RET gives the series of returns for the equities in SMALLCAP. We chose to work with BKE, FCEL, GG, OII and SEB. Execute:

```
> names(SMALLCAP)  #to see the contents of the data set
> Data = SMALLCAP.RET #to get returns
> Data = Data[, c("BKE","FCEL","GG","OII","SEB")]
> covData <- covEstimator(Data) ; covData
```

The covEstimator function from fPortfolio gives the sample mean of the returns of each stock and the sample covariance matrix. The results are shown in Table 8.1. The sample mean $\widehat{\mu}$ is shown in percentage units.

Thus, for example, BKE has mean 0.0274 (or 2.74 %) and variance 0.0225 (hence, standard deviation $\sqrt{0.0225} = 0.15$). Each stock is plotted in Fig. 8.1 above.

Now compute the unrestricted (long-short) Mean-Variance (MV) portfolio for the given set of stocks. This is done using an exact solver for the system (8.8)–(8.9) which returns 50 points in the efficient frontier.

```
> ##MV|solveRshortExact:
> shortSpec <- portfolioSpec()
> setSolver(shortSpec) <- "solveRshortExact"
> shortFrontier <- portfolioFrontier(Data,spec=shortSpec,
+    constraints="Short")
> print(shortFrontier) #report results for portfolio:1,13,25,
37,50
> Frontier <- shortFrontier  ##Plot the Efficient Frontier
> frontierPlot(Frontier,frontier="both",risk="Sigma",type="l")
> ## Plot some portfolios
> minvariancePoints(Frontier,pch=19,col="red") #the MVP point
> ##Position of each asset in the sigma-mu plane
> singleAssetPoints(Frontier,risk="Sigma",pch=19,cex=1.5,
+    col=topo.colors(6))
```

To compute the minimum variance portfolio (MVP), and a particular efficient portfolio for a given target return:

Table 8.1 Covariance matrix and sample mean of five stocks

Stock	Covariance					$\widehat{\mu}$
	BKE	FCEL	GG	OII	SEB	
BKE	0.0225	0.0026	−0.0035	−0.0010	0.0073	2.74
FCEL	−0.0026	0.0732	0.0072	0.0112	−0.0052	6.88
GG	−0.0034	0.0072	0.0360	0.0126	−0.0016	2.19
OII	−0.0010	0.0112	0.0126	0.0269	−0.0040	1.84
SEB	0.0073	−0.0052	−0.0016	−0.0040	0.0155	1.01

```
> ##MVP: the minimum Variance.
> minvariancePortfolio(Data)
> ##EP(mu): an efficient portfolio for given target return
> mu = 0.05; Spec = portfolioSpec()
> setSolver(Spec) = "solveRshortExact"
> setTargetReturn(Spec) = mu
> efficientPortfolio(Data, Spec)
```

To compute the global maximum return portfolio we use the linear programming solver lp to resolve the optimization problem (8.11) for a vector of 5 unknown weights. Note that lp only works with the standard form of a linear program that requires all variables to be positive; hence if we want to allow unrestricted solutions (i.e. the possibility of long and short positions) we need to use the trick of replacement of variables by the difference of two new positive variables. We show how to do the unrestricted version in the notes for this chapter (Note 8.5.7), and now show the standard solution requiring all weights to be positive (a long only portfolio).

```
> ##MaxR: Global maximum return portfolio
> ## maximize: (w1,w2,w3,w4,w5)*covData$mu
> ## subject to: w1+w2+w3+w4+w5 = 1
> ##Use the linear programming solver lp from lpSolve:
> f.obj <- covData$mu
> f.con <- matrix(c(1,1,1,1,1), nrow=1, byrow=TRUE)
> f.dir <- "="
> f.rhs <- 1
> lp ("max", f.obj, f.con, f.dir, f.rhs)
   Success: the objective function is 0.06882631
> lp ("max", f.obj, f.con, f.dir, f.rhs)$solution
 [1] 0 1 0 0 0
> ## portfolio containing only FCEL
```

The weights defining the portfolios just computed, together with their target return and standard deviation are shown in Table 8.2. Note that portfolio 37 is one of the 50 efficient portfolios computed with portfolioFrontier; it contains negative weights which implies short selling of the corresponding stocks. The maximum return portfolio is attained by investing all our budget in only one company, FCEL; hence its standard deviation is that of FCEL. This is not a wise investment strategy as it goes against diversification.

Table 8.2 Four portfolios for the given set of stocks, their expected returns (in percentage) and standard deviations

Stock	Portfolios			
	MVP	37	EP(0.05)	MaxR
BKE	0.1818	0.6178	0.5737	0
FCEL	0.0766	0.5354	0.4890	1
GG	0.1201	0.1854	0.1788	0
OII	0.2121	−0.1454	−0.1093	0
SEB	0.4094	−0.1932	−0.1323	0
$\widehat{\mu}$ (%)	2.10	5.33	5.00	6.88
$\widehat{\sigma}$	0.0791	0.1670	0.1540	

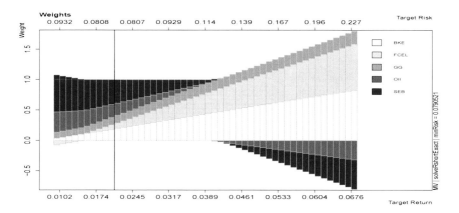

Fig. 8.2 Weights composition along the efficient frontier for the data of R Example 8.1

We can produce a graphical description of all portfolios in the efficient frontier with the command `weightsPlot(Frontier)`. The picture of different weights for unrestricted compositions of portfolios along the efficient frontier can be seen in Fig. 8.2. The bold vertical line marks the MVP. ☐

8.2 Portfolios with a Risk-Free Asset

We have so far assumed that the assets available for constituting a portfolio are all risky. In this section we deal with the case of adding a risk-free asset and the pricing model obtained as consequence. Adding a risk free asset to a portfolio corresponds to lending or borrowing cash at a known interest rate r_0 and with zero risk. Lending corresponds to the risk free asset having a positive weight, and borrowing corresponds to its having a negative weight.

Let $r_f = r_0\tau$ be the risk free rate, or return, over the time period τ of the risk free asset. A portfolio consisting solely of this risk free asset has mean value r_f and variance 0. This risk free portfolio is represented in the *risk-mean* plane by the point $(0, r_f)$.

Now consider a portfolio consisting of the risk free asset, with mean value r_f, plus N risky assets with aggregated mean and variance given by (cf. Eqs. (8.2) and (8.3))

$$\mu_{\mathbf{w}} = E(R^{\mathbf{w}}) \quad \text{and} \quad \sigma_{\mathbf{w}}^2 = Var(R^{\mathbf{w}})$$

where $\mathbf{w} = (w_1, \ldots, w_N)$ is the vector of weights of the N risky assets and $R^{\mathbf{w}}$ the return of the portfolio. Let w_f be the weight of the risk free asset in the portfolio; therefore,

$$1 - w_f = w_1 + \cdots + w_N$$

which implies that the total wealth has been distributed in two parts: one for investing in the risk free asset and the rest for investing among N risky assets. We can then view the portfolio $(\mathbf{w}, w_f) = (w_1, \ldots, w_N, w_f)$ as consisting of one risk free asset, with weight w_f, mean r_f and zero standard deviation, together with a risky asset, which is the aggregation of the N risky assets, with weight $1 - w_f$, mean $\mu_{\mathbf{w}}$ and standard deviation $\sigma_{\mathbf{w}}$. Note that this pair of risky and a risk free asset has covariance equal to zero. Then the expected return and standard deviation of this combined portfolio $\omega = (\mathbf{w}, w_f)$ are

$$\mu_\omega = w_f r_f + (1 - w_f)\mu_{\mathbf{w}} \tag{8.12}$$

$$\sigma_\omega = (1 - w_f)\sigma_{\mathbf{w}} \tag{8.13}$$

We see in these equations that the mean and the standard deviation of the portfolio ω depend linearly on w_f. Therefore, varying w_f we get that the different portfolios represented by Eqs. (8.12) and (8.13) trace a straight line from the point $(0, r_f)$ and passing through the point $(\sigma_{\mathbf{w}}, \mu_{\mathbf{w}})$ in the risk-mean plane. Furthermore, varying the weights of the N risky assets, i.e. building another risky portfolio $\mathbf{w}'$, and combining it with the risk free asset, we get another straight line describing all possible portfolios which are combination of these two. We conclude that the feasible region of mean-variance portfolios obtained from N risky assets and one risk free asset is a triangle with a vertex in the point representing the risk free asset and enveloping the hyperbola containing all feasible portfolios on the N risky assets. This region is shown in Fig. 8.3.

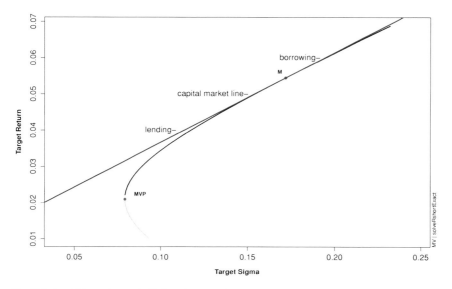

Fig. 8.3 Feasible region and efficient frontier of portfolios with a risk free asset with $r_f = 2\%$. Capital Market Line (*blue*); Market Portfolio (M). Efficient portfolios above (resp. *below*) M need borrowing (resp. *lending*) at the risk free rate

8.2.1 The Capital Market Line and the Market Portfolio

The efficient frontier for a portfolio of risky assets and one risk-free asset is now a straight line with intercept point $(0, r_f)$ and tangent to the efficient frontier of risky portfolios (denoted from now on EF_r) in the risk-mean plane. This tangent line describing all efficient portfolios is named the *Capital Market Line* (CML), and the point where the CML makes contact with the curve EF_r has as coordinates the standard deviation and expected return of a particular portfolio named the *Market Portfolio*. The Market Portfolio is the best portfolio with respect to the (excess return)/(risk) ratio, and it is the best representative of the market for it contains shares of every stock in proportion to the stock's weight in the market. Let us see why is this so.

Let θ be the angle between the horizontal axis and a line passing through $(0, r_f)$ and a point $(std(R^w), E(R^w))$ corresponding to some feasible portfolio of risky assets only. Then

$$\tan \theta = \frac{E(R^w) - r_f}{std(R^w)} \tag{8.14}$$

The Market Portfolio is the point that maximizes $\tan \theta$, for this gives the slope of the CML computed at the point tangent to the risky efficient frontier (and this is the reason why the Market Portfolio is also known as *the tangency portfolio*). From Eq. (8.14) it is clear that the Market Portfolio gives the maximum (excess return)/(risk) ratio; and the weights w that are solution of the problem of maximizing $\tan \theta$ are in proportions to the stocks' market weights. We leave the verification of this last assertion as exercise (see Note 8.5.3).

R Example 8.2 (Continuation of R Example 8.1) To compute the Market Portfolio for the stocks treated in R Example 8.1 and a risk free asset earning a risk free rate of 1.2 % type the commands:

```
> RiskF = portfolioSpec()
> setRiskFreeRate(RiskF) = 0.012 ##at 1.2% risk-free rate
> setSolver(RiskF) <- "solveRshortExact"
> ##Market portfolio for this Risk free rate
> M = tangencyPortfolio(Data,spec=RiskF,constraints="Short"); M
```

This gives the following portfolio

BKE	FCEL	GG	OII	SEB
0.6342	0.5526	0.1879	−0.1589	−0.2159

with target return 0.0545, and target risk (standard deviation) 0.1719. To draw the Capital Market Line and plot the Market Portfolio type the commands:

```
> ##recompute the efficient frontier to consider risk-free
asset
> Frontier=portfolioFrontier(Data,spec=RiskF,constraints=
"Short")
> frontierPlot(Frontier,risk="Sigma",type="l",lwd=3)
> ##Capital Market Line
> tangencyLines(Frontier,risk= "Sigma",col ="blue",lwd=3)
> ##plot Market Portfolio
> cmlPoints(Frontier,pch=19,col="purple")
```

8.2.2 The Sharpe Ratio

Let $(std(R_M), E(R_M))$ be the point describing the Market Portfolio, i.e. the tangency portfolio or contact point between the CML and the EF_r. As before, $(0, r_f)$ represents the risk-free asset. Then any point $(std(R^w), E(R^w))$ on the CML, that is, any efficient portfolio, is such that

$$E(R^w) = w_f \cdot r_f + (1 - w_f)E(R_M) \text{ and}$$
$$std(R^w) = (1 - w_f) \cdot std(R_M)$$

where w_f is the fraction of wealth invested in the risk-free asset. If $w_r \geq 0$ we are lending at the risk-free rate; whereas if $w_r < 0$ we are borrowing at the risk-free rate, in order to increase our investment in risky assets. We can rewrite the CML as

$$E(R^w) = r_f + (1 - w_f)(E(R_M) - r_f)$$

$$= r_f + \frac{(E(R_M) - r_f)}{std(R_M)} std(R^w) \qquad (8.15)$$

(We could have also arrived to this point-slope equation for the CML from Eq. (8.14).)

The quantity $SR_M = (E(R_M) - r_f)/std(R_M)$ is known as the *Sharpe ratio* of the Market Portfolio. In general, the Sharpe ratio of any portfolio is the number

$$SR^w = \frac{E(R^w) - r_f}{std(R^w)} \qquad (8.16)$$

which gives a measure of the portfolio reward to variability ratio, and has become standard for portfolio evaluation. The higher the Sharpe ratio the better the investments, with the upper bound being the Sharpe ratio of the Market Portfolio, namely SR_M. By Eq. (8.15) all efficient portfolios, with risky assets and one risk-free asset, should have same Sharpe ratio and equal to the Sharpe ratio of the market:

$$SR^w = \frac{E(R^w) - r_f}{std(R^w)} = SR_M$$

Thus, summarizing, the best an investor can do in a mean-variance world is to allocate a proportion of his investment money in a risk-free asset and the rest in the Market Portfolio. This guarantees the best possible ratio of excess return to variability.

8.2.3 The Capital Asset Pricing Model and the Beta of a Security

The Capital Market Line shows an equilibrium between the expected return and the standard deviation of an efficient portfolio consisting of a risk free asset and a basket of risky assets (Eq. (8.15)). It would be desirable to have a similar equilibrium relation between risk and reward of an individual risky asset with respect to an efficient risky portfolio, where it could be included. This risk-reward equilibrium for individual assets and a generic efficient portfolio (i.e., the Market Portfolio), is given by the *Capital Asset Pricing Model* (CAPM).

Theorem 8.1 (**CAPM**) *Let $E(R_M)$ be the expected return of the Market Portfolio, $\sigma_M = std(R_M)$ its standard deviation, and r_f the risk-free return in a certain period of time τ. Let $E(R_i)$ be the expected return of an individual asset i, σ_i its standard deviation, and $\sigma_{iM} = Cov(R_i, R_M)$ be the covariance of the returns R_i and R_M. Then*

$$E(R_i) = r_f + \beta_i(E(R_M) - r_f) \qquad (8.17)$$

where $\beta_i = \dfrac{\sigma_{iM}}{\sigma_M^2} = \dfrac{Cov(R_i, R_M)}{Var(R_M)}$.

Proof Exercise (see Note 8.5.4).   □

The CAPM states that the expected excess rate of return of asset i, $E(R_i) - r_f$, also known as the asset's *risk premium*, is proportional by a factor of β_i to the expected excess rate of return of the Market Portfolio, $E(R_M) - r_f$, or the *market premium*. The coefficient β_i, known as the *beta* of asset i, is then the degree of the asset's risk premium relative to the market premium. We shall soon get back to discussing more on the meaning and use of beta, but before doing that let us see how the CAPM serves as a pricing model.

Suppose that we want to know the price P of an asset whose payoff after a period of time τ is set to be some random value P_τ. Then the rate of return of the asset through the period τ is $R_\tau = (P_\tau/P) - 1$, and by the CAPM the expected value of R_τ relates to the expected rate of return of the market, on the same period of time, as follows:

$$E(R_\tau) = \frac{E(P_\tau)}{P} - 1 = r_f + \beta(E(R_M) - r_f)$$

where β is the beta of the asset. Solving for P we get the pricing formula

$$P = \frac{E(P_\tau)}{1 + r_f + \beta(E(R_M) - r_f)} \qquad (8.18)$$

This equation is a natural generalization of the discounted cash flow formula for risk free assets (Sect. 1.2.2) to include the case of risky assets, where the present value of a risky asset is its expected future value discounted back by a risk-adjusted interest rate given by $r_f + \beta(E(R_M) - r_f)$. Indeed, observe that if the asset is risk free, e.g. a bond, then $E(P_\tau) = P_\tau$ and $\beta = 0$, since the covariance with any constant return is null; hence, $P = P_\tau/(1 + r_f)$.

On the meaning of beta. Let's get back to discussing beta. Formally the beta of an asset measures the linear dependence of the asset's return and the return of the market in proportion to the asset to market volatility ratio. To better see this rewrite β_i in terms of correlations

$$\beta_i = \frac{Cov(R_i, R_M)}{Var(R_M)} = \rho(R_i, R_M)\frac{\sigma_i}{\sigma_M} \qquad (8.19)$$

Using this equation we produce Table 8.3 containing interpretations of the values of an asset's beta as a measure of its co-movement with the market. Let $\beta = \beta_i$ and $\rho = \rho(R_i, R_M)$ the correlation coefficient of the asset and the market.

Thus, for example, a $\beta = 0$ means that the asset and the market are uncorrelated, since the standard deviation ratio σ_i/σ_M is always positive. In this case, the CAPM states that $E(R_i) = r_f$, or that the risk premium is zero. The explanation for this surprising conclusion is that the risk of an asset which is uncorrelated with an efficient

Table 8.3 Interpretation of beta

Value of beta	Effect on correlation and volatility ratio	Interpretation
$\beta < 0$	$\rho < 0$, $\frac{\sigma_L}{\sigma_M} > 0$	Asset moves in the opposite direction of the movement of the market
$\beta = 0$	$\rho = 0$	Movements of the asset and the market are uncorrelated
$0 < \beta \leq 1$	$\rho > 0$, $0 < \frac{\sigma_L}{\sigma_M} \leq 1/\rho$	Asset moves in the same direction as the market, volatility of asset can be < or > volatility of market
$\beta > 1$	$\rho > 0$, $\frac{\sigma_L}{\sigma_M} > 1/\rho > 1$	Asset moves in the same direction as the market but with greater volatility

portfolio gets neutralize by the risk compounded from the different positions of the portfolio, and hence, one should not expect greater benefits than those that could be obtained at the risk free rate.

A different situation is presented by $\beta > 1$. This implies that the asset and the market are positively correlated; but since $\rho \leq 1$ always, the risk of the asset is greater than the risk of the market. This extra risk might have its compensation since from the CAMP follows that $E(R_i) > E(R_M)$, i.e. the asset could beat the market. The reader can work out from these examples other consequences to the asset's risk premium for other values of its beta.

However we should caution the reader that our interpretation of beta differs from mainstream interpretation of this statistic, and as it is frequently used by most financial institutions and investors, and as explained in various financial reports, text books, and many sources of popular knowledge (e.g. *Wikipedia*). The most common usage of beta is as some measure of risk relative to the market. But this is only one factor in the equation, the volatility ratio, which is then confounded with the correlation factor. Then so often it seems that mainstream interpretation wrongly considers the product of these two statistics as if it were a conjunction, as it is the case when $0 < \beta \leq 1$ which is taken to mean that *jointly* $0 < \rho \leq 1$ *and* $0 < \frac{\sigma_L}{\sigma_M} \leq 1$. This leads to interpretations such as "*movement of the asset is in the same direction as, but less than the movement of the benchmark*", or equivalently that "*the asset and the market are correlated but the asset is less volatile*". This is not only wrong but dangerous, for it could lead to bad investment decisions (and surely have done so in the past). Strictly speaking, from Eq. (8.3), the conclusion that can be drawn for a positive beta but less than one is that there is some correlation between the return of the asset and the return of the market, but the relative volatility can be any value in the interval $(0, 1/\rho)$, which leaves open the possibility for the volatility of the asset being greater than the volatility of the market. Let us illustrate this with a numeric example. Assume $\beta = 0.4$. This number can be decomposed in the following fractions:

$$0.4 = \frac{10}{25} = \frac{1}{5} \cdot \frac{10}{5}$$

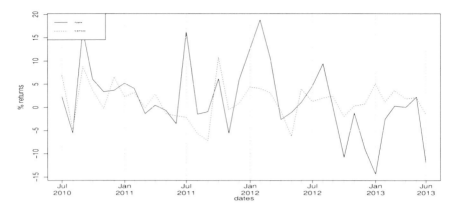

Fig. 8.4 Monthly excess returns of Apple (*solid line*) and S&P500 (*dashed line*) over a period of three years

so that $\rho = \frac{1}{5}$ and $\frac{\sigma_i}{\sigma_M} = \frac{10}{5}$; hence, the standard deviation of the asset doubles the standard deviation of the market. Hence, small positive beta does not necessarily mean that the asset is less riskier than the market. We can further check this matter with real data.

R Example 8.1 Yahoo Finance uses as beta the *beta of equity*. This is "*the monthly price change of a particular company relative to the monthly price change of the S&P 500. The time period for beta is 3 years (36 months) when available*"[3]. By the end of July 2013 the beta listed in *finance.yahoo.com* for Apple (AAPL) was of 0.76, which is less than 1, and according to popular interpretation this means that the stock should be less volatile than the reference index, the S&P 500. We have calculated and plot the monthly excess return of Apple and the monthly excess return of the S&P 500, for the time period from July 1, 2010 to July 1, 20013 (3 years). The resulting plot can be seen in Fig. 8.4, and it is clear from the picture that the variance of Apple's returns (solid black line) is higher than the variance of the S&P 500 (dashed red line), contradicting the common interpretation attached to a beta below 1.

One can further compute the correlation for the pair of returns in question, and will find that $\rho = \rho(\text{Apple}, \text{S\&P500}) = 0.3927$. Using Eq. (8.19) we can estimate the relative volatility for the two assets, and obtain $\beta/\rho = 0.76/0.3927 = 1.93$, a number that quantifies better the phenomenon observed in the picture: it says that the volatility of Apple almost doubles the volatility of the S&P 500. □

The widespread wrong use and abuse of beta have been reported decades ago (see, for example, Lindahl-Stevens (1978) and the review on this topic by Tofallis (2006)), but beta is still today interpret in a manner inconsistent with its mathematical formula. There are several academic proposals for adjusting beta to fit with the desire use as a measure of market risk. One such alternative estimator, due to Tofallis (2006), is

---

[3] source: help.yahoo.com, Understanding the key statistics page.

precisely to consider, as motivated by the previous example, a beta as the relative volatility with the sign for correlation; that is, $\beta^* = (\text{sign } \rho)\sigma_i/\sigma_M$. This β^* has the computational advantage of being easier to calculate, as it is the quotient of two standard deviations, and it does measures the risk of a stock against a portfolio, or the market.

Estimating beta from sample. Given n pairs of sample returns of stock i, R_i, and the market, R_M, over the same time period, the beta of stock i with respect to the market can be estimated by using the unbiased estimators for the covariance and the variance statistics, and by Eq. (8.19), we have[4]

$$\widehat{\beta}_i = \frac{\sum_{t=1}^{n}(R_{i,t} - \widehat{\mu}(R_i))(R_{M,t} - \widehat{\mu}(R_M))}{\sum_{t=1}^{n}(R_{M,t} - \widehat{\mu}(R_M))^2} \qquad (8.20)$$

For this estimator we've assumed that the risk free rate r_f remains constant through the time period considered. If we have a variable risk free rate, then instead of the returns $R_{i,t}$ and $R_{M,t}$, we should consider the excess returns $R_{i,t} - r_f$ and $R_{M,t} - r_f$ in Eq. (8.20).

R Example 8.3 A suite of functions for computing beta of returns and other statistics related to the CAPM can be found in the R package Performance Analytics. Load the package by typing library(PerformanceAnalytics). Beta is computed by the function CAPM.beta(Ra, Rb, Rf = 0), where Ra is a vector of an asset's returns or a matrix of various returns, Rb is the return vector of benchmark asset (e.g. the Market Portfolio), and Rf is the risk free rate. If the risk free rate is constant then we can set Rf to 0 (default value); otherwise for a variable risk free rate then set Rf to the series of interests for the same period of time as the returns considered. The CAPM.beta( ) applies the numerical computation given by Eq. (8.20), applied to excess returns. In SMALLCAP we have a MARKET index and T90 interest rates; we use those for testing. Also we compute the beta of some stocks against the Market Portfolio M build in R Example 8.2.

```
> ##Beta of BKE w.rto MARKET index and T90 interest rates
> CAPM.beta(SMALLCAP.RET[,"BKE",drop=false],SMALLCAP.RET
[,"MARKET",
+   drop=false],Rf=SMALLCAP[,"T90",drop=false])
> #Some betas w.r.to  portfolio M of Data. 1) Get returns of M
> Mw = c(0.6342,0.5526,0.1879,-0.1589,-0.2159 ) ##weights of M
> ##build returns in same period as assets BKE, ...
> RM = Return.portfolio(Data,weights=Mw)
> ##Beta of SEB wrto M and T90 interests
> CAPM.beta(SMALLCAP.RET[,"SEB",drop=false],Rb=RM,
```

---

[4] The reader acquainted with linear regression analysis will note that this $\widehat{\beta}_i$ is the least squares estimation of the slope of the linear regression of $R_{i,t}$ on $R_{M,t}$. This gives a geometrical interpretation for beta.

```
+   Rf=SMALLCAP[,"T90",drop=false])
> ##same but for a fixed interest rate of 3%
> CAPM.beta(SMALLCAP.RET[,"SEB",drop=false],Rb=RM,Rf=0.03)
> ##Note that you get same result with Rf=0
> ##Beta of M wrto to MARKET index and any fixed interest rate
> CAPM.beta(RM,SMALLCAP.RET[,"MARKET",drop=false])
```

8.3 Optimization of Portfolios Under Different Constraint Sets

Let us go back to our general mean-variance portfolio model. We have N assets with expected return vector $\boldsymbol{\mu} = (\mu_1, \ldots, \mu_N)$, where $\mu_i = E(R_i)$ is the expected return of asset i, and covariance matrix $\mathbf{C} = [\sigma_{ij}]_{1 \leq i,j \leq N}$, where $\sigma_{ij} = Cov(R_i, R_j)$. We choose a level of risk $\gamma \in [0, 1]$, and the problem is to find a vector of weights $\boldsymbol{w} = (w_1, \ldots, w_N)$ that maximizes

$$\gamma \boldsymbol{w}' \boldsymbol{\mu} - (1 - \gamma) \boldsymbol{w}' \mathbf{C} \boldsymbol{w} \tag{8.21}$$

$$\text{subject to:} \quad \sum_{i=1}^{N} w_i = 1 \tag{8.22}$$

Equation (8.22) is the budget constraint and is a necessary restriction to norm the solution to the mean-variance problem. There are several other constraints that can be added to the model to adapt it to more realistic scenarios. We can add some or all of the following constraints (Perold 1984; Crama and Schyns 2003):

Upper and lower bounds in holdings. These constraints limit the proportions of each asset that can be held in the portfolio, and model situations where investors require to have only long positions ($w_i \geq 0$), or specific proportions of certain assets. The constraints are formalized by the equations

$$L_i \leq w_i \leq U_i, \quad \text{for each } i = 1, \ldots, N, \tag{8.23}$$

where L_i, U_i are given real numbers. Observe that if $L_i = 0$ and $U_i = \infty$, for all i, then we are in the case of building portfolios with long positions only.

Turnover limits in purchase or sale. These constraints impose upper bounds on the variation of the holdings from one period to the next (e.g. to limit proportions added or subtracted when rebalancing, or to model periodic administrative commissions and taxes). These are expressed by the equations:

$$\max[w_i - w_i^{(0)}, 0] \leq \overline{B}_i \quad \text{for } i = 1, \ldots, N \quad \text{(purchase)} \tag{8.24}$$

$$\max[w_i^{(0)} - w_i, 0] \leq \overline{S}_i \quad \text{for } i = 1, \ldots, N \quad \text{(sale)} \tag{8.25}$$

where $w_i^{(0)}$ denotes the weight of asset i in the initial portfolio, $\overline{B}_i$ and $\overline{S}_i$ are the maximum allowed purchase and sale of asset i during the current period.

Trading limits. These constraints impose lower bounds on the variation of the holdings from one period to the next (e.g., if modifications to the portfolio should be enforced to trade a minimum volume for each position). These are expressed by the disjunctions:

$$w_i = w_i^{(0)} \text{ or } w_i \geq w_i^{(0)} + \underline{B_i} \text{ or } w_i \leq w_i^{(0)} - \underline{S_i}, \text{ for } i = 1, \ldots, N \quad (8.26)$$

where $w_i^{(0)}$ is the initial weight, $\underline{B_i}$ and $\underline{S_i}$ are minimal quantities of purchase and sell.

Size of portfolio. This refers to imposing a limit in the number of assets that can be included in the portfolio. Formally, for a given integer $K \leq N$, it is asked that

$$|\{i \in \{1, \ldots, N\} : w_i \neq 0\}| \leq K \quad (8.27)$$

The addition of these constraints turns the optimization problem (8.21) computationally more harder to resolve, and for that we would have to recur to optimization heuristics. We discuss some of these constrained problems in the following sections.

8.3.1 Portfolios with Upper and Lower Bounds in Holdings

If the weights of assets in the portfolio are constrained to be within some limits then the quadratic programming problem given by Eq. (8.21), subject to constraints (8.22) and (8.23), cannot always be reduced to a linear system and has to be solved numerically. In R, solve.QP() in the package quadprog is a suitable solver for quadratic programming problems of the form

$$\min_{\mathbf{w}} \frac{1}{2}\mathbf{w}'\mathbf{C}\mathbf{w} - \mathbf{w}'\mathbf{u}$$
$$\text{subject to:} \quad A'\mathbf{w} \geq b$$

In fact, the fPortfolio function portfolioFrontier() uses as default the solver solve.QP(); hence, if we do not specify the solver and constraints parameters, it will be solving the long only portfolio optimization problem (i.e. $b = 0$).

R Example 8.4 Compute the long only portfolio problem for the stocks analyzed in R Example 8.1.

```
> Data = SMALLCAP.RET
> Data = Data[, c("BKE","FCEL","GG","OII","SEB")]
> ##MV|solveRquadprog:
> ##Compute the long-only MV model for 50 points (default)
> longFrontier = portfolioFrontier(Data);  longFrontier
```

Continue with plotting the long only efficient frontier, minimum variance portfolio, etc.   □

The portfolio selection problem with the turnover limits in purchase and sale, or trading constraints can also be posed and solved as a quadratic programming problem. However, the problem becomes more demanding of computational resources as the number of assets increases, and as alternatives some ad hoc methods that exploit the special structure of the covariance matrix have been developed and applied successfully (Perold 1984).

8.3.2 Portfolios with Limited Number of Assets

Consider the situation where an investor can only hold at most K different assets from the N available, for some given $K < N$. This is the case, more often than not, when an investor does not have enough money to invest in all available securities. This portfolio optimization problem with cardinality constraints is formally expressed by Eqs. (8.21), (8.22) and (8.27). We add the positivity condition on weights (i.e. forbid short selling), only to simplify the exposition, and consequently restate the problem as follows: Given N different assets, a number $K < N$, and $\gamma \in [0, 1]$, find weights $w = (w_1, \ldots, w_N)$ satisfying

$$\max_{w} \gamma w' \mu - (1 - \gamma) w' C w \qquad (8.28)$$

subject to the constraints

$$\sum_{i=1}^{N} w_i = 1 \qquad (8.29)$$

$$w_i \geq 0 \quad \text{for all } i = 1, \ldots, N \qquad (8.30)$$

$$\sum_{i=1}^{N} b_i \leq K \quad \text{where } b_i = 1 \text{ (if } w_i > 0\text{), } 0 \text{ (otherwise)} \qquad (8.31)$$

Observe that if we have N assets to choose from but only K assets can be included in the portfolio, then there are $\binom{N}{K}$ possible portfolios. This means, for example, that if we have $N = 50$ stocks to choose from (e.g. those from the Eurostoxx 50), and $K = 10$, we have $\binom{50}{10}$, which is around ten billion possible combinations. Thus, finding optimal solutions by enumerating all possibilities is impracticable. As a matter of fact, one can see that the problem (8.28) with constraints (8.29)–(8.31) is a form of the Knapsack problem (Garey and Johnson 1979), which is computationally NP-hard. We shall study a solution with simulated annealing.

8.3.3 Simulated Annealing Optimization of Portfolios

The simulated annealing (SA) solution presented below for the portfolio optimization problem with cardinality constraint is a simple and direct application of the method studied and applied more throughly in Crama and Schyns (2003); therefore ours may not be the best algorithmic version as it might be prone to get stuck in a local optimum. Our purpose is to present a basic SA approach to this problem of portfolio selection, where from the reader can design better variants by tuning the appropriate parameters following the discussion on the subject by Crama and Schyns (2003), or at a more profound level by Maringer 2005.

We have N different assets to choose from, but can only take $K < N$ to build an optimum portfolio. To achieve this goal we implement a SA scheme (Algorithm 8.1) which initially chooses K assets out of the N available (represented by a N-vector of weights $w = (w_1, \ldots, w_N)$, where $N - K$ are set to 0), and progressively updates the assets by exchanging some of the previously chosen by new others within the neighbor of current solution, so that the updated vector of weights w^u improves (maximizes) the objective function

$$L(w) = \gamma w' \mu - (1 - \gamma) w' C w .$$

Algorithm 8.1 SA solution to portfolio optimization with cardinality constraints

---

1. **initialization**: generate $w = (w_1, \ldots, w_N)$ with random values ≥ 0,
        so that $N - K$ are 0, and $\sum_{i=1}^{N} w_i = 1$
2. $w^* = w$, $T = T_0$ {*initial temperature*}
3.   cooling parameter γ; neighborhood range u_0
4. **for** $i = 1$ **to** I **do**
5. $w^u = w$
        {*perform neighborhood search:* }
6.   choose randomly two assets i and j with $w_i > 0$ and $j \neq i$.
7.   **if** $w_j = 0$ **then** $w_j^u = w_i$ and $w_i^u = 0$
8.   **else** $w_j^u = w_j + u_i$ and $w_i^u = w_i - u_i$ $(0 < u_i \leq w_i)$
9. $\Delta L = L(w^u) - L(w)$
10.   **if** $(\Delta L > 0)$ **then** {$w = w^u$
11.   **if** $(L(w) > L(w^*))$ **then** $w^* = w$ }
12.   **else** {$p = randomNumber \in [0, 1]$
13.   **if** $p \leq exp(\Delta L / T_{i-1})$ **then** $w = w^u$ }
14. $T_i := \gamma \cdot T_{i-1}$ {*lower the temperature*}
15.   **if** applicable adjust neighborhood range: $u_i = \gamma_u \cdot u_{i-1}$
16. **end for**
17. **output**: w^* {*best solution*}

Note that lines 6–8 (neighborhood search) looks for candidate uphill moves, keeping the vector of weights with at most K non null entries, and enforcing the constraints (8.29)–(8.31). Lines 9–14 realize the move update, and apply a geometrical cooling scheme to control next uphill moves in accordance with the dynamics of an annealing process. The parameters of the algorithm are set according to Crama and Schyns (2003) as follows. The initial temperature $T_0 = -\Delta/\ln(0.8)$, where Δ is average ΔL through several runs. The parameter γ, which reflects the level of risk can be taken arbitrarily with $0 < \gamma < 1$, and the authors suggest to use $\gamma = 0.95$. The number of iterations is estimated experimentally as $I \approx \binom{N}{3}$, but this value may produce too long computations, so for practical purposes it is sufficient to use a value proportional to N, that is $I \approx c \cdot N$, for some constant $c > 1$. Crama and Schyns report results of the SA algorithm applied to $N = 151$ US stocks to build optimal portfolios of 20 assets. For $I = 2N$ the algorithm achieved near-optimal solutions, a hypothesis corroborated by performing further experiments with larger values of I ($I = N^2$), and finding no improvement. Moreover, the solutions described a smooth mean-variance frontier, which was obtained in less than 5 min of computation. For smaller instances (e.g., $N = 30$ and $K = 5$), the SA algorithm computed the full mean-variance frontier perfectly and fast (i.e., perfectly meaning that the resulting feasible portfolios coincide with solutions obtained by a commercial package that handles nonlinear programming problems and with high precision when solving small instances of the portfolio problem).

8.4 Portfolio Selection

When a portfolio deviates from the investor's expected return, its positions must be revised by either increasing or decreasing (possibly down to zero) their weights in the portfolio. All in order to align the portfolio back to the investor's financial objectives. The problem we are posed with is how to do this reallocation of wealth effectively, increasing future gains, through a finite sequence of trading periods, that conforms an investor's schedule for revising the portfolio positions. To model solutions to this problem, we focus on a simple situation of an investor who is long in a number m of securities and with a determined investment horizon. We assume that there are no transaction costs, and that the only limitation for an investor to trade any amount of each security at any time is the amount of money he possesses at the time. The time between the present and the ending date of the life of the portfolio is divided into n trading periods, limited by time instants $t = 0, \ldots, n$, with $t = 0$ representing the initial time and $t = n$ the final time. At the end of each period $[t - 1, t], t = 1, \ldots, n$, the proportion of wealth invested in each position is revised, being the overall goal to maximize the total wealth at the final date. The change in the market price for a given period is represented by the *market vector* $x_t = (x_{t1}, \ldots, x_{tm})$, a sequence of price relatives simple gross returns) for the given trading period of the m securities; that is, each $x_{ti} = P_i(t)/P_i(t - 1)$ is the ratio of closing to opening price (price relative) of the i-th security for the period $[t - 1, t]$. The distribution of wealth among the m securities for the given period is represented by the portfolio vector $\boldsymbol{w}_t = (w_{t1}, \ldots, w_{tm})$ with non negative entries summing to one; that is $\boldsymbol{w}_t \in \mathcal{W}$, where $\mathcal{W} = \{\boldsymbol{w} : \boldsymbol{w} \in \mathbb{R}_+^m, \sum_{j=1}^m w_j = 1\}$. Thus, an investment according to portfolio $\boldsymbol{w}_t$ produces an increase of wealth, with respect to market vector $\boldsymbol{x}_t$ for the period $[t-1, t]$, by a factor of

$$\boldsymbol{w}_t' \cdot \boldsymbol{x}_t = \sum_{j=1}^m w_{tj} x_{tj}.$$

A sequence of n investments according to a selection of n portfolio vectors $\boldsymbol{w}^n = (\boldsymbol{w}_1, \ldots, \boldsymbol{w}_n)$ results in a wealth increase of

$$S_n(\boldsymbol{w}^n, \boldsymbol{x}^n) = \prod_{t=1}^n \boldsymbol{w}_t' \cdot \boldsymbol{x}_t = \prod_{t=1}^n \sum_{j=1}^m w_{tj} x_{tj} \tag{8.32}$$

where $\boldsymbol{x}^n = (\boldsymbol{x}_1, \ldots, \boldsymbol{x}_n)$ is the sequence of price relative vectors corresponding to the n trading periods considered. The quantity $S_n(\boldsymbol{w}^n, \boldsymbol{x}^n)$ is the *wealth factor* achieved by $\boldsymbol{w}^n$ with respect to the sequence of market vectors $\boldsymbol{x}^n$.

Example 8.2 At all times of investment, a j-th stock ($1 \le j \le m$) is represented by the portfolio vector $\boldsymbol{e}_j = (0, \ldots, 1, \ldots, 0) \in \mathcal{W}$ (where 1 is in the j-th coordinate). It has a wealth factor for n trading periods equal to the n-period simple gross return:

$$S_n(\boldsymbol{e}_j, \boldsymbol{x}^n) = \prod_{t=1}^n x_{tj} = \prod_{t=1}^n \frac{P_j(t)}{P_j(t-1)} = R_j(n) + 1.$$

Definition 8.1 A portfolio selection strategy for n trading periods is a sequence of n choices of portfolio vectors $\boldsymbol{w}^n = (\boldsymbol{w}_1, \ldots, \boldsymbol{w}_n)$, where each $\boldsymbol{w}_t \in \mathcal{W}$. A portfolio selection algorithm is an algorithm that produces a portfolio selection strategy. □

If *ALG* is a portfolio selection algorithm, by identifying it with its output (a selection strategy $\boldsymbol{w}^n$), we can also use $S_n(ALG, \boldsymbol{x}^n)$ to denote the wealth factor of *ALG* with respect to a sequence of market vectors $\boldsymbol{x}^n$.

To evaluate the performance of a portfolio selection strategy (or algorithm), independently of any statistical property of returns, the common procedure is to compare its wealth factor against the wealth factor achieved by the best strategy in a class of reference investment strategies (a benchmark strategy). An alternatively is to compare their *exponential growth rate*, which for a selection algorithm *ALG* is defined as

$$W_n(ALG, \boldsymbol{x}^n) = \frac{1}{n} \ln S_n(ALG, \boldsymbol{x}^n) \tag{8.33}$$

We can rewrite the wealth factor $S_n(ALG, \boldsymbol{x}^n)$ in terms of the exponential growth rate as

$$S_n(ALG, \boldsymbol{x}^n) = \exp(nW_n(ALG, \boldsymbol{x}^n)). \tag{8.34}$$

Example 8.3 (**Buy and hold**) This is the simplest strategy where an investor initially allocates all his wealth among m securities according to portfolio vector $\boldsymbol{w}_1 \in \mathcal{W}$, and does not trade anymore. The wealth factor of this strategy after n trading periods is

$$S_n(\boldsymbol{w}_1, \boldsymbol{x}^n) = \sum_{j=1}^{m} w_{1j} \prod_{t=1}^{n} x_{tj}$$

Let $\boldsymbol{e}^* \in \mathcal{W}$ represent (as in Example 8.2) the best performing stock for the sequence of market vectors $\boldsymbol{x}^n$. This stock has a wealth factor of $S_n(\boldsymbol{e}^*, \boldsymbol{x}^n) = \max_{1 \le j \le m} \prod_{t=1}^{n} x_{tj}$. If $BH = \boldsymbol{w}_1$ is the buy and hold strategy, then $S_n(BH, \boldsymbol{x}^n) \le S_n(\boldsymbol{e}^*, \boldsymbol{x}^n)$, and equality holds (i.e. BH achieves the wealth of the best performing stock) if the investors allocates all his wealth in $\boldsymbol{e}^*$ from the beginning. The problem is, of course, how could he know?   □

Example 8.4 (**Constant Rebalanced Portfolios**) A constant rebalanced portfolio (CRP) is a market timing strategy that uses the same distribution of weights throughout all trading periods. Let $CRP_{\boldsymbol{w}}$ be the CRP strategy with fixed weights $\boldsymbol{w} = (w_1, \dots, w_m)$. The wealth factor achieved by applying this strategy for n trading periods is

$$S_n(CRP_{\boldsymbol{w}}, \boldsymbol{x}^n) = \prod_{t=1}^{n} \sum_{j=1}^{m} w_j x_{tj}.$$

Cover Gluss (1986) gave the following illustration of the power of the CRP strategy. In a market consisting of cash and one stock, consider the sequence of market vectors $(1, 1/2), (1, 2), (1, 1/2), (1, 2), \dots$, where the first entry corresponds to cash (so its value is invariant through time) and the second entry corresponds to the stock (on odd days its value reduces one half, while on even days it doubles). Consider a constant rebalancing with $\boldsymbol{w} = (1/2, 1/2)$. Then, on each odd day the simple gross return of this $CRP_{(1/2, 1/2)}$ is $\frac{1}{2}1 + \frac{1}{2}\frac{1}{2}$ and on each even day it is $\frac{3}{2}$. Therefore the wealth achieved over n days is $(9/8)^{n/2}$. This means a 12.5% of wealth growth every two investment periods. Observe that no buy-and-hold will yield any gains.

For a sequence of market vectors $\boldsymbol{x}^n$, the best constant rebalanced portfolio is given by the solution $\boldsymbol{w}^*$ of the optimization problem:

$$\max_{\boldsymbol{w} \in \mathcal{W}} S_n(CRP_{\boldsymbol{w}}, \boldsymbol{x}^n)$$

that is, $\boldsymbol{w}^*$ maximizes the wealth $S_n(CRP_{\boldsymbol{w}}, \boldsymbol{x}^n)$ over all portfolios $\boldsymbol{w}$ of m fixed real positive values applied to the sequence of n market vectors $\boldsymbol{x}^n$.

The optimal strategy $CRP_{\boldsymbol{w}^*}$ outperforms the following common portfolio strategies:

(1) Buy-and-Hold (BH): The wealth factor of $CRP_{\boldsymbol{w}^*}$ is at least the wealth factor of the best performing stock (i.e., $S_n(CRP_{\boldsymbol{w}^*}, \boldsymbol{x}^n) \ge S_n(\boldsymbol{e}^*, \boldsymbol{x}^n)$), since each stock $e_j \in \mathcal{W}$ and $S_n(CRP_{\boldsymbol{w}^*}, \boldsymbol{x}^n)$ is the maximum over all $\mathcal{W}$.

(2) Arithmetic mean of stocks: $S_n(\mathrm{CRP}_{\mathbf{w}^*}, \mathbf{x}^n) \geq \sum_{j=1}^{m} \alpha_j S_n(\mathbf{e}_j, \mathbf{x}^n)$, for $\alpha_j \geq 0$, $\sum_{j=1}^{m} \alpha_j = 1$. In particular, the $\mathrm{CRP}_{\mathbf{w}^*}$ strategy outperforms the DJIA index. (*Prove as exercise.*)

(3) Geometric mean of stocks: $S_n(\mathrm{CRP}_{\mathbf{w}^*}, \mathbf{x}^n) \geq \left(\prod_{j=1}^{m} \alpha_j S_n(\mathbf{e}_j, \mathbf{x}^n) \right)^{1/m}$. In particular, the $\mathrm{CRP}_{\mathbf{w}^*}$ strategy outperforms the geometric Value Line (VLIC) (*Exercise.*). □

The best constant rebalanced portfolio strategy $\mathrm{CRP}_{\mathbf{w}^*}$ is an extraordinary profitable strategy by the above properties; however, it is unrealistic in practice because it can only be computed with complete knowledge of future market performance. We shall explore more practical alternative strategies, and performing as well as $\mathrm{CRP}_{\mathbf{w}^*}$, in Chap. 9.

8.5 Notes, Computer Lab and Problems

8.5.1 Bibliographic remarks: A clear and comprehensible introduction to portfolio management is given by Luenberger(1998). Much of the material for the first section of this chapter is derived from this reference. Brealey et al. (2011) give a more practical, although less mathematical, presentation of portfolio theory. For a mathematical discussion of the Markowitz portfolio model as a quadratic programming problem see Vanderbei (2008). The original paper markowitz, (1952) is an excellent lecture, together with criticism by Samuelson (1967), who discusses distributional conditions that will make the optimal solution not to be within the efficient frontier, and a proof that under the general assumptions on independence of the mean-variance model, the weights that give an optimal portfolio of N assets can be all taken equal to $1/N$. The Sharpe ratio is due to William Sharp (1964). The CAPM is due to Sharpe, Lintner, Mossin and Treynor (see Rubinstein 2006 for a historical account). To learn the full capabilities of R/RMetrics for portfolio optimization see Würtz et al. (2009) For more heuristic optimization schemes, such as ant colony, genetic methods, and memetic algorithms, applied to portfolio management see (Maringer 2005).

8.5.2: Solve the linear system for the Lagrangian (8.7) to obtain the analytic solutions for the weights for a portfolio with mean r^* and corresponding minimum variance.

8.5.3: Find the formulae for the weights that maximizes $\tan \theta$ in Eq. (8.14). (Hint: write

$$\tan \theta = \frac{E(R^{\mathbf{w}}) - r_f}{std(R^{\mathbf{w}})} = \frac{\displaystyle\sum_{i=1}^{N} w_i (E(R_i) - r_f)}{\left(\displaystyle\sum_{1 \leq i,j \leq N} w_i w_j \sigma_{ij} \right)^{1/2}}$$

Then set the derivatives of $\tan \theta$ with respect to each w_i equal to zero, and solve the resulting system of linear equations.)

8.5.4: Prove Theorem 8.1, i.e., show the CAPM formula, Eq. (8.17). (Hint: Consider a portfolio consisting of asset i, with weight ω, and the Market Portfolio M, with weight $1 - \omega$. This portfolio has expected return $E(R^{\omega}) = \omega E(R_i) + (1 - \omega)E(R_M)$ and standard deviation $\sigma_{\omega} = (\omega^2 \sigma_i^2 + 2\omega(1 - \omega)\sigma_{iM} + (1 - \omega)^2 \sigma_M^2)^{1/2}$. As ω varies, the points $(\sigma_{\omega}, E(R^{\omega}))$ trace a curve C in the σ-μ plane. For $\omega = 0$ we have the Market Portfolio M, and at this point the Capital Market Line must be tangent to C. Hence, the slope of C is equal to the slope of the CML at $\omega = 0$. Use this equality to obtain Eq. (8.17).)

8.5.5 (Linearity of the CAPM pricing formula): Show that the pricing formula given by Eq. (8.18) is linear; i.e. for any real number λ and two asset prices P_1 and P_2, $F(P_1 + \lambda P_2) = F(P_1) + \lambda F(P_2)$, where $F(P) = \dfrac{E(P_\tau)}{1 + r_f + \beta(E(R_M) - r_f)}$.
This is a nice property, which among other things, allow us to find the price of a portfolio as the linear combination of the prices of its positions. (Hint: substitute R_τ by $(P_\tau/P) - 1$ in $\beta = Cov(R_\tau, R_M)/\sigma_M^2$ and put the resulting term in the place of β in Eq. (8.18) to obtain the required formula on P_τ.)

8.5.6 R Lab: To get pictures fancier than the ones computed in R Example 8.1, containing labels and other information, the reader should study the code of the fPortfolio function `tailoredFrontierPlot`. In it you will find how to set up the boundaries of the picture, write labels for points and axis, and other tricks.

8.5.7 R Lab: To resolve the optimization problem (8.11) with the solver `lp` allowing unrestricted solutions (i.e. long and short positions) we need to use the trick of replacement of variables by the difference of two new positive variables. There is a caveat with this trick in that it may produce unbounded solutions and the solver fails to converge, outputting `Error: status 3`. The fix is to add extra constraints. We explain the trick for a weight vector of 2 variables. We want to

$$\text{maximize: } w_1\mu_1 + w_2\mu_2$$
$$\text{subject to: } w_1 + w_2 = 1 \tag{8.35}$$

Setting each $w_i = w_{i1} - w_{i2}$, $i = 1, 2$, Eq. (8.35) extends to

$$\text{maximize: } w_{11}\mu_1 + w_{21}\mu_2 - (w_{12}\mu_1 + w_{22}\mu_2)$$
$$\text{subject to: } w_{11} + w_{21} - (w_{12} + w_{22}) = 1 \tag{8.36}$$

This extended system has unbounded solutions and the `lp` will fail to give an answer; hence we place as extra restrictions the inequalities:

$$|w_{11} + w_{21} + w_{12} + w_{22}| \le C$$

where C is some bound (e.g. for $C = 1$ we get the long-only solution; hence consider $C > 1$). For the stocks in the R Example 8.1, the R commands for computing the maximum return portfolio with unrestricted weights are:

```
> f.obj <- c(covData$mu,-covData$mu)
> f.con <-matrix(c(1,1,1,1,1,-1,-1,-1,-1,-1,
    +              1,1,1,1,1,1,1,1,1,1,
    +              1,1,1,1,1,1,1,1,1,1),nrow=3, byrow=TRUE)
> f.dir <- c("=",">=","<=")
> bound <- 2 ##alternatives 3, 1
> f.rhs <- c(1,-bound,bound)
> lp ("max", f.obj, f.con, f.dir, f.rhs)
> lp ("max", f.obj, f.con, f.dir, f.rhs)$solution
```

Chapter 9
Online Finance

Many financial problems require making decisions based on current knowledge and under uncertainty about the future. This is the case, for example, of problems such as:

- to decide when to buy or sell a stock by observing its price daily quotes;
- to find a killing price for an American put option;
- to rebalance a portfolio for maximizing future returns.

We have so far given algorithmic solutions to some of these problems, where the uncertainty is filled in with expectations (as for the first two problems of the list), or complete information is assumed (as in the best constant rebalancing portfolio strategy). These solutions are termed *offline*. We are now concerned with developing *online* solutions to financial problems that are natural online problems; that is, where decisions have to be made with the partial information at hand, and without knowledge about the future. The performance of online algorithms is analyzed from the perspective of *competitive analysis* (a worst-case paradigm), where the online solutions are compared against that of the optimum offline algorithm. This type of analysis can lead to different solutions with improved performance, and from different perspectives.

The broad subject of online financial problems is described by Ran El-Yaniv (1998) in terms of a one player game against an adversary, pictured as the online player against nature, where each game takes place during a time horizon, which can be either continuous or discrete, divided in time periods. In all the games, the objective of the online player is to minimize a cost function, while the adversary goal is to maximize it. Under this perspective, El-Yaniv classifies online financial problems as variants or applications of one of the following four elementary online problems: (1) online searching; (2) online portfolio selection; (3) online replacement; (4) online leasing. In this chapter we shall study online financial problems belonging to the first two classes.

A. Arratia, *Computational Finance*, Atlantis Studies in Computational Finance
and Financial Engineering 1, DOI: 10.2991/978-94-6239-070-6_9,
© Atlantis Press and the authors 2014

9.1 Online Problems and Competitive Analysis

We begin outlining the general formalities of online computation. Let $\mathscr{P}$ be an optimization problem, $\mathscr{S}$ be the set of all input instances of $\mathscr{P}$, and F the cost or objective function, defined for all instances $I \in \mathscr{S}$ and onto a set of feasible outputs. We assume that the cost function F is bounded.

$\mathscr{P}$ is an *online optimization problem* if every input instance $I \in \mathscr{S}$ is given as a finite sequence $I = \{i_1, \ldots, i_n\}$, which is incrementally revealed by stages, such that at stage $j \in \{1, \ldots, n\}$ only the input i_j is made known and a decision o_j has to be made with that partial information. The cost function F has its value updated after each decision and the final cost is the aggregation of the successive partial valuations obtained at the end of the finite sequence that constitute the input. By contrast, in an *offline* problem the input I is fully given at once, and a decision is made with the full information.

An *online algorithm* is a decision making procedure for an online optimization problem. Let $OALG$ be an online algorithm for online problem $\mathscr{P}$, and denote by $OALG(I)$ the total cost attained by algorithm $OALG$ for its sequence of outputs $\{o_1, o_2, \ldots, o_n\}$ obtained when processing the input sequence $I = \{i_1, i_2, \ldots, i_n\}$. Let OPT be an optimum offline algorithm for $\mathscr{P}$, and $OPT(I)$ be the optimal cost attained by OPT on the input I.

The *competitiveness* of the online algorithm $OALG$ with respect to the optimum offline algorithm OPT, in accordance with $\mathscr{P}$ being a maximization or minimization problem, is defined as follows.

- If $\mathscr{P}$ is a minimization problem then $OALG$ is c-competitive if for all $I \in \mathscr{S}$

$$OALG(I) \leq c \cdot OPT(I) \tag{9.1}$$

- If $\mathscr{P}$ is a maximization problem then $OALG$ is c-competitive if for all $I \in \mathscr{S}$

$$c \cdot OALG(I) \geq OPT(I) \tag{9.2}$$

The least c such that $OALG$ is c-competitive is the algorithm's *competitive ratio*.

The analysis of performance (or competitiveness) of an online algorithm with respect to the competitive ratio measure is called *competitive analysis*. To do this type of analysis it is convenient to view the process of finding an online algorithmic solution to an online problem as a two-player game. In this game one of the players, known as the online player, chooses an online algorithm $OALG$ and reveals it to his adversary, the offline player, who in turn chooses an input I and feeds it to the algorithm. The goal of the online player is to minimize the competitive ratio, so he must choose $OALG$ wisely, whereas the goal of the offline player is to maximize the competitive ratio, so once he knows $OALG$ chooses the input according to this objective. This means that

- if $\mathscr{P}$ is a minimization problem (Eq. (9.1)) offline chooses input I that maximizes

$$c = \frac{OALG(I)}{OPT(I)};$$

- if $\mathscr{P}$ is a maximization problem (Eq. (9.2)) offline chooses input I that maximizes

$$c = \frac{OPT(I)}{OALG(I)}.$$

In the following sections we review some online solutions to online financial problems pertaining to price searching, trading and portfolio management.

9.2 Online Price Search

9.2.1 Searching for the Best Price

The goal of this online problem is to select the minimum (resp. maximum) price to buy (resp. sell) some asset from a series of prices observed through time. We assume prices are quoted through a finite sequence of time instances $t_1, \ldots t_n$, and are taken from a finite interval of values $[m, M]$. At each time instant t_i a price is quoted and the online player must decide to take it or not, and once taken that price is the player's payoff. If the online player doesn't select a price throughout the n time instances, then at the end of the series he must accept the last quoted price. Let $\varphi = M/m$ be the global fluctuation ratio. If m and M are known we have the following deterministic online algorithm, known as the Reservation Price Policy (RPP), which we split in two versions according to the goal of the online player being to buy or to sell.

RESERVATION PRICE POLICY (RPP).
RPP_{buy}: Buy at the first price less than or equal to $\sqrt{M \cdot m}$.
RPP_{sell}: Sell at the first price greater than or equal to $\sqrt{M \cdot m}$.

Theorem 9.1 RPP *attains a competitive ratio of* $\sqrt{\phi} = \sqrt{M}/\sqrt{m}$.

Proof We analyze the case where the online player wants to buy and is searching for a minimum price (i.e. RPP_{buy}). The inputs are price quotations given at times $t_1, t_2, \ldots, t_n$. There are basically two strategies that the offline player has to offset the online player's profits, which depends on the online player buying at some given price or not:

Case 1: at some instant t_i the price p quoted is the lowest so far, but the online player decides not to buy. Then, immediately after, the offline player raises the price to M and keeps it at that value until the end, which forces the online player to buy at the highest price possible, i.e. M. The player's performance ratio is $\dfrac{online}{offline} = \dfrac{M}{p}$.

Case 2: at some instant t_i the price p quoted is the lowest so far, and the online player decides to buy. Then, immediately after, the offline player lowers the price to m and keeps it at that value until the end. The player's performance ratio for this case is $\dfrac{online}{offline} = \dfrac{p}{m}$.

Equating both performance ratios for a unified strategy, we find

$$\frac{M}{p} = \frac{p}{m} \Rightarrow p = \sqrt{M \cdot m}$$

and the competitive ratio

$$c = \frac{OALG(I)}{OPT(I)} = \frac{\sqrt{M}}{\sqrt{m}} = \sqrt{\varphi}.$$

The argument for the sell objective (i.e. RPP_{sell}) is analogous.   □

If we only know φ then it can be shown that the competitive ratio attainable by any deterministic price search strategy is trivially φ (see Exercise 9.5.2).

9.2.2 Searching for a Price at Random

An important improvement to the Reservation Price Policy to solve more efficiently the price search problem is to allow the online player to make his transaction decisions randomly. In a randomized algorithm the players may know the algorithm but can not know the random numbers that such algorithm will generate upon execution to make some decisions. This makes it harder for the offline player to select an appropriate input to maximize his competitiveness. The following randomized version of RPP is due to Leonid Levin,[1] and we will see that attains a competitive ratio in the order of $O(\log \varphi)$. We state the algorithm for the case of selling, being the case of buying very similar. We suppose $\varphi = M/m = 2^k$ for some integer $k > 0$; hence $M = m2^k$.

---

[1] Communicated in El-Yaniv (1998).

RANDOMIZED RPP (RRPP).
At each instant of time choose an integer $i \in \{0, 1, \ldots, k - 1\}$ with probability $1/k$ and sell when the price is $\geq 2^i$.

Theorem 9.2 RRPP *attains a* $O(\log \varphi)$–*competitive ratio.*

Proof Let p^* be the maximum price that can be possibly obtained in a given input sequence. There exists a positive integer j such that

$$m2^j \leq p^* < m2^{j+1} \tag{9.3}$$

Note that $j \leq k$, since all prices are in $[m, M]$, and $j = k$ if and only if $p^* = M$. The offline player has control over the input and can choose p^* to his advantage. Nonetheless, choosing $p^* = M$ (and consequently, $j = k$) will be seen to be not an optimal choice; hence, we can assume that $j \leq k - 1$. For any j satisfying Eq. (9.3), the expected return that the online player can obtain applying algorithm RRPP is

$$\frac{m}{k}\left(k - j + \sum_{i=0}^{j} 2^i\right) = \frac{m}{k}(2^{j+1} + k - j - 2).$$

For any such j the optimal choice of p^* for the offline player is a value close to $m2^{j+1}$. Then the player's performance ratio, offline to online, for a j satisfying Eq. (9.3) is given by the following function $R(j)$

$$R(j) = k\frac{2^{j+1}}{2^{j+1} + k - j - 2}$$

As a simple calculus exercise one can see that the function $R(j)$ attains its maximum at $j^* = k - 2 + 1/\ln 2$ (and this is why it is optimal for offline to choose $j \leq k - 1$). Also one can see that the coefficient of k in $R(j)$ is almost 1, resulting in a competitive ratio of the order of $O(k) = O(\log \varphi)$. $\qquad\square$

Remark 9.1 If m and M are not known but φ is known, then the same competitive ratio of $O(\log \varphi)$ holds. This is better than the deterministic strategy RPP where the competitive ratio is φ. Even further the randomized strategy RRPP can be modified to work when there is no knowledge of φ. This enhanced version of RRPP can attain a competitive ratio of $O((\log \varphi) \cdot \log^{1+\varepsilon}(\log \varphi))$, for arbitrary $\varepsilon > 0$ and where φ is the posterior global function ratio. For details see El-Yaniv et al. (2001).

9.3 Online Trading

We are now concern with using the price search strategies of the Sect. 9.2 for online trading, that is, to implement them for consecutive transactions of buying and selling, with possible repetitions, through a given investment horizon divided into trading periods marked by the time instants when a transaction is made. We will see first that the problem of searching for a price can be seen as a particular online trading problem.

9.3.1 One-Way Trading

The one-way trading game consists of a trader who needs to exchange some initial wealth given in some currency to some other currency.[2] To fix ideas let's assume the trader wants to exchange dollars to euros. The trader can partition his initial wealth and trade the parts through a sequence of time instants, each part at a different exchange rate offered at the time. Once all the dollars have been exchange the total return of the trader is the amount of euros he accumulated. The problem is then how to do this sequential exchange to maximize total return. Some variants of this problem include: knowledge of a minimum m and a maximum M prices within which exchange rates fluctuates, or the number n of time instants for making the trade.

Note that the one-way trading problem resembles the price search problem, where the online player has the apparent advantage of partition his wealth and select different prices to buy sequentially. Nevertheless, it can be shown that the total return is the same as the price obtained by an optimal randomized search algorithm. This is the content of the following theorem of El-Yaniv et al. (2001)

Theorem 9.3 *For any deterministic one-way trading algorithm ALG, there exists a randomized algorithm ALG_r that makes a single trade, and such that for any exchange rate sequence σ, $E(ALG_r(\sigma)) = ALG(\sigma)$. Conversely, for any randomized single trade algorithm ALG_r, there exists a one-way trading deterministic algorithm ALG, such that for any sequence of exchange rates σ, $E(ALG_r(\sigma)) = ALG(\sigma)$.*

Proof Let the budget for investment be of $1. Suppose a deterministic algorithm trades a fraction s_i of the budget at the i-th period, $i = 1, \ldots, I$. Each $0 < s_i < 1$ and $\sum_{i \in I} s_i = 1$; hence, s_i can be interpreted as the probability of trading the entire budget at the i-th period. The average return of this randomized algorithm equals the return of the deterministic algorithm. Conversely, any randomized single trade algorithm can be considered as a *mixed strategy* (i.e., a probability distribution over

---

[2] Here the trader is selling an asset -a currency- and buying another asset. The one-way refers to the fact that the original asset is not gotten back, for it is completely exchange by the other asset.

deterministic algorithms),[3] and by linearity of expectation, and that the sum of all traded amounts is 1, this mixed strategy is equivalent to a deterministic algorithm that trades the initial investment budget through a sequence σ of fractions.   □

Can we make profits in the stock markets with online trading algorithms?
Theorem 9.3 assure us that on average it is possible to obtain the same profit applying either a trading strategy that at the time of trade uses all our budget, or a strategy consisting in trading one-way using fragments of our budget sequentially until it is all used up. However, a particular online trading strategy might be more profitable than others, at a certain periods of time, depending on the conditions of the market and the nature of the assets (e.g. their volatility, industrial sectors, economic and social factors, etc.), and so it is of interest to evaluate the performance on real data of different online trading strategies typified by either way of using the initial budget for investment, i.e., by using it all in each transaction, or by fractions. Online trading strategies that use all the budget in each transaction can be defined from the online price search problems of the Sect. 9.2; other online strategy of this type is the *difference maximization* in Kao and Tate (1999). On the other hand, online strategies that trade with fractions of the allocated budget for investments are to some extend versions of the *thread-based policy* described in El-Yaniv (1998); others in this family are the *max-min search* in Loren et al. (2009), the *dollar-cost averaging* and the *value averaging*, these last two coming from the folklore in financial markets (but see Marshall (2000) for a statistical comparison and El-Yaniv (op. cit.) for performance bounds under certain restricted scenario). We propose as a research problem the question posed at the beginning of this paragraph.

9.4 Online Portfolio Selection

In Section 8.4 we approached the problem of portfolio selection from an offline perspective. The downside with that view is that the best performing offline strategies, as the best constant rebalancing portfolio strategy CRP_{w^*}, are not realistically attainable since it requires full knowledge of the future. Thus, we frame here the portfolio selection problem as a online problem, which is in fact a more natural model since decisions about portfolio management are made in practice only with current information. Nonetheless the offline strategies we have already studied for the portfolio selection problem will be important benchmarks for testing the quality of online solutions through competitive analysis.

   Recall that the space of portfolio vectors is $\mathcal{W} = \{w : w \in \mathbb{R}_+^m, \sum_{j=1}^m w_j = 1\}$, and that $S_n(w^n, x^n)$ denotes the wealth factor achieved by the sequence of n portfolio vectors, w^n, with respect to the sequence of market vectors x^n. An online portfolio selection (or rebalancing) strategy is defined as follows.

---

[3] This follows from a theorem of Kuhn in game theory: Kuhn, H. W. (1953) Extensive games and the problem of information. In *Contribution to the Theory of Games*, vol. II, H. W. Kuhn and A. W. Tucker, eds. Princeton Univ. Press, 193–216.

Definition 9.1 An online portfolio selection strategy is a sequence of maps

$$\widehat{w}_t : \mathbb{R}_+^{m(t-1)} \to \mathscr{W}, \quad t = 1, 2, \dots$$

where $\widehat{w}_t = \widehat{w}_t(x_1, \dots, x_{t-1})$ is the portfolio used at trading time t given only past securities price relatives vectors (i.e. market vectors) $x_1, \dots, x_{t-1}$. An online portfolio selection algorithm is an algorithm that produces an online portfolio selection strategy. □

The performance of an online portfolio selection algorithm $OALG$ can be measured with the same worst-case measures used for offline algorithms, namely:

- by comparing its compounded return to that of a benchmark offline algorithm, e.g., the best performing stock or the best constant rebalanced strategy CRP_{w^*};
- using the exponential growth rate of wealth $W_n(OALG, x^n) = \frac{1}{n} \ln S_n(OALG, x^n)$.

This is considered in a competitive analysis of performance, where the competitive ratio of $OALG$ is given by

$$\sup_{x^n} \frac{S_n(OPT, x^n)}{S_n(OALG, x^n)} \tag{9.4}$$

where OPT is an optimal offline portfolio selection algorithm, which is usually taken as CRP_{w^*}. Recall that the CRP_{w^*} is a very high benchmark, since as an offline strategy it outperforms: the best performing stock (and hence, buy and hold); the DJIA index (or any arithmetic market index); and the Value Line (see Example 8.4). Thus, designing an online portfolio selection strategy with a wealth factor similar to the wealth factor achieved by the CRP_{w^*} algorithm, for any sequence of market vectors x^n, would be a much desirable strategy. That an online portfolio selection strategy performing as well as CRP_{w^*} is possible is what we shall see in the Sect. 9.4.1.

9.4.1 The Universal Online Portfolio

Cover (1991) proposed an online portfolio selection strategy that performs asymptotically as well as the best constant rebalanced portfolio CRP_{w^*}. Cover's *Universal Online Portfolio* (UOP) algorithm assigns a uniform distribution of the initial wealth to the initial portfolio, at the beginning of the first trading period, and for subsequent trading periods selects the performance weighted average of all portfolios $w = (w_1, \dots, w_m) \in \mathscr{W}$ over the partial sequence of price relative information known up to the current period. Formally it is defined as the following strategy:

$$\widehat{w}_1 = \left(\frac{1}{m}, \dots, \frac{1}{m} \right), \quad \widehat{w}_{k+1} = \frac{\int_{\mathscr{W}} w \, S_k(w, x^k) dw}{\int_{\mathscr{W}} S_k(w, x^k) dw} \tag{9.5}$$

where

$$S_k(\boldsymbol{w}, \boldsymbol{x}^k) = \prod_{t=1}^{k} \boldsymbol{w}' \cdot \boldsymbol{x}_t = \prod_{t=1}^{k} \sum_{j=1}^{m} w_j x_{tj} \tag{9.6}$$

and the integration is over the set of $(m-1)$-dimensional portfolios $\mathcal{W}$. The initial wealth $S_0 = S_0(\boldsymbol{w}, \boldsymbol{x}^0) \equiv 1$, for it corresponds to $\widehat{\boldsymbol{w}}_1$.

To have a real sense of the computations involved in this strategy, rewrite the formulas as follows:

$$\begin{aligned}
\widehat{\boldsymbol{w}}_{k+1} &= \int_{\mathcal{W}} (w_1, \dots, w_m) \left(\frac{S_k(\boldsymbol{w}, \boldsymbol{x}^k)}{\int_{\mathcal{W}} S_k(\boldsymbol{w}, \boldsymbol{x}^k) d\boldsymbol{w}} \right) d\boldsymbol{w} \\
&= \left(\int_{\mathcal{W}} w_1 M_k(\boldsymbol{w}) d\boldsymbol{w}, \dots, \int_{\mathcal{W}} w_m M_k(\boldsymbol{w}) d\boldsymbol{w} \right)
\end{aligned}$$

where $M_k(\boldsymbol{w}) = S_k(\boldsymbol{w}, \boldsymbol{x}^k) / \int_{\mathcal{W}} S_k(\boldsymbol{w}, \boldsymbol{x}^k) d\boldsymbol{w}$.

Note also that $S_k(\boldsymbol{w}, \boldsymbol{x}^k) = S_{k-1}(\boldsymbol{w}, \boldsymbol{x}^{k-1}) \sum_{j=1}^{m} x_{kj} w_j$, and by the definition of this wealth factor, the strategy is performing a constant rebalance up to time $t = k$, since the distribution of weights $\boldsymbol{w}$ does not changes. Thus, using the notation of Example 8.4 we could write $S_k(\boldsymbol{w}, \boldsymbol{x}^k) = S_k(\text{CRP}_{\boldsymbol{w}}, \boldsymbol{x}^k)$. And now, let us look at a small specific case.

Example 9.1 Consider a market with three stocks; that is, $m = 3$ and the set of possible portfolios $\mathcal{W}$ is a 2-dimensional space, where each portfolio $\boldsymbol{w} = (w_1, w_2, 1 - w_1 - w_2)$, with $0 \le w_1 \le 1$ and $0 \le w_2 \le 1 - w_1$.

The simplex that these set of points describe in $\mathbb{R}^3$ is shown in the figure. Then the universal online portfolio strategy consists of

$$\widehat{\boldsymbol{w}}_1 = (1/3, 1/3, 1/3)$$
$$\widehat{\boldsymbol{w}}_{k+1} = (\widehat{w}_{k+1,1}, \widehat{w}_{k+1,2}, \widehat{w}_{k+1,3})$$
$$= (\widehat{w}_{k+1,1}, \widehat{w}_{k+1,2}, 1 - \widehat{w}_{k+1,1} - \widehat{w}_{k+1,2})$$

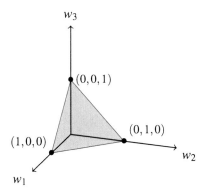

where

$$\widehat{w}_{k+1,i} = \int_0^1 \int_0^{1-w_1} w_i \left(\frac{S_k(w_1, w_2, \boldsymbol{x}^k)}{\int_0^1 \int_0^{1-w_1} S_k(w_1, w_2, \boldsymbol{x}^k) dw_2 dw_1} \right) dw_2 dw_1, \quad \text{for } i = 1, 2.$$

and $S_k(w_1, w_2, \boldsymbol{x}^k) = \prod_{i=1}^k (w_1 x_{i1} + w_2 x_{i2} + (1 - w_1 - w_2) x_{i3})$ is a polynomial of degree at most k in the variables w_1 and w_2. $\qquad\square$

The wealth factor resulting from applying the universal online portfolio strategy UOP is given by

$$S_n(\text{UOP}, \boldsymbol{x}^n) = \prod_{k=1}^n \widehat{\boldsymbol{w}}_k' \cdot \boldsymbol{x}_k = \prod_{k=1}^n \sum_{j=1}^m \widehat{w}_{k,j} x_{k,j}$$

This can be seen to be the average of the wealths achieved by the individual strategies in the simplex $\mathscr{W}$.

Lemma 9.1

$$S_n(UOP, \boldsymbol{x}^n) = \int_{\mathscr{W}} S_n(\boldsymbol{w}, \boldsymbol{x}^n) d\boldsymbol{w} \Big/ \int_{\mathscr{W}} d\boldsymbol{w}. \tag{9.7}$$

Proof

$$S_n(\text{UOP}, \boldsymbol{x}^n) = \prod_{k=1}^n \sum_{j=1}^m \widehat{w}_{k,j} x_{k,j} = \prod_{k=1}^n \frac{\int_{\mathscr{W}} \sum_{j=1}^m x_{k,j} w_j S_{k-1}(\boldsymbol{w}, \boldsymbol{x}^{k-1}) d\boldsymbol{w}}{\int_{\mathscr{W}} S_{k-1}(\boldsymbol{w}, \boldsymbol{x}^{k-1}) d\boldsymbol{w}}$$

$$= \prod_{k=1}^n \frac{\int_{\mathscr{W}} S_k(\boldsymbol{w}, \boldsymbol{x}^k) d\boldsymbol{w}}{\int_{\mathscr{W}} S_{k-1}(\boldsymbol{w}, \boldsymbol{x}^{k-1}) d\boldsymbol{w}} \qquad \text{(product telescopes and } S_0 = 1\text{)}$$

$$= \frac{\int_{\mathscr{W}} S_n(\boldsymbol{w}, \boldsymbol{x}^n) d\boldsymbol{w}}{\int_{\mathscr{W}} d\boldsymbol{w}} \qquad\qquad\qquad\qquad\qquad\qquad \square$$

From previous remark, $S_n(\boldsymbol{w}, \boldsymbol{x}^n) = S_n(\text{CRP}_{\boldsymbol{w}}, \boldsymbol{x}^n)$, and hence, an immediate consequence of Eq. (9.7) is

$$S_n(\text{UOP}, \boldsymbol{x}^n) = E(\{S_n(\text{CRP}_{\boldsymbol{w}}, \boldsymbol{x}^n) : \boldsymbol{w} \in \mathscr{W}\})$$

under uniform distribution of all portfolios. Other interesting properties of the wealth achieved by the universal online portfolio strategy can be derived from (9.7).

Theorem 9.4 (Properties of UOP)

(1) $S_n(UOP, x^n)$ is invariant under permutations of the sequences of stock market
vectors $x_1, \ldots, x_n$.
(2) The Universal Portfolio exceeds the Value Line index (VLIC):

$$S_n(UOP, x^n) \geq \left(\prod_{j=1}^{m} S_n(e_j) \right)^{1/m}$$

Proof (1) In Eq. (9.7) the integrand

$$\int_{\mathscr{W}} S_n(w, x^n)dw = \int_{\mathscr{W}} \prod_{k=1}^{n} w' \cdot x_k \, dw$$

and not matter how you permute the stock vectors $x_1, \ldots, x_n$, the product $\prod_{k=1}^{n} w' \cdot x_k$
yields the same result.
(2) We sketch the main idea of the proof, and leave as an exercise for the reader to
complete the proof (or read through (Cover 1991, Prop. 2.5)). Rewrite $S_n(UOP, x^n)$
in terms of exponential growth rate

$$S_n(UOP, x^n) = \frac{1}{\int_{\mathscr{W}} dw} \int_{\mathscr{W}} S_n(w, x^n)dw = \frac{1}{\int_{\mathscr{W}} dw} \int_{\mathscr{W}} \exp(n W_n(w, x^n))dw$$

Next, use *Jensen's inequality* to push everything inside the exponential, do the appropriate cancelations and further manipulations (e.g. recover $\ln S_n(w, x^n)$ and apply
Jensen's inequality a second time) to get the desired result.   □

The first property assess for the robustness of the universal portfolio: it implies
that violent periods of changes in the market prices have no worse consequences for
the wealth $S_n(UOP, x^n)$ of the universal portfolio, that if the days comprising those
unsteady times are dispersed among the regularly behaved price times. The second
property shows that the universal portfolio performs as well as the best constant
rebalanced portfolio (offline) strategy, CRP_{w^*}, at least with respect to the VLIC
index.

Furthermore, a much stronger property of the wealth factor function of the universal online portfolio strategy states that this is similar (or asymptotically close) to
the wealth factor of CRP_{w^*}; that is

$$\lim_{n \to \infty} \frac{1}{n} \left(\ln S_n(CRP_{w^*}, x^n) - \ln S_n(UOP, x^n) \right) = 0 \tag{9.8}$$

for arbitrary sequence of stock vectors x^n. Note that the difference is between
the exponential growth rates of the two algorithms: Eq. (9.8) is equivalent to
$\lim_{n \to \infty} (W_n(CRP_{w^*}, x^n) - W_n(UOP, x^n)) = 0$. Hence, this equation says that

the strategy UOP should perform as well as the best constant rebalanced portfolio CRP_{w^*} to the first order in the exponent.

The limit in (9.8) is a consequence of the following result.

Theorem 9.5 *For all markets with m stocks, for all n periods and all securities price relatives sequence x^n,*

$$\frac{S_n(UOP, x^n)}{S_n(CRP_{w^*}, x^n)} \geq \left(\frac{1}{(n+1)^{m-1}}\right) \cdot \frac{1}{e} \tag{9.9}$$

Proof The idea is that portfolios "near" to each other perform similarly, and that a large fraction of portfolios are near the optimal one. Suppose, in hindsight, that w^* is an optimal CRP for the market of m stocks. A portfolio near to w^* is given by $w = (1 - \alpha)w^* + \alpha z$, for some $z \in \mathcal{W}, 0 \leq \alpha < 1$. Then, a single period's gain of CRP_w is at least $(1 - \alpha)$ times the gain of CRP_{w^*}, and so, over n periods

$$S_n(CRP_w, x^n) \geq (1 - \alpha)^n S_n(CRP_{w^*}, x^n) \tag{9.10}$$

A considerable large volume of portfolios is sufficiently near w^*, because the set of near portfolios is a shrunken simplex $\alpha\mathcal{W}$, translated from the origin to $(1 - \alpha)w^*$. The volume of simplex $\mathcal{W}$ is $Vol(\mathcal{W}) = \int_{\mathcal{W}} dw$, and

$$Vol(\{(1 - \alpha)w^* + \alpha z : z \in \mathcal{W}\}) = Vol(\{\alpha z : z \in \mathcal{W}\}) = \alpha^{m-1} Vol(\mathcal{W})$$

From these equalities we get $Vol(\alpha\mathcal{W})/Vol(\mathcal{W}) = \alpha^{m-1}$. Now, using equations (9.7) and (9.10), together with previous observations, we get

$$S_n(UOP, x^n) = \frac{1}{\int_{\mathcal{W}} dw} \int_{\mathcal{W}} S_n(CRP_w, x^n) dw = \frac{1}{Vol(\mathcal{W})} \int_{\mathcal{W}} S_n(CRP_w, x^n) dw$$

$$\geq \frac{1}{Vol(\mathcal{W})} \int_{\alpha\mathcal{W}} (1 - \alpha)^n S_n(CRP_{w^*}, x^n) dw$$

$$= \left(\frac{Vol(\alpha\mathcal{W})}{Vol(\mathcal{W})}\right) (1 - \alpha)^n S_n(CRP_{w^*}, x^n)$$

$$= \alpha^{m-1} (1 - \alpha)^n S_n(CRP_{w^*}, x^n)$$

Take $\alpha = 1/(n + 1)$. Then

$$\frac{S_n(UOP, x^n)}{S_n(CRP_{w^*}, x^n)} \geq \left(\frac{1}{(n+1)^{m-1}}\right) \cdot \left(\frac{n}{n+1}\right)^n \geq \left(\frac{1}{(n+1)^{m-1}}\right) \cdot \frac{1}{e}$$

$\square$

All of the previously stated properties of the universal portfolio strategy makes it as profitable as the best constant rebalanced portfolio CRP_{w^*}, and yet realistic due to its online character, so what's the catch (if any)? Well, indeed, there is a catch, and it is its computational complexity. Observe that the computation of the universal portfolio involves solving an integral over an m-dimensional simplex. This has an exponential cost relative to m, and even the numerical approach to compute these integrals proposed by Cover in its original work (Cover 1991, §8), turns intractable beyond nine assets. Cover proposes to sample portfolios at each trading period and approach the integral by averaging on the wealth of CRP on these samples; more precisely, his suggestion is to fix some level of precision $1/N$, and take a finite sum over all possible portfolios w of the form $(a_1/N, a_2/N, \ldots, a_m/N)$, where each a_i is a nonnegative integer and $a_1 + a_2 + \cdots + a_m = N$. Helmbold et al. (1998) tried this approximation method with real data to select portfolios consisting of nine stocks, and the computing time exceeded 9.5 hours to calculate the universal portfolio updates over a 22-year trading period. Blum and Kalai (1999) suggested to do randomized sampling to approximate the universal portfolio selection algorithm, and possibly alleviate the computational costs at the sacrifice of precision. We shall review this and other efficient approximation algorithms for UOP in the Sect. 9.4.2.

9.4.2 Efficient Universal Online Portfolio Strategies

Blum and Kalai (1999) uniform randomized approximation. Because the universal portfolio strategy is a weighted average of all CRPs, Blum and Kalai suggested to do the following natural uniform randomized approximation.

(1) Choose N portfolios uniformly at random (i.e., N many weight vectors $w \in \mathscr{W}$).
(2) Invest a $1/N$ fraction of the money in each of the N CRPs and let it sit within these CRPs (i.e. do not transfer between CRPs) all through the n periods of trading.

Note that this is like Cover's approximation that we mentioned before, but instead of considering all possible portfolios, just considers N portfolios randomly chosen. How good is this strategy? If the best constant rebalanced portfolio achieves a wealth in the order of $R \cdot S_n(\text{UOP}, x^n)$, for some real number $R > 0$, then by Chebyshev's inequality one can guarantee that using $N = (R-1)/\varepsilon^2 \delta$ random portfolios, the approximation achieves a wealth at least $(1-\varepsilon) \cdot S_n(\text{UOP}, x^n)$, with probability $1 - \delta$. For a given market one can determine in hindsight the optimal CRP, and then estimate R (Helmbold et al. (1997) give a method to do this). However, note that by Theorem 9.5 the value of R could be as big as n^{m-1}. Nonetheless, experiments on stock market data by Cover (1991) and Helmbold et al. (1998), all give a ratio of $R < 2$, for various combinations of two stocks, hence making this approximation particularly promising.

Kalai and Vempala (2003) non-uniform randomized approximation. The key idea of Kalai and Vempala is to sample N portfolios on a discretization of the simplex $\mathscr{W}$, and construct each sample as a random walk. In order to do this $\mathscr{W}$ is divided into grids of size δ, and the wealth function S_k is defined on integral multiples of δ. Then all portfolios begin at the center of the grid, the point $(1/m, \ldots, 1/m)$, and continues through a random walk, each step taken according to their wealth performance, as determined by $\dfrac{S_k(\boldsymbol{w}, \boldsymbol{x}^k)}{\int_{\mathscr{W}} S_k(\boldsymbol{v}, \boldsymbol{x}^k) d\boldsymbol{v}}$. The details of Kalai and Vempala algorithm, which they named R-Universal, follows below. Its parameters are: δ, the spacing of the grid; ω_0, minimum coordinate; N, the number of samples; K, the number of steps in the random walk.

Algorithm 9.1 R-Universal (δ, ω_0, N, K)

On each trading time, take the average of N sample portfolios obtained as follows:
1. Start each portfolio at the point $\boldsymbol{w} = (1/m, \ldots, 1/m)$
2. **for each** portfolio take K steps of the following random walk:
   (a) Choose $1 \le j \le m - 1$ at random
   (b) Choose $D \in \{-1, 1\}$ at random.
   if $(\omega_0 \le w_j + D\delta$ **and** $\omega_0 \le w_m - D\delta)$ **set** $prev = S_k(w_1, \ldots, w_m, \boldsymbol{x}^k)$,
   and $next = S_k(w_1, \ldots, w_j + D\delta, \ldots, w_{m-1}, w_m - D\delta, \boldsymbol{x}^k)$
   with probability $\min(1, prev/next)$:
   set $w_j := w_j + D\delta, w_m := w_m - D\delta$

Kalai and Vempala showed that, with this non-uniform sampling, the R-Universal algorithm has a performance guarantees of at least $(1 - \varepsilon) \cdot S_n(\text{UOP}, \boldsymbol{x}^n)$, with probability $1 - \eta$, and a computational cost bounded by a polynomial in $1/\varepsilon, \log(1/\eta)$, m, and n.

Other approximation algorithms. Helmbold et al. (1998) give an experts-based approximation algorithm for the UOP, the *EG* investment strategy, which uses a weight assignment based on a multiplicative update rule, as those employed in on-line regression; that is, to find a wealth distribution vector $\boldsymbol{w}^k$ that maximizes $F(\boldsymbol{w}^k) = \eta \log(\boldsymbol{w}^k \cdot \boldsymbol{x}^{k-1}) - d(\boldsymbol{w}^k, \boldsymbol{w}^{k-1})$, where $\eta > 0$ is a parameter representing the learning rate, and d is a distance measure that serves as a penalty term. This *EG* investment strategy has a computational cost linear in m, but performance guarantees inferior to the one given by Theorem 9.5. Agarwal et al. (2006) give an algorithm based on the Newton method for offline optimization. It has a computational cost of order $O(m^2)$, but better performance guarantees than the *EG* strategy.

What lies beyond? In this presentation of the problem of portfolio selection we have at all times assumed that there are zero transaction costs, nor commissions for trading. The analysis of the universal portfolio strategy in the presence of transaction costs is treated by Blum and Kalai (1999). Another enhancement to the model of universal portfolio selection is to consider side information. This was proposed and analyzed first by Cover and Ordentlich (1996). For further details on these extensions and other related subjects see El-Yaniv (1998) and Cesa-Bianchi and Lugosi (2006), and references therein.

9.5 Notes, Computer Lab and Problems

9.5.1 Bibliographic remarks: A general introduction to online computation and competitive analysis is the textbook by Borodin and El-Yaniv (1998), with specific applications to finance in Chapter 14. El-Yaniv (1998) is an excellent survey of online computation in finance, and together with El-Yaniv et al. (2001) constitute the basic material we use for writing Sects. 9.2 and 9.3. The universal portfolio is due to Cover (1991), and we have based our presentation mostly on that article. Theorem 9.5 is due to Ordentlich and Cover (1996), but the proof we presented is a simplified version from one given by Blum and Kalai (1999). In this last reference you will find a more refined argument that removes the $1/e$ in Eq. (9.9); moreover, you will find an analysis of the wealth of UOP when transaction costs are considered. The subject of portfolio selection, and extensions to using side information, is treated from the perspective of prediction with expert advice in the textbook by Cesa-Bianchi and Lugosi (2006).

9.5.2: Show that if we only know $\varphi = M/m$, but we don't know m and M, then the best competitive ratio that can be attained by a deterministic online price search is φ.

9.5.3: Complete the proof of Theorem 9.4, part (2): The Universal Portfolio exceeds the Value Line.

9.5.4: Show that Theorem 9.5 implies Eq. (9.8).

9.5.5: Consider a portfolio with one risk-free asset and one stock. The stock is highly volatile having a price relative factor of $a > 1$ on odd days, and of $1/a$ on even days. Thus,

$$x_1, x_2, x_3, x_4, \ldots = (1, a), (1, 1/a), (1, a), (1, 1/a), \ldots$$

is the sequence of price relative vectors, where the first component x_{i1} corresponds to the risk-free asset and the second component x_{i2} to the stock.

(i) Show that a constant rebalanced strategy with weights $w = (b, 1 - b)$, $0 \le b \le 1$, on the above sequence x^n of market vectors gives a wealth, for n even,

$$S_n(w, x^n) = (b + (1 - b)a)^{n/2} \left(b + \frac{1 - b}{a}\right)^{n/2} \tag{9.11}$$

(ii) Using Eq. (9.11) show that the best constant rebalanced portfolio is obtained with $w^* = (1/2, 1/2)$, and compute the value of $S_n(w^*, x^n)$, for n even. (Hint: recall that w^* is the value that maximizes (9.11)).
(iii) What is the value of $S_n(\text{UOP}, x^n)$, for n even? Try at least an approximate solution.

9.5.6: Implement in R the R-Universal approximation algorithm for UOP, and test it with twenty real stocks with data downloaded with `getSymbols` methods. Consider, for example, computing the price relatives on monthly periods, and feed the algorithm with these numbers to progressively construct the random walks. Note that this construction can be parallelized, thus improving computational time. Compare the performance with other strategies, e.g., buy-and-hold.

Appendix A
The R Programming Environment

This a brief introduction to R, where to get it from and some pointers to documents that teach its general use. A list of the available packages that extends R functionality to do financial time series analysis, and which are use in this book, can be found at the end. In between, the reader will find some general programming tips to deal with time series. Be aware that all the R programs in this book (the R Examples) can be downloaded from http://computationalfinance.lsi.upc.edu.

A.1 R, What is it and How to Get it

R is a free software version of the commercial object-oriented statistical programming language S. R is free under the terms of GNU General Public License and it is available from the R Project website, http://www.r-project.org/, for all common operating systems, including Linux, MacOS and Windows. To download it follow the CRAN link and select the mirror site nearest to you. An appropriate IDE for R is R Studio, which can be obtained for free at http://rstudio.org/ under the same GNU free software license (there is also a link to R Studio from the CRAN). By the way, CRAN stands for Comprehensive R Archive Network.

Although R is defined at the R Project site as a "language and environment for statistical computing and graphics", R goes well beyond that primary purpose as it is a very flexible and extendable system, which has in fact been greatly extended by contributed "packages" that adds further functionalities to the program, in all imaginable areas of scientific research. And it is continuously growing. In the realm of Econometrics and Quantitative Finance there is a large number of contributed packages, where one can find all the necessary functions for the analysis of financial instruments. Many of these packages are used for the financial analytics experiments performed throughout the book, and so it is necessary to install them in order to reproduce all the R Examples. There are instructions on how to do this in the next section.

A. Arratia, *Computational Finance*, Atlantis Studies in Computational Finance
and Financial Engineering 1, DOI: 10.2991/978-94-6239-070-6,
© Atlantis Press and the authors 2014

The best place to start learning R is at the home page of the R Project. Once there, go to Manuals and download *An Introduction to R*; afterwards, go to Books and browse through the large list of R related literature. Each package comes also with an interactive Help detailing the usage of each function, together with examples.

A.2 Installing R Packages and Obtaining Financial Data

After downloading and installing R, tune it up with the contributed packages for doing serious financial time series analysis and econometrics listed in Sect. A.4. Except for RMetrics, to install a package XX, where XX stands for the code name within brackets in the package references list, from the R console enter the command:

```
install.packages("XX")
```

and once installed, to load it into your R workspace, type:

```
library(XX)
```

For example, to install and load the package quantmod type:

```
install.packages("quantmod")
library(quantmod)
```

RMetrics is a suite of packages for financial analytics. It contains all the package prefixed with f (e.g. fArma, fGarch, etc.) and others; each of the f packages can be installed by its own using the previous installation procedure, or to install all these packages within the RMetrics set, type:

```
source("http://www.rmetrics.org/Rmetrics.R")
install.Rmetrics( )
```

Then load each package with the library( ) command (e.g., library(fArma)). For more details and tutorials on RMetrics go to www.rmetrics.org.

Installing packages from RStudio is easier: just click on the "Packages" window, then on "Install Packages" and in the pop-up window type the code name of the package. Make sure to select "Install all dependencies".

To get acquainted with all existing R packages for empirical work in Finance, browse through the list of topics at the CRAN Task View in Empirical Finance: http://cran.r-project.org/web/views/Finance.html.

Sources of Financial Data

Financial data can be freely obtained from various sources. For example: Yahoo finance (http://finance.yahoo.com/), Google finance (www.google.com/finance), MSN Moneycentral (http://moneycentral.msn.com), and the Federal Reserve Bank of St. Louis - FRED (http://research.stlouisfed.org/fred2/). One can obtain the data manually by directly accessing these sites, or access the server through various functions built in some of the R packages. In the package quantmod there is the getSymbols method, and in Rmetrics there is the fImport package for importing market data from the internet. To obtain US company's filings (useful for evaluating a company's bussiness) the place to go is EDGAR at http://www.sec.gov/.

Also, there are various R packages of data, or that includes some data sets for self-testing of their functions. One R package of data used in this book for testing econometric models is the AER; the fOptions package comes with some markets data to test the portfolios analytic methods.

A.3 To Get You Started in R

Getting ready to work with R. The first thing you should do at your R console is to set your working directory. A good candidate is the directory where you want to save and retrieve all your data. Do this with setwd("<path-to-dir>"), where <path-to-dir> is the computer address to the directory. Next step is to load necessary libraries with library( ).

Obtaining financial data, saving and reading. The most frequent method that we use to get financial data from websites like Yahoo or the Federal Reserve (FRED) is

library(quantmod)

getSymbols("<ticker>",src='yahoo')

where <ticker> is the ticker symbol of the stock that we want (to learn the ticker look up the stock in finance.yahoo.com). The getSymbols retrieves and converts financial data in xts format. This is an "extensible time-series object", and among other amenities it allows to handle the indices of the time series in Date format. We need to load library(xts) to handle xts objects.

Based on our own experience, it seems that the only way to save into disk all the information contained in an xts object is with saveRDS() which saves files in binary format. This in fact makes the data files portable across all platforms. To reload the data from your working directory into the R working environment use readRDS(). Here is a full example of obtaining the data for Apple Inc. stock from Yahoo, saving it to a working directory, whose path is in a variable string wdir, and later retrieving the data with readRDS:

```
wdir = "<path-to-working-directory>"
setwd(wdir)
library(quantmod); library(xts)
getSymbols("AAPL",src='yahoo')
saveRDS(AAPL, file="Apple.rds")
Apple = readRDS("Apple.rds")
```

The suffix .rds identifies the file of type RDS (binary) in your system. Sometimes we also get files from other sources in .csv format. In that case it is save in disk with save command, and read from disk with read.csv() or read.zoo().

Some manipulations of xts objects. With the data in the xts format it can be explored by dates. Comments in R instructions are preceded by #. The command names() shows the headers of the column entries of the data set:

```
appl = AAPL['2010/2012']  #consider data from 2010 to 2012
names(appl)
```

The output shows the headers of the data. These are the daily Open, High, Low, Close, Volume and Adjusted Close price:

```
[1] "AAPL.Open"      "AAPL.High"      "AAPL.Low"
[4] "AAPL.Close"     "AAPL.Volume"    "AAPL.Adjusted"
```

We choose to plot the Adjusted Close. To access the column use $:

```
plot(appl$AAPL.Adjusted, main="Apple Adj. Close")
```

Handling missing or impossible values in data. In R, missing values (whether character or numeric) are represented by the symbol NA (Not Available). Impossible values (e.g., dividing by zero) are represented by the symbol NaN (Not a Number). To exclude missing values from analysis you can either delete it from the data object X with na.omit(X), or set the parameter na.rm=TRUE in the function (most functions have this feature). Note that na.omit() does a listwise deletion and this could be a problem for comparing two time series, for their length might not be equal. You can check length of a list object with length(X).

This is enough to get you started. The R Examples in the book will teach you further programming tricks. Enjoy!

A.4 References for R and Packages Used in This Book

[R]   R Core Team (2013). *R: A Language and Environment for Statistical Computing*, R Foundation for Statistical Computing, Vienna, Austria.

[AER]   Kleiber, C. and Zeileis, A. (2008). *Applied Econometrics with R*, (Springer, New York).

[caret]   Kuhn, M. and various contributors (2013). *caret: Classification and Regression Training*.

[e1071]   Meyer, D., Dimitriadou, E., Hornik, K., Weingessel, A. and Leisch, F. (2012). *e1071: Misc. Functions of the Dept. of Statistics (e1071), TU Wien*.

[fArma]   Würtz, D. et al (2012). *fArma: ARMA Time Series Modelling*.

[fExoticOptions]   Würtz, D. et al (2012). *fExoticOptions: Exotic Option Valuation*.

[fGarch]   Würtz, D. and Chalabi, Y. with contribution from Miklovic, M., Boudt, C., Chausse, P. and others (2012). *fGarch: Rmetrics - Autoregressive Conditional Heteroskedastic Modelling*.

[fOptions]   Würtz, D. et al (2012). *fOptions: Basics of Option Valuation*.

[kernlab]   Karatzoglou, A., Smola, A., Hornik, K. and Zeileis, A. (2004). *kernlab – An S4 Package for Kernel Methods in R*, Journal of Statistical Software **11**, 9, pp. 1–20.

[Metrics]   Hamner, B. (2012). *Metrics: Evaluation metrics for machine learning.*

[PerformanceAnalytics]   Carl, P. and Peterson, B. G. (2013). *PerformanceAnalytics: Econometric tools for performance and risk analysis.*

[quantmod]   Ryan, J. A. (2013). *quantmod: Quantitative Financial Modelling Framework.*

[RMetrics]   Würtz, D., Chalabi, Y., Chen, W., and Ellis, A. (2009). *Portfolio Optimization with R / RMetrics*, RMetrics Pub., Zurich.

[tseries]   Trapletti, A. and Hornik, K. (2013). *tseries: Time Series Analysis and Computational Finance.*

[TTR]   Ulrich, J. (2013). *TTR: Technical Trading Rules.*

[xts]   Ryan, J. A. and Ulrich, J. M. (2013). *xts: eXtensible Time Series.*

(A package url in the CRAN.R-project is http://CRAN.R-project.org/package=XX, where XX is the name of the package within the brackets in the list above.)

References

Aarts, E. H. L., & Lenstra, J. K. (2003). *Local search in combinatorial optimization.* Princeton: Princeton University Press.

Achelis, S. B. (2001). *Technical analysis from A to Z* (2nd ed.). New York: McGraw Hill.

Agarwal, A., Hazan, E., Kale, S., & Schapire, R. (2006). Algorithms for portfolio management based on the newton method. In *Proceedings of the 23rd International Conference on Machine Learning, ICML '06* (pp. 9–16). ACM.

Aggarwal, C., Hinneburg, A., & Keim, D. (2001). On the surprising behavior of distance metrics in high dimensional space. In J. V. den Bussche & V. Vianu (Eds.), *Proceedings 8th International Conference Database Theory (ICDT 2001)* (Vol. 1973, pp. 420–434). Springer.

Alizadeh, S., Brandt, M. W., & Diebold, F. X. (2002). Range-based estimation of stochastic volatility models. *The Journal of Finance, 57*(3), 1047–1091.

Allen, F., & Karjalainen, R. (1999). Using genetic algorithms to find technical trading rules. *Journal of Financial Economics, 51*, 245–271.

Arratia, A. (2010). *Reconocimiento automático de patrones candlestick, in i-Math Jornadas sobre Matemática de los Mercados Financieros.* Spain: Universidad de Murcia.

Arratia, A., & Cabaña, A. (2013). A graphical tool for describing the temporal evolution of clusters in financial stock markets. *Computational Economics, 41*(2), 213–231.

Arthur, D., & Vassilvitskii, S. (2007). k-means++: The advantages of careful seeding. In *Proceedings 18th Annual ACM-SIAM Symposium on Discrete Algorithms (SIAM)* (pp. 1027–1035).

Baker, J. E. (1985). Adaptive selection methods for genetic algorithms. In *Proceedings of the 1st International Conference on Genetic Algorithms* (pp. 101–111). Hillsdale, NJ: Lawrence Erlbaum Associates Incorporates.

Bélisle, C. J. P. (1992). Convergence theorems for a class of simulated annealing algorithms on $\mathbb{R}^d$. *Journal of Applied Probability, 29*, 885–895.

Bernstein, P. L. (1992). *Capital ideas: The improbable origins of modern wall street.* New York: Free Press.

Black, F., & Scholes, M. (1973). The pricing of options and corporate liabilities. *Journal of Political Economy, 81*(3), 637–654.

Blum, A., & Kalai, A. (1999). Universal portfolios with and without transaction costs. *Machine Learning, 35*(3), 193–205.

Bollerslev, T., Engle, R. F., & Nelson, D. B. (1994). Arch models. In R. F. Engle & D. L. McFadden (Eds.), *Handbook of Econometrics* (Vol. IV, Chap. 49, pp. 2959–3032). Elsevier

Borodin, A., & El-Yaniv, R. (1998). *Online computation and competitive analysis.* New York: Cambridge University Press.

Box, G. E. P., & Jenkins, G. M. (1976). *Time series analysis: forecasting and control* (Revised ed.). San Francisco: Holden-Day.

Boyle, P., Broadie, M., & Glasserman, P. (1997). Monte carlo methods for security pricing. *Journal of Economic Dynamics and Control, 21*(8), 1267–1321.

Boyle, P. P. (1977). Options: A monte carlo approach. *Journal of Financial Economics, 4*(3), 323–338.

Brabazon, A., & O'Neill, M. (2006). *Biologically inspired algorithms for financial modelling*. UK: Springer.

Brealey, R. A., Myers, S. C., & Allen, F. (2011). *Principles of corporate finance* (10th ed.). New York: McGraw-Hill.

Breiman, L. (1992). *Probability* (Vol. 7). Addison-Wesley: Philadelphia. (SIAM, Reprinted from Addison-Wesley 1968 ed.).

Brockwell, P. J., & Davis, R. A. (1991). *Time series: Theory and methods*. New York: Springer.

Brockwell, P. J., & Davis, R. A. (2002). *Introduction to time series and forecasting* (2nd ed.). New York: Springer.

Brown, S. J., Goetzmann, W. N., & Kumar, A. (1998). The dow theory: William Peter Hamilton's track record reconsidered. *The Journal of Finance, 53*(4), 1311–1333.

Brucker, P. (1977). On the complexity of clustering problems. In R. Henn, B. Korte & W. Oletti (Eds.), *Optimizing and Operations Research, Kecture Notes in Economics and Mathematical Systems*. Springer.

Burges, C. (1998). A tutorial on support vector machines for pattern recognition. *Data Mining and Knowledge Discovery, 2*(2), 121–167.

Campbell, J. Y. (2000). Asset pricing at the millennium. *Journal of Finance LV, 4*, 1515–1567.

Campbell, J. Y., Lo, A. W., & MacKinlay, A. C. (1997). *The econometrics of financial markets*. Princeton: Princeton University Press.

Campbell, J. Y., & Shiller, R. J. (2005). Valuation ratios and the long-run stock market outlook: an update. *Advances in Behavioral Finance, 2*, 173–201.

Černý, V. (1985). Thermodynamical approach to the traveling salesman problem: An efficient simulation algorithm. *Journal of Optimization Theory and Applications, 45*(1), 41–41.

Cesa-Bianchi, N., & Lugosi, G. (2006). *Prediction, learning, and games*. Cambridge: Cambridge University Press.

Cohen, J. (1960). A coefficient of agreement for nominal scales. *Educational and psychological measurement, 20*(1), 37–46.

Colorni, A., Dorigo, M., & Maniezzo, V. (1992a). Distributed optimization by ant colonies. In F. J. Varela & P. Bourgine (Eds.), *Towards a Practice of Autonomous Systems: Proceedings of the First European Conference on Artificial Life* (pp. 134–142). Cambridge, MA: MIT Press.

Colorni, A., Dorigo, M., & Maniezzo, V. (1992b). An investigation of some properties of an ant algorithm. In R. Männer & B. Manderick (Eds.), *Proceedings 2nd Conference Parallel Problem Solving from Nature—PPSN II: International Conference on Evolutionary Comput.* (pp. 509–520). North-Holland.

Cont, R. (2001). Empirical properties of asset returns: Stylized facts and statistical issues. *Quantitative Finance, 1*(2), 223–236.

Cormen, T. H., Leiserson, C. E., & Rivest, R. L. (1990). *Introduction to algorithms*. Cambridge: MIT Press.

Cover, T. M. (1991). Universal portfolios. *Mathematical Finance, 1*(1), 1–29.

Cover, T. M., & Gluss, D. H. (1986). Empirical bayes stock market portfolios. *Advances in Applied Mathematics, 7*(2), 170–181.

Cover, T. M., & Ordentlich, E. (1996). Universal portfolios with side information. *IEEE Transactions on Information Theory, 42*, 2.

Cox, J., Ross, S., & Rubinstein, M. (1979). Option pricing: A simplified approach. *Journal of Financial Economics, 7*(3), 229–263.

Cox, J., & Rubinstein, M. (1985). *Options markets*. Prentice Hall: Englewood Cliffs.

Crama, Y., & Schyns, M. (2003). Simulated annealing for complex portfolio selection problems. *European Journal of Operational Research, 150*, 546–571.

Deza, M., & Deza, E. (2009). *Encyclopedia of distances*. Boca Raton: Springer.

Diks, C., & Panchenko, V. (2006). A new statistic and practical guidelines for nonparametric granger causality testing. *Journal of Economic Dynamics and Control, 30*, 1647–1669.

Doherty, G. C. (2003). Fundamental analysis using genetic programming for classification rule induction. In J. R. Koza (Ed.), *Genetic Algorithms and Genetic Programming at Stanford* (pp. 45–51).

Dorigo, M. (1992). Optimization, learning and natural algorithms. Ph.D. Thesis, Politecnico di Milano, Milan, Italy.

Dorigo, M., & Stützle, T. (2004). *Ant colony optimization*. Cambridge: MIT Press.

Drineas, P., Frieze, A., Kannan, R., Vempala, S., & Vinay, V. (2004). Clustering large graphs via the singular value decomposition. *Machine Learning, 56*(1–3), 9–33.

Edwards, R. D., & Magee, J. (1966). *Technical analysis of stock trends* (5th ed.). Boston: John Magee.

Eiben, A. E., Aarts, E. H., & Van Hee, K. M. (1991). Global convergence of genetic algorithms: A Markov chain analysis. In H. P. Schwefel & R. Männer (Eds.), *Parallel Problem Solving From Nature, Lecture Notes in Computer Science* (Vol. 496, pp. 3–12). Springer.

El-Yaniv, R. (1998). Competitive solutions for online financial problems. *ACM Computing Surveys, 30*(1), 28–69.

El-Yaniv, R., Fiat, A., Karp, R. M., & Turpin, G. (2001). Optimal search and one-way trading online algorithms. *Algorithmica, 30*(1), 101–139.

Embrechts, P., McNeil, A., & Straumann, D. (2002). Correlation and dependence in risk management: Properties and pitfalls. In M. Dempster & H. Moffatt (Eds.), *Risk Management: Value at Risk and Beyond* (pp. 176–223).

Fama, E. F. (1965). The behavior of stock-market prices. *The Journal of Business, 38*(1), 34–105.

Fama, E. F. (1970). Efficient capital markets: a review of theory and empirical work. *The Journal of Finance, 25*(2), 383–417.

Fama, E. F. (1991). Efficient capital markets: ii. *The Journal of Finance, 46*(5), 1575–1617.

Feller, W. (1968). *An introduction to probability theory and its applications* (3rd ed.). New York: Wiley.

Figlewski, S. (1994). *Forecasting volatility using historical data*, (Working Paper FIN-94-032). New York University: Stern School of Business.

Fogel, L. J., Owens, A. J., & Walsh, M. J. (1966). *Artificial intelligence through simulated evolution*. New York: Wiley.

Forbes, C. S., Evans, M., Hastings, N. A. J., & Peacock, J. B. (2011). *Statistical distributions* (4th ed.). New York: Wiley.

Garey, M. R., & Johnson, D. S. (1979). *Computers and intractability: A guide to the theory of NP-completeness*. San Francisco: W. H. Freeman.

Geweke, J., Meese, R., & Dent, W. (1983). Comparing alternative tests of causality in temporal systems: analytic results and experimental evidence. *Journal of Econometrics, 21*(2), 161–194.

Gibbons, J. D., & Chakraborti, S. (2003). *Nonparametric statistical inference* (4th ed., Vol. 168). New York: Marcel Dekker.

Gilli, M., & Winker, P. (2008). *A review of heuristic optimization methods in econometrics*, Research paper 08–12. Geneva: Swiss Finance Institute.

Glasserman, P. (2004). *Monte Carlo methods in financial engineering* (Vol. 53). New York: Springer.

Goldberg, D. E., & Deb, K. (1991). A comparative analysis of selection schemes used in genetic algorithms. In *Foundations of genetic algorithms* (pp. 69–93). San Mateo: Morgan Kaufmann.

Gonzalez, T. F. (1985). Clustering to minimize the maximum intercluster distance. *Theoretical Computer Science, 38*, 293–306.

Gooijer, J. G. d., & Sivarajasingham, S. (2008). Parametric and nonparametric granger causality testing: Linkages between international stock markets.*Physica A, 387*, 2547–2560.

Graham, B. (2003). *The intelligent investor* (Revised ed.). New York: Harper Business Essentials.

Graham, B., & Dodd, D. L. (1934). *Security analysis*. New York: Whittlesey House. (Whittlesey House, updated 2009 ed. by McGraw-Hill).

Granger, C. W. J. (1969). Investigating causal relations by econometric models and cross-spectral methods. *Econometrica, 37*(3), 424–438.

Granger, C. W. J., & Ding, Z. (1995). Some properties of absolute return: An alternative measure of risk. *Annales d'Économie et de Statistique, 40*, 67–91.

Granger, C. W. J., & Morris, M. J. (1976). Time series modelling and interpretation. *Journal of the Royal Statistical SocietySeries A, 139*(2), 246–257.

Grant, D., Vora, G., & Weeks, D. (1997). Path-dependent options: extending the monte carlo simulation approach. *Management Science, 43*(11), 1589–1602.

Greenwald, B. C. N. (2001). *Value investing: From Graham to Buffet and beyond*. New York: Wiley.

Grossman, S., & Stiglitz, J. (1980). On the impossibility of informationally efficient markets. *American Economic Review, 70*, 393–408.

Hald, A. (1999). On the history of maximum likelihood in relation to inverse probability and least squares. *Statistical Science, 14*(2), 214–222.

Hamilton, W. P. (1922). *The stock market barometer: A study of its forecast value based on charles H. Dow's theory of the price movement*. New York: Harper & brothers. (Harper & brothers, Republished by Wiley in 1998).

Hardin, C. D. (1982). On the linearity of regression. *Zeitschrift für Wahrscheinlichkeitstheorie und verwandte Gebiete, 61*(3), 293–302.

Hartigan, J. A., & Wong, M. A. (1979). Algorithm as 136: A k-means clustering algorithm. *Journal of the Royal Statistical Society Series C (Applied Statistics), 28*(1), 100–108.

Hasanhodzic, J., Lo, A. W., & Viola, E. (2011). A computational view of market efficiency. *Quantitative Finance, 11*(7), 1043–1050.

Hastie, T., Tibshirani, R., & Friedman, J. H. (2009). *The elements of statistical learning: Data mining, inference, and prediction* (2nd edn.). Springer Series in Statistics. New York: Springer.

Haug, E. G. (2007). *The complete guide to otpon pricing formulas* (2nd ed.). New York: McGraw-Hill.

Helmbold, D. P., Schapire, R. E., Singer, Y., & Warmuth, M. K. (1997). A comparison of new and old algorithms for a mixture estimation problem. *Machine Learning, 27*(1), 97–119.

Helmbold, D. P., Schapire, R. E., Singer, Y., & Warmuth, M. K. (1998). On-line portfolio selection using multiplicative updates. *Mathematical Finance, 8*(4), 325–347.

Hiemstra, C., & Jones, J. (1994). Testing for linear and nonlinear granger causality in the stock price- volume relation. *The Journal of Finance, 49*(5), 1639–1664.

Holland, J. H. (1975). *Adaptation in natural and artificial systems: An introductory analysis with applications to biology, control, and artificial intelligence*. Ann Arbor: University of Michigan Press.

Hornik, K., Stinchcombe, M., & White, H. (1989). Multilayer feedforward networks are universal approximators. *Neural Networks, 2*(5), 359–366.

Hull, J. (2009). *Options, futures and other derivatives* (7th ed.). Upper Saddle River: Prentice Hall.

Hyndman, R. J., & Koehler, A. B. (2006). Another look at measures of forecast accuracy. *International Journal of Forecasting, 22*(4), 679–688.

Jain, A., Murty, M., & Flynn, P. (1999). Data clustering: a review. *ACM Computing Surveys, 31*(3), 264–323.

Jain, A. K., & Dubes, R. C. (1988). *Algorithms for clustering data*. New Jersy: Prentice Hall.

Jensen, M. C. (1978). Some anomalous evidence regarding market efficiency. *Journal of Financial Economics, 6*, 95–101.

Joag-dev, K. (1984). Measures of dependence. In P. R. Krishnaiah (Ed.) *Handbook of Statistics* (Vol. 4, Chap. 6, pp. 79–88). Elsevier.

Kalai, A., & Vempala, S. (2003). Efficient algorithms for universal portfolios. *Journal of Machine Learning Research, 3*, 423–440.

Kao, M.-Y., & Tate, S. R. (1999). On-line difference maximization. *SIAM Journal of Discrete Mathematics, 12*(1), 78–90.

Karatzoglou, A., Meyer, D., & Hornik, K. (2006). Support vector machines in R. *Journal of Statistical Software*, *15*, 9.

Katz, J., & Mccormick, D. (2000). *The encyclopedia of trading strategies*. New York: McGraw-Hill.

Keber, C., & Schuster, M. G. (2002). Option valuation with generalized ant programming. In *Proceedings of the Genetic and Evolutionary Computation Conference* (pp. 74–81). Morgan Kaufmann.

Keith, M. J., & Martin, M. C. (1994). Genetic programming in C++: Implementation issues. In *Advances in Genetic Programming* (Chap. 13, pp. 285–310). Cambridge, MA: MIT Press.

Kirkpatrick, S., Gelatt, C. D., & Vecchi, M. P. (1983). Optimization by simulated annealing. *Science*, *220*(4598), 671–680.

Kloeden, P. E., & Platen, E. (1995). *Numerical solution of stochastic differential equations* (2nd ed., Vol. 23). Berlin: Springer.

Korczak, J., & Roger, P. (2002). Stock timing using genetic algorithms. *Applied Stochastic Models in Business and Industry*, *18*(2), 121–134.

Koza, J. R. (1992). *Genetic programming: on the programming of computers by means of natural selection*. Cambridge: MIT Press.

Kuhn, M. (2008). Building predictive models in r using the caret package. *Journal of Statistical Software*, *28*(5), 1–26.

Laarhoven, P. J. M. v., & Aarts, E. H. L. (1987). *Simulated annealing:Theory and applications, mathematics and Its applications*. Dordrecht: D. Reidel.

Lewis, H. R., & Papadimitriou, C. H. (1998). *Elements of the theory of computation* (2nd ed.). Upper Saddle River: Prentice-Hall.

Liao, T. W. (2005). Clustering of time series data—a survey. *Pattern Recognition*, *38*, 1857–1874.

Lindahl-Stevens, M. (1978). Some popular uses and abuses of beta. *Journal of Portfolio Management*, *4*(2), 13–17.

Llorente-Lopez, M., & Arratia, A. (2012). *Construcción automática de reglas de inversión utilizando programación genética*. BSc Thesis, Universitat Politècnica de Catalunya.

Lorenz, J., Panagiotou, K., & Steger, A. (2009). Optimal algorithm for k-search with application in option pricing. *Algorithmica*, *55*(2), 311–328.

Luenberger, D. G. (1998). *Investment science*. New York: Oxford University Press.

Lundy, M., & Mees, A. (1986). Convergence of an annealing algorithm. *Mathematical Programming*, *34*(1), 111–124.

Maller, R. A., Müller, G., & Szimayer, A. (2009). Ornstein-Uhlenbeck processes and extensions. In *Handbook of Financial Time Series* (pp. 421–438). Springer.

Mandelbrot, B. (1963). The variation of certain speculative prices. *The Journal of Business*, *36*(4), 394–419.

Maringer, D. (2005). Portfolio management with heuristic optimization. In *Advances in Computational Management Science* (Vol. 8). Heidelberg: Springer.

Markowitz, H. (1952). Portfolio selection. *The Journal of Finance*, *7*(1), 77–91.

Marshall, P. S. (2000). A statistical comparison of value averaging versus dollar cost averaging and random investment techniques. *Journal of Financial and Strategic Decisions*, *13*(1), 87–99.

Maymin, P. Z. (2011). Markets are efficient if and only if p = np. *Algorithmic Finance*, *1*(1), 1–11.

McLeod, A. I., Yu, H., & Mahdi, E. (2012). Time Series analysis with R. In T. S. Rao, S. S. Rao, & C. Rao (Eds.), *Handbook of statistics 30: Time series analysis and applications* (Vol. 30, Chap. 23, pp. 661–712). Amsterdam: Elsevier.

McNelis, P. D. (2005). *Neural networks in finance: Gaining predictive edge in the market*. Burlington, MA: Elsevier Academic Press.

Merton, R. (1973). Theory of rational option pricing. *The Bell Journal of Economics and Management Science*, *4*(1), 141–183.

Metropolis, N., Rosenbluth, A. W., Rosenbluth, M. N., Teller, A. H., & Teller, E. (1953). Equation of state calculations by fast computing machines. *The Journal of Chemical Physics*, *21*(6), 1087–1092.

Neely, C., Weller, P., & Dittmar, R. (1997). Is technical analysis in the foreign exchange market profitable? a genetic programming approach. *The Journal of Financial and Quantitative Analysis*, *32*(4), 405–426.

Neftci, S. (1991). Naive trading rules in financial markets and wiener-kolmogorov prediction theory: a study of "technical analysis". *Journal of Business*, *64*(4), 549–571.

Neftci, S. N. (2000). *An Introduction to the mathematics of financial derivatives* (2nd ed.). San Diego: Academic Press.

Nison, S. (1991). *Japanese candlestick charting techniques: A contemporary guide to the ancient investment techniques of the far east*. New York: New York Institute of Finance.

O'Neill, M., Dempsey, I., Brabazon, A., & Ryan, C. (2003). Analysis of a digit concatenation approach to constant creation. In C. Ryan, T. Soule, M. Keijzer, E. Tsang, R. Poli, & E. Costa (Eds.), *Genetic programming, Lecture notes in computer Science* (Vol. 2610, pp. 173–182). Berlin: Springer.

Ordentlich, E., & Cover, T. M. (1996). On-line portfolio selection. In *Proceedings of the 9th Annual Conference on Computational Learning Theory*, COLT '96 (ACM), pp. 310–313.

Papadimitriou, C. H. (1994). *Computational complexity*. Reading: Addison-Wesley.

Papadimitriou, C. H., & Steiglitz, K. (1998). *Combinatorial optimization: Algorithms and complexity*. New York: Dover.

Park, C.-H., & Irwin, S. H. (2007). What do we know about the profitability of technical analysis? *Journal of Economic Surveys*, *21*(4), 786–726.

Paulos, J. A. (2007). *A mathematician plays the stock market*. New York: MJF Books/Fine Communications.

Perold, A. F. (1984). Large-scale portfolio optimization. *Management Science*, *30*, 1143–1160.

Rechenberg, I. (1973). *Evolutionsstrategie: Optimierung technischer Systeme nach Prinzipien der biologischen Evolution*, Problemata, 15 (Frommann-Holzboog).

Reeves, C. R., & Rowe, J. E. (2003). *Genetic algorithms. Principles and perspectives: A guide to GA theory*. Boston: Kluwer Academic Publishers.

Rhea, R. (1932). *The Dow theory: An explanation of its development and an attempt to define its usefulness as an aid in speculation*. New York:Barron's.

Ripley, B. D. (1994). Neural networks and related methods for classification. *Journal of the Royal Statistical Society. Series B (Methodological)*, *56*, 409–456.

Ripley, B. D. (1995). Statistical ideas for selecting network architectures. In B. Kappen & S. Gielen (Eds.), *Neural networks: Artificial intelligence and industrial applications* (pp. 183–190). Berlin: Springer.

Rizzo, M. L. (2008). *Statistical computing with R*. London: Chapman & Hall/CRC.

Rodgers, J. L., & Nicewander, W. A. (1988). Thirteen ways to look at the correlation coefficient. *The American Statistician*, *42*(1), 59–66.

Rubinstein, M. (2006). *A history of the theory of investments: My annotated bibliography*. New York: Wiley.

Ryan, C., Collins, J. J., & O'Neill, M. (1998). Grammatical evolution: Evolving programs for an arbitrary language. In *Proceedings of the First European Workshop on Genetic Programming, Lecture Notes in Computer Science* (Vol. 1391, pp. 83–95). Berlin: Springer.

Samuelson, P. A. (1967). General proof that diversification pays. *The Journal of Financial and Quantitative Analysis*, *2*(1), 1–13.

Sharpe, W. (1964). Capital asset prices: a theory of market equilibrium under conditions of risk. *The Journal of Finance*, *19*(3), 425–442.

Shilling, A. G. (1992). Market timing: better than buy-and-hold strategy. *Financial Analysts Journal*, *48*(2), 46–50.

Shiryaev, A. N. (2007). *Optimal stopping rules* (Vol. 8). Berlin: Springer.

Sims, C. A. (1972). Money, income, and causality. *The American Economic Review*, *62*(4), 540–552.

Stirzaker, D. (2005). *Stochastic processes and models*. Oxford: Oxford University Press.

Stützle, T., López-Ibáñez, M., Dorigo, M., Cochran, J., Cox, L., Keskinocak, P., Kharoufeh, J., & Smith, C. (2011). A concise overview of applications of ant colony optimization. In J. J. Cochran et al. (Eds.), *Wiley encyclopedia of operations research and management science*. New York: Wiley.

Talbi, E. G. (2002). A taxonomy of hybrid metaheuristics. *Journal of Heuristics, 8*, 541–564.

Taylor, S. J. (1986). *Modelling financial time series*. New York: Wiley.

The Options Institute (Ed.). (1995). *Options: Essential concepts and trading strategies* (2nd ed.) Irwin Chicago: Professional Publishing (Educational Division of the Chicago Options Exchange).

Thruman, W., & Fisher, M. (1988). Chickens, eggs, and causality, or which came first? *American Journal of Agricultural Economics, 70*(2), 237–238.

Toda, H., & Yamamoto, T. (1995). Statistical inference in vector autoregressions with possibly integrated processes. *Journal of Econometrics, 66*(1–2), 225–250.

Tofallis, C. (2006). Investment volatility: a critique of standard beta estimation and a simple way forward. *European Journal of Operational Research, 187*, 1358–1367.

Tsay, R. S. (2010). *Analysis of financial time series* (3rd ed.). New York: Wiley.

Vanderbei, R. J. (2008). *Linear programming: Foundations and extensions* (3rd ed.). Berlin: Springer.

Vapnik, V. N. (2000). *The nature of statistical learning theory* (2nd ed.). New York: Springer.

Vidyamurthy, G. (2004). *Pairs trading, quantitative methods and analysis*. New York: Wiley.

Wagner, D., & Wagner, F. (1993). Between min cut and graph bisection. In *Proceedings of the 18th International Symposium on Mathemetical Foundations of Computer Science, Lecture Notes in Computer Science* (Vol. 711, pp. 744–750). New York: Springer.

Wang, J. (2000). Trading and hedging in s&p 500 spot and futures markets using genetic programming. *Journal of Futures Markets, 20*(10), 911–942.

Ward, J. H. (1963). Hierarchical grouping to optimize an objective function. *Journal of the American Statistical Association, 58*, 236–244.

Williams, J. B. (1938). *The theory of investment value*. Cambridge: Harvard University Press

Würtz, D., Chalabi, Y., Chen, W., & Ellis, A. (2009). *Portfolio optimization with R/RMetrics* (1st ed.). Zurich: Rmetrics Publishing.

Zivot, E., & Wang, J. (2006). *Modeling financial time series with S-plus* (2nd ed.). New York: Springer.

Index

A

AIC (Akaike information criterion), 122
Ant colony, 226
   algorithm, 228
AR(p), 112, 113
   autocovariances, 116
   best linear forecaster, 118
   transformed into MA(∞), 114
Arbitrage, 20
   principle, 25
Arbitrageur, 20
ARCH effect, 129
ARCH model, 124
   ARCH(p) h-step ahead prediction, 126
   likelihood function, 125
ARMA(p, q), 112, 113
   autocovariances, 120
Autocorrelation, 73
   estimate, 74
   function (ACF), 74
   lag-k, 73
   partial (PACF), 118
Autocovariance, 58
Autoregressive
   and moving average, *see* ARMA(p, q)
   conditional heteroscedastic, *see* ARCH
     model
   order p, 112

B

Bear
   investor, 21
   market, 21, 178
Beta
   estimation from sample, 255
   interpretation, 252

of equity, 254
of security, 252
Black, Fisher, 153, 172
Black-Scholes formula, 154
Bollerslev, Tim, 127
Bond, 2
   coupons, 2
   payoff, 4
   principal, 2
   valuation, 2, 4
Brownian motion, 146
   arithmetic, 146
   geometric, 149
Bull
   investor, 21
   market, 21, 178
Buy and hold, 21, 262

C

Capital Asset Pricing Model (CAPM), 251
   linearity, 264
Capital Market Line, 249, 251
Causality, 78
   Granger, 79
   nonparametric, 82
Chart
   bars, 180
   candlesticks, 180
   line, 180
Clustering, 85
   k-means, 91
   agglomerative, 85
   agglomerative hierarchical algorithm, 89
   divisive, 85
   evaluation, 94
   graph–based, 93

hierarchical, 88
partitional, 90
Competitive analysis, 268
Conditional expectation, 60
Constant Rebalanced Portfolio (CRP), 262
Correlation
    linear (Pearson), 72
    rank (Kendall), 77
Covariance, 57

D

Derivative, 1
Digit concatenation, 231
Discounted cash flow (DCF), 23, 197
Distribution
    binomial, 68
    conditional, 45
    cumulative function, 43
    empirical function, 43
    log-normal, 53
    normal or Gaussian, 49
Dow, Charles, 178
    investment principles, 178

E

Efficient frontier, 243
Efficient Market Hypothesis, 31
    semi–strong, 31
    strong, 31
    weak, 31, 148
Empirical risk minimization, 134
Engle, Robert, 124
Equity, 1
ETF, 18
    inverse, 20
Exchange market, 2
Extended no arbitrage, 26

F

Fama, Eugene, 31, 173
Forward contract, 17
    payoff, profit, 17
    price formula, 28
Forward price, 17
Function
    logistic, 132
    threshold, 133
Fundamental analysis, 196
    basic principles, 196
    indicators, *see* Value indicators
Future contract

payoff, profit, 17
Future price, 17
Futures contract, 17

G

GARCH model, 127
    and simulated annealing, 212
    GARCH(p, q), 127
    GARCH(1,1) h-step ahead prediction,
        128
    unconditional variance, 128
Genetic programming, 213
    algorithm, 216
    default parameters, 226
    generating trading rules, 218
Graham, Benjamin, 196
    investment principles, 196
    stock valuation guide, 203
Granger, Clive, 71, 79

H

Hansen, Lars Peter, 173
Hedger, 20
Hybrid heuristic, 233
    flat, 233
    hierarchical, 233

I

Interest
    compounding, 2
    continuously compounded, 3

K

Kurtosis, 46

L

Level line
    resistance, 181
    support, 181
Likelihood
    function, 63
    log function, 63
    maximum estimate (MLE), 63
Limit order, 6
Ljung–Box
    statistic, 74
    test, 74, 129

M

M-dependent, 56, 194
MA(q), 112, 113
   transformed into AR(∞), 119
Market capitalization, 5
Market order, 5
Market portfolio, 249
Markov time, 190
Markowitz, Harry, 239
   mean-variance model, 242
   portfolio selection problem, 241
Maximum return portfolio, 244
Mean, 46
   of portfolio, 240
Merton, Robert, 153, 172
Minimum variance portfolio, 243
MLE, *see* likelihood,maximum estimate
Monte Carlo method, 152, 164
   convergence, 167
   Euler approximation, 165
   for path dependent options, 169
   Milstein approximation, 168
Moving average
   order q, *see* MA(q)
   technical rule, 183
Mutual fund, 18

N

Neglected nonlinearity, 137
Neural networks
   feed-forward, 131
   model estimation, 133

O

Online
   algorithm, 268
   Blum-Kalai randomized algorithm, 279
   Kalai-Vempala randomized algorithm (R-Universal), 279
   one-way trading, 272
   optimization problem, 268
   portfolio selection strategy, 274
   randomized RPP, 270
   Reservation Price Policy (RPP), 269
   universal portfolio, 274
Option, 12
   call, 12
   Greeks, 156
   payoff, 13
   pricing problem (defn.), 16
   profit, 14
   put, 12

Option styles, 12
   American, 12, 161, 164
   Asian, 12, 164, 169
   Barrier, 12, 164
   Bermudan, 12, 161, 164
   European, 12, 156, 161, 164, 166
   path dependent, 13, 169
   vanilla, 13

P

Portfolio, 18
   constant rebalanced, *see* Constant Rebalanced Portfolio
   insurance strategy, 19
   replicating, 27
   value, 18
Portfolio selection, 260
   algorithm, 261
   exponential growth rate, 261
   strategy, 261
   wealth factor, 261
Positions
   long, 19
   of portfolio, 18
   short, 19
Process
   ARCH, *see* ARCH model
   ARMA, *see* ARMA(p, q)
   GARCH, *see* GARCH model
   Gaussian, 59, 61, 196
   generalized Wiener, 146
   linear, 111
   nonlinear, 124
   Ornstein-Uhlenbeck, 174
   stationary, *see* stationary
   Wiener, 146
Profit graph, 15
Put-call parity, 29

R

Random walk, 59, 148
Return, 38
   annualized, 39
   compounded, 39
   excess, 42
   log, 40
   of portfolio, 42, 240
   simple, 38
   with dividends, 42
Risk-neutral valuation, 30, 155

S

Scholes, Myron, 153, 172
Security, 1
    payoff, 4
    profit, 4
Shares outstanding, 5
Sharpe ratio, 251
Shiller, Robert, 122, 140, 173
    PE 10 series, 122, 123, 129, 137
Simulated annealing, 209
    algorithm, 210
    foundations, 210
    GARCH estimation, 212
    portfolio optimization, 259
Skewness, 46
Speculator, 20
Stable clusters, 97
    algorithm, 99
Standard deviation, 46
Stationary, 56
    covariance, 58
    strictly, 56, 194
    weakly, 58
Stock, 4
    common, 4
    payoff, 9
    preferred, 4
    profit, 9
    ticker, 7
    trading order forms, 5
Stock cover, 99
    algorithm, 100
Stock indices, 9
    capitalization weighted, 10
    DJIA (defn.), 10
    IBEX35 (defn.), 10
    price weighted, 10
    Value Line Arithmetic, VLIA (defn.), 11
    Value Line Geometric,VLIC (defn.), 11
Stock quotes, 7
    ticker, open, close, high, low, volume,
        adjusted close, 7
Stop gains, 6
Stop loss, 6
Stopping time, 192
Structural risk minimization, 134
Support vector machines
    for regression, 134
    model estimation, 136

T

Tangency portfolio, 249

Technical analysis, 177
    basic principles, 178, 179
    mathematical foundation, 190
    predictive power assessment, 195
Technical trading rule, 183
    Bollinger bands, 185
    candlesticks evening tar, 206
    candlesticks Hammer, 189
    candlesticks morning star, 190
    gaps, 188
    head and shoulders, 186, 205
    high–low envelopes, 185
    in R, 204
    moving averages crossover, 183
    occurring times, 191
    Relative Strength Index (RSI), 206
    trend lines crossing, 184
    triangles, 188
    volume-to-price, 205
Temporal Graph of clusters, 95
Trend
    lines, 182
    type (TA), 178

V

Value indicators
    asset turnover, 201
    cash-coverage ratio, 201
    current ratio, 202
    debt ratio, 201
    earnings per Share (EPS), 199
    inventory turnover, 201
    price to book (P/B), 200
    price to earnings (P/E), 199
    profit margin, 201
    Return on assets (ROA), 201
    Return on capital (ROC), 201
    Return on equity (ROE), 200
    ROE, ROC, ROA revisited, 205
    times-interest-earned-ratio, 201
    working capital, 202
Variance, 46
    of portfolio, 240
Volatility, 64
    EWMA model, 67
    historical, 65
    implied, 173
    ranged-based, 65
    spillover, 84

W

White noise, 56

Wold's decomposition, 111   in R, 118

Y
Yule-Walker equations, 116

Printed by Publishers' Graphics LLC
ICISO140521.23.35.3